Molecular Endocrinology

genetic analysis of hormones and their receptors

The HUMAN MOLECULAR GENETICS series

Series Advisors

D.N. Cooper, *Institute of Medical Genetics, University of Wales College of Medicine, Cardiff, UK*

S.E. Humphries, *Division of Cardiovascular Genetics, University College London Medical School, London, UK*

T. Strachan, *Department of Human Genetics, University of Newcastle upon Tyne, Newcastle upon Tyne, UK*

Human Gene Mutation
From Genotype to Phenotype
Functional Analysis of the Human Genome
Molecular Genetics of Cancer
Environmental Mutagenesis
HLA and MHC: Genes, Molecules and Function
Human Genome Evolution
Gene Therapy
Molecular Endocrinology

Forthcoming titles

Venous thrombosis: from Genes to Clinical Medicine
Protein Dysfunction in Human Genetic Disease
Molecular Genetics of Early Human Development

Molecular Endocrinology

genetic analysis of hormones and their receptors

Gill Rumsby
Department of Molecular Pathology, Windeyer Institute of Medical Sciences, London, UK

Sheelagh M. Farrow
Department of Medicine, UCL Medical School, Middlesex Hospital, London, UK

First published in 1997

A CIP catalogue record for this book is available from the British Library.

ISBN 1 859962 35 1

BIOS Scientific Publishers Ltd
9 Newtec Place, Magdalen Road, Oxford OX4 1RE, UK.
Tel. +44 (0) 1865 726286. Fax. +44 (0) 1865 246823
World-Wide Web home page: http://www.Bookshop.co.uk/BIOS/

DISTRIBUTORS

Australia and New Zealand
DA Information Services
648 Whitehorse Road, Mitcham
Victoria 3132

India
Viva Books Private Limited
4346/4C Ansari Road
New Delhi 110002

Singapore and South East Asia
Toppan Company (S) PTE Ltd
38 Liu Fang Road, Jurong
Singapore 2262

USA and Canada
BIOS Scientific Publishers
PO Box 605, Herndon,
VA 20172-0605

Production Editor: Priscilla Goldby
Typeset by Saxon Graphics Ltd, Derby, UK.
Printed by Biddles Ltd, Guildford, UK.

Contents

Contributors

Barker, S. The William Harvey Research Institute, St Bartholomew's Medical College, Charterhouse Square, London EC1M 6BQ, UK

Baumbach, W.R. American Cyanamid Company, Box 400, Princeton, NJ 08543, USA

Bingham, B. Wyeth-Ayerst Research, CN 8000, Princeton, NJ 08543, USA

Brickell, P.M. Molecular Haematology Unit, Institute of Child Health, 30 Guilford Street, London WC1N 1EH, UK

Chatterjee, V.K.K. Department of Medicine, University of Cambridge, Level 5, Addenbrooke's Hospital, Hills Road, Cambridge CB2 2QQ, UK

Clifton-Bligh, R.J. Department of Medicine, University of Cambridge, Level 5, Addenbrooke's Hospital, Hills Road, Cambridge CB2 2QQ, UK

Conway, G.S. Department of Endocrinology, The Middlesex Hospital, Mortimer Street, London W1N 8AA, UK

Farrow, S.M. Department of Medicine, University College, London Medical School, Middlesex Hospital, Mortimer Street, London W1N 8AA, UK

Gillespie, M.T. St Vincent's Institute of Medical Research, 41 Victoria Parade, Fitzroy 3065, Australia

Martin, T.J. St Vincent's Institute of Medical Research, 41 Victoria Parade, Fitzroy 3065, Australia

Matthews, C. Department of Medicine, University of Cambridge, Level 5, Addenbrooke's Hospital, Hills Road, Cambridge CB2 2QQ, UK

Rumsby, G. Department of Molecular Pathology, Windeyer Institute of Medical Sciences, Windeyer Building, Cleveland Street, London W1P 6DB, UK

Shepherd, P.R. Department of Biochemistry and Molecular Biology, University College London, Gower St, London WC1E 6BT, UK

White, P.C. Department of Paediatrics, University of Texas SW Medical Center, 5323 Harry Hines Blvd, Dallas, TX 75235-9063, USA

Zysk, J.R. Cephalon, Inc., 145 Brandywine Parkway, West Chester, PA 19380, USA

Abbreviations

AC	adenylyl cyclase
ACC	acetyl CoA carboxylase
ACTH	adrenocorticotrophic hormone
AD	activation domain
AHC	adrenal hypoplasia congenita
AME	apparent mineralocorticoid excess
Ang II	angiotensin II
AP-PCR	arbitrarily primed PCR
AR	androgen receptor
AVP	vasopressin
AZF	azoospermia factor
β-AR	β-adrenergic receptor
βARK	β-AR kinase
CAT	chloramphenicol acetyltransferase
CGRP	calcitonin gene-related peptide
CMD	camptomelic dwarfism
CNS	central nervous system
COUP-TF	chick ovalbumin upstream promoter transcription factor
CPT	carnitine palmitoyl CoA transferase
CRAC	calcium release-activated calcium (channel)
CRE	cAMP response element
DAG	diacylglycerol
DAX1	dosage-sensitive sex-reversal–AHC critical region of the X chromosome gene 1
DAZ	deleted in azoospermia
DBD	DNA-binding domain
DGGE	denaturing gradient gel electrophoresis
DHT	dihydrotestosterone
DNase I	deoxyribonuclease I
DSS	dosage-sensitive sex
EGF	epidermal growth factor
ER	oestrogen receptor
ES	embryonic stem
FHH	familial hypocalciuric hypercalcaemia
FSH	follicle-stimulating hormone
FTZ-F1	fushi-tarazu factor 1
G-protein	guanine nucleotide-binding protein
GAP	GTPase-activating protein
GFP	green fluorescent protein
GH	growth hormone
GHBP	GH-binding protein
GHR	GH receptor
GHRP	GH-releasing peptide
GLP	glucagon-like peptide
GPLR	G-protein-linked receptor
GnRH	gonadotrophin-releasing hormone

GR	glucocorticoid receptor
GRF	GH-releasing factor
GRK	GPLR kinase
GST	glutathione *S*-transferase
hCG	human choriogonadotrophin
HHM	humoral hypercalcaemia of malignancy
HLA	human leukocyte antigen
HMG	high mobility group
hnRNA	heterogeneous nuclear RNA
HSD	hydroxysteroid dehydrogenase
HTLV	human T-cell leukaemia virus
IDDM	insulin-dependent diabetes
IGF	insulin-like growth factor
IGFBP	IGF-binding protein
IL	interleukin
IP_3	inositol (1,4,5) triphosphate
IRE	iron-responsive element
IRF	iron regulatory factor
IRMA	immunoradiometric assay
IRS	insulin receptor substrate
JAK	Janus kinase
LBD	ligand-binding domain
LC-CoA	long-chain fatty-acyl-CoA ester
LDL	low density lipoprotein
LH	luteinizing hormone
LPS	lipopolysaccharide
m^7G	methylguanine cap
MAP	mitogen-activated protein
MIS	Müllerian inhibiting substance
MR	mineralocorticoid receptor
mRNP	messenger ribonucleoprotein
MSH	melanocyte-stimulating hormone
NGF	nerve growth factor
NIDDM	non-insulin-dependent diabetes
NYP	neuropeptide Y
PABP	poly(A)-binding protein
PBR	peripheral benzodiazepine receptor
PCR	polymerase chain reaction
PDE	phosphodiesterase
PDH	pyruvate dehydrogenase
PEPCK	phosphoenolpyruvate carboxykinase
PI3-kinase	phosphatidylinositol-3-kinase
PKA	A protein kinase
PKC	protein kinase C
PLC	phospholipase C
PMDS	persistent Müllerian duct syndrome
POF	premature ovarian failure
POMC	proopiomelanocortin
PP-1	protein phosphatase-1
PPAR	peroxisome-proliferator-activated receptor
PR	progesterone receptor
PRL	prolactin

PTH	parathyroid hormone
PTHrP	parathyroid hormone-related protein/peptide
RAPD	random amplified polymorphic DNA
RAR	retinoic acid receptor
RBM	RNA-binding motif
RIA	radioimmunoassay
ROR	retinoid orphan receptor
RT–PCR	reverse transcriptase–polymerase chain reaction
RXR	retinoid X receptor
SCAD	short-chain alcohol dehydrogenase
SCP	sterol carrier protein
SF-1	steroidogenic factor 1
SH	*src* homology
SH-PTP2	SH2-containing protein tyrosine phosphatase
snRNP	small nuclear ribonucleoprotein
SOX9	SRY-related, HMG-box, gene 9
SRIF	somatostatin release-inhibiting factor
SRP	signal recognition particle
SRY	sex-determining region Y chromosome gene
SSCP	single-stranded conformational polymorphism
SSTR	somatostatin receptor
StAR	steroidogenic acute regulatory protein
STAT	signal transducers and activators of transcription
TGF	transforming growth factor
TGGE	temperature gradient gel electrophoresis
7TMDR	seven-transmembrane-domain receptor
TNF	tumour necrosis factor
TR	thyroid hormone receptor
TSH	thyroid-stimulating hormone
UCP	uncoupling protein
UTR	untranslated region
VDR	vitamin D receptor
X-Gal	5-bromo-4-chloro-3-indolyl-β-D-galactopyranoside
ZFX	zinc finger gene on X chromosome

Single-letter codes for amino acids

A alanine	**C** cysteine	**D** aspartic acid	**E** glutamic acid	**F** phenylalanine
G glycine	**H** histidine	**I** isoleucine	**K** lysine	**L** leucine
M methionine	**N** asparagine	**P** proline	**Q** glutamine	**R** arginine
S serine	**T** threonine	**V** valine	**W** tryptophan	**Y** tyrosine

X means a stop codon

Preface

The pace of research in endocrinology has increased greatly with the use of modern techniques which have led to the characterization of many of the genes involved in endocrine homeostasis. However, it is clear that the identification of a gene leads to a great many other questions, such as the nature and role of the protein product and the regulation of its synthesis. Consequently, much of the current research in many endocrine systems is directed at providing answers to these questions.

Our aim in this book is to provide a flavour of the progress in this area and we hope that it will be of particular interest to scientists and clinicians embarking on a research career in endocrinology. By assuming the reader to have some knowledge of molecular biology, we have been able to focus on the less commonly described mechanisms of gene regulation and the techniques used to investigate them. For example, the first chapters include descriptions of RNA processing and some of the powerful techniques which have been developed to study protein function, such as the production of transgenic mice and the yeast two-hybrid system. Subsequent chapters detail our current understanding of the principles of hormone action at peptide and steroid hormone receptors since, as can be seen in later chapters, this is fundamental to endocrinology.

Rather than follow a disease-oriented course, we have selected various systems in which significant progress has been made in recent years and which illustrate the successful application of modern technologies. The expansion in our knowledge of endocrine physiology is illustrated by chapters on growth, glucose homeostasis and the regulation of blood pressure. The complexity of the steroid biosynthetic pathway with its multiple isoenzymes is now largely unravelled due to the application of molecular biology.

The potential for rapid progress in research is exemplified by the characterization of a newly discovered hormone, parathyroid hormone-related protein, and the advances in our understanding of the regulation of sexual differentation and development.

It has not been possible to cover all areas of endocrinology in this book and we apologize to those researchers who feel that their subject area has been overlooked. We would like to thank all the authors for their contributions together with their patience and good humour with regard to editorial changes.

G. Rumsby (*London*)

S.M. Farrow (*London*)

1

Basis of gene structure and function

Sheelagh M. Farrow

1.1 Introduction

This chapter is not intended to be a complete guide to the molecular biology of the genome, since this information is widely available in many textbooks (see for example Lewin, 1994; Strachan and Read, 1996). Therefore, it will be assumed that the reader is familiar with the structure and biochemistry of DNA and RNA, the organization of the genome, the basic mechanisms of transcription and translation, and the structure of proteins. Instead, the purpose is to concentrate on those areas in molecular biology, and particularly the recent advances in our knowledge of the mechanisms of gene regulation, which are less well-addressed in other sources. It will focus on developments in our understanding of promoters and enhancers, RNA processing (particularly with reference to the mechanisms of alternate splicing and differential polyadenylation), and the role and function of the untranslated regions in mRNA.

1.2 Gene transcription to polypeptide synthesis

The stages in the protein synthetic pathway are shown schematically in *Figure 1.1*. Initially, the gene is transcribed to produce the primary transcript, heterogeneous nuclear RNA (hnRNA). This RNA is subsquently processed, during which a methylguanine cap (m^7G) is added to the 5′ end of the molecule and a poly(A) tail is added to the 3′ end. At this point, the RNA undergoes splicing to excise the introns, and yield the mature mRNA which is transported from the nucleus into the cytoplasm. After this step, the mRNA may be translated immediately, it may be degraded or, alternatively, it may become a cytosolic, untranslated 'store' of mRNA. The latter usually exists as mRNA associated with proteins forming messenger ribonucleoproteins.

If the mRNA is translated, initiation of polypeptide chain synthesis follows the interaction of the 40S ribosomal subunit with the cap site at the 5′ end of the RNA, recognition of the AUG codon and the association of the 60S subunit.

Molecular Endocrinology, edited by G. Rumsby and S.M. Farrow.

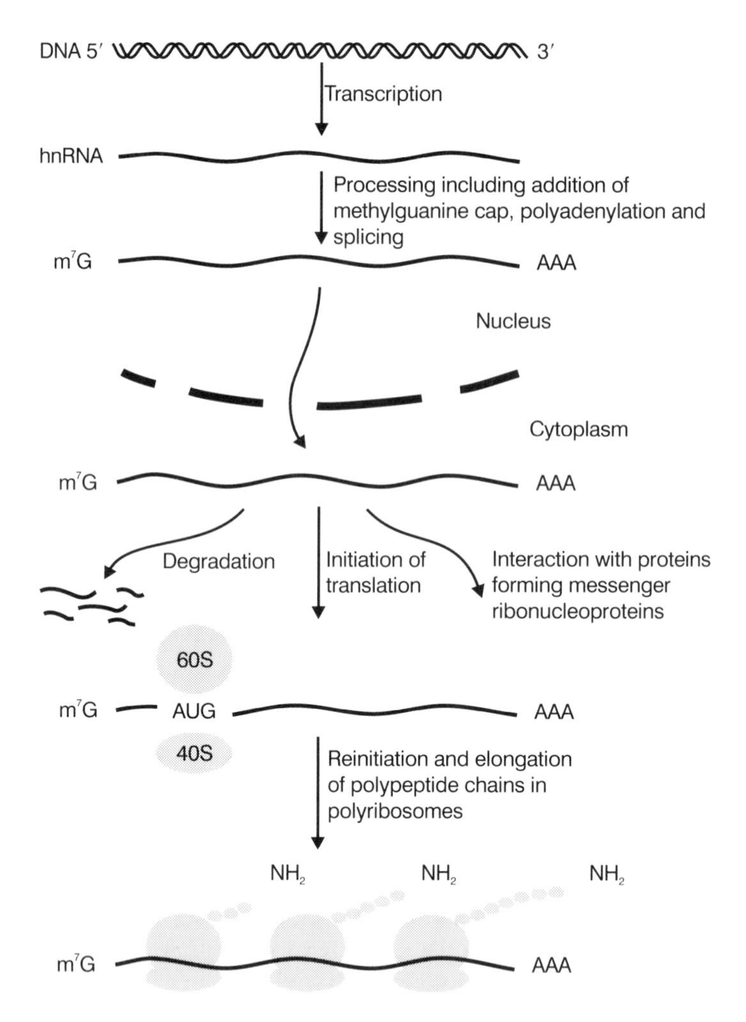

Figure 1.1. Regulatory steps in protein synthesis from gene transcription to mRNA translation. Reproduced from Farrow (1993) Post-transcriptional regulation of hormone synthesis. *J. Endocrinol.* **138**: 363–369, by permission of the Journal of Endocrinology Ltd.

Concomitantly with synthesis and elongation of the polypeptide chain, multiple ribosomal units may associate with mRNA and initiate additional polypeptide chain production, so forming polyribosomes or polysomes. Proteins destined for export are generally synthesized containing an amino terminal leader sequence. This acts as a signal sequence to direct the nascent polypeptide to interact with the endoplasmic reticulum. Following transport through the endoplasmic reticulum membrane, the signal sequence is cleaved and peptides or proteins are subsequently modified through, for example, glycosylation and sulphation.

It is clear from this brief schema that there are multiple stages in the process of protein synthesis at which gene expression may be regulated. These can be divided broadly into regulation occurring at transcription, RNA processing and translation, and will be discussed in the following sections.

1.3 Gene transcription

The initiation of transcription of eukaryotic genes is dependent on a number of defined sequences upstream of the start site. These include those elements which are required for constitutive or basal expression and those which modulate transcription. The former group consists of, for example, the TATA and CCAAT boxes whereas the second group contains the enhancers, silencers and hormone response elements. A schematic representation of these elements is shown in *Figure 1.2*.

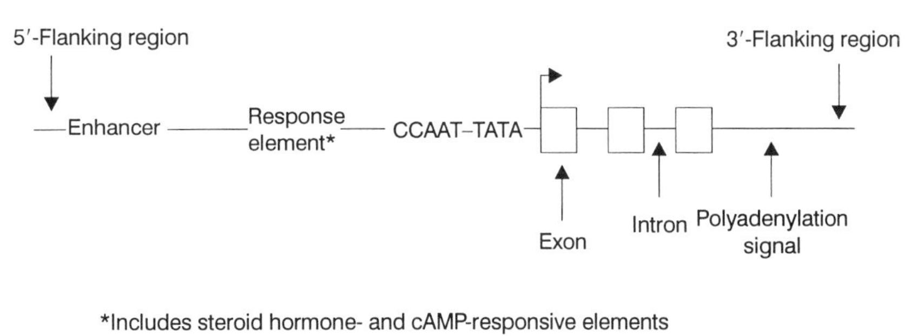

Figure 1.2. A schematic representation of some regulatory elements in a gene.

1.3.1 Constitutive gene transcription

The elements involved in this process are components of the promoter region of genes. One of these moieties is the TATA box: a *cis*-acting motif which appears to act as a signal to define the start of transcription, which usually occurs about 25–30 bp downstream of the box. During transcriptional initiation, a sequence of events occurs beginning with binding of the TATA-binding protein (a component of the transcription factor, TFIID) to the TATA box region. This process is followed by a cascade of protein–protein interactions between a family of general transcription factors resulting in the assembly of a large complex which includes RNA polymerase II and leads to the initiation of transcription. A second consensus sequence in this region is the CCAAT box which is located upstream of the TATA box. The role of the CCAAT box during transcription is likely to be as a modulator of the frequency of initiation.

Not all genes contain TATA and/or CCAAT boxes. In these cases, the process of transcriptional initiation is mediated through other factors and elements. One common feature of such genes is that the promoter regions tend to be GC-rich, and often contain one or more binding sites for the Sp1 transcription factor, that is, GGGCGG. It appears that this element, like the TATA box, acts as a determinant of the start site of transcription. Such properties tend to be characteristic of genes which exhibit multiple promoter regions and initiation start sites (see Section 1.4.1).

1.3.2 Modulators of transcription

Elements which regulate the rate of gene transcription are often gene specific and include the enhancers and silencers. These moieties exhibit several notable characteristics. That is, they may be located up- or downstream of, and often at great distance from, the transcription start site; they modulate the rates of transcription but cannot initiate transcription; they function independently of orientation. Another important characteristic of enhancer elements is that the realization of their effects is often dependent on cell type and thus provides a means by which cell-specific gene expression is achieved. Therefore, unlike the promoter elements, the enhancer sequences often interact with tissue-specific proteins. Other motifs which mediate the regulation of gene expression include the cAMP response elements and steroid hormone response elements (discussed in detail in Chapter 4).

Whilst it is easy to understand how proteins interacting with sequences in the gene such as the TATA box can influence processes 30 bases downstream, it is more difficult to envisage the mechanisms by which proteins binding to enhancers several kilobases upstream of the initiation site influence transcription. Several models have been proposed to explain how this may occur and are described by Dynan and Tjian (1985). One of these proposes that the protein interacts with the specific enhancer site and 'slides' along the DNA until coming into close proximity to the initiation site. In a second model, the binding of the first protein stimulates a cascade of proteins binding along the gene. Finally, a third model proposes that the bound protein causes 'looping out' of the DNA. This conformational change brings the enhancer physically close to the initiation site at which point the bound protein may interact with general transcription factors.

An essential component of the transcriptional apparatus is the interaction between protein and DNA and also protein–protein associations, whether between general transcription factors in the initiation complex, or other regulatory proteins. A great many studies have been performed on the nature of these interactions and a number of binding motifs have been described. Proteins can be grouped according to which of these are used in their association with the gene or other proteins. These elements include the leucine zipper, helix–loop–helix and zinc finger structures and the reader is referred to other sources in which a comprehensive description is provided (Latchman, 1990; Lewin, 1994).

1.4 Regulation of gene transcription

1.4.1 The use of alternate promoters

Regulation at the transcriptional level is considerably more complex when a gene contains multiple possible start sites for transcription, that is, several promoter elements. The use of different promoters may be tissue specific or several may be used within the same cell. Furthermore, switching between promoters in response to a stimulus is an important component in the process of gene regulation. The consequences of using different transcription start sites within a gene may be severalfold. For instance, the resultant variation in the length and sequence of RNA may not alter the coding region; that is, the same protein is synthesized but there may be differences in processing of mRNA (see Section 8.2.8). However, in other cases, the change in the sequence of the RNA causes a different protein to be produced.

An important example of the latter is provided by the α2(1) collagen gene. Early studies demonstrated that, although the gene is transcribed in both bone and cartilage cells, no α2(1) collagen is produced in cartilage. Subsequently, it was found that this difference was due to the use of alternate promoters directing transcription initiation from different start sites (Bennett *et al.*, 1989). The cartilage-specific start site arises from sequences within intron 2 and is preceded by a CCAATT sequence and a TGTAAA sequence which, given its proximity to the former sequence, and that transcription begins about 25 bases downstream of this element, is likely to have basal promoter activity (Bennett and Adams, 1990).

The use of the alternate promoter in cartilage alters the 5′ sequence of the mRNA such that the transcript does not encode collagen. Instead, there is evidence that this mRNA is translated to yield a non-collagenous protein of low molecular weight with an as yet unknown function. Since the use of the alternate promoter is restricted to chondrocytes, this protein will also be tissue specific. A particularly significant characteristic of this gene is that the promoter used can be 'switched'. Thus, the cessation of α2(1) collagen production during differentiation of prechondrogenic mesenchymal cells into chrondrocytes is due to a developmental-dependent change in the promoter used.

Alternate promoters are also present in the gene encoding the parathyroid hormone-related protein (PTHrP) receptor (see Section 6.4). A TATA-less promoter region was initially identified which is GC-rich and contains a number of Spl binding sites (McCuaig *et al.*, 1994). A second promoter has since been described which is more than 3 kb upstream of the original promoter, is also TATA-less but is not GC-rich, although a CCAAT box is present (McCuaig *et al.*, 1995). Unlike the GC-rich promoter, which is active in all tissues, this promoter is particularly tissue specific, with strong activity in kidney and weak activity in liver. The significance of the differential use of these promoters is yet to be ascertained. However, some indication of the consequences can be inferred since the use of the alternate promoters yields mRNAs which differ in the 5′-untranslated region (5′-UTR). As the nature of this region affects translation (see Section 1.6.2), it is possible that the alternate mRNA species are translated with different efficiencies, with the consequent effects on protein synthesis.

Whilst there may be tissue-dependent usage of promoters, there are also instances in which their use alters in response to stimuli. For example, the promoter region of the rat mitochondrial aspartate aminotransferase gene is GC-rich with the first 200 bases of the 5′-flanking region of the gene containing 67% of these nucleotides, no TATA box but three putative Spl binding sites (Juang *et al.*, 1995). Seven transcription start sites are present within this gene and are all active in liver and prostate. However, testosterone induction of enzyme synthesis is confined to the prostate epithelium, and changes in androgen levels cause a change in the pattern of start site usage in this tissue. Thus, it appears that some of the promoters have constitutive activity whereas others contain putative androgen response elements (see Chapter 4) and are hormonally regulated in a tissue-specific manner.

1.4.2 Modulation of the rate of transcription

A major part of the modulation of protein synthesis is mediated not by changes in promoter usage but through other regulatory elements. These include the enhancers and hormone response elements and are usually (although not exclusively) in the 5′-flanking regions of genes. Interactions through these sequences may determine tissue-specific expression of proteins or may provide a means by which transcription of the gene is up- or downregulated. Thus, enhancer elements may increase the basal promoter activity of a gene to different extents in different tissues and even determine whether that gene is expressed or not.

For example, the expression of growth hormone receptors is low in fetal tissues but increases rapidly postnatally (Mathews *et al.*, 1988). There is evidence that this postnatal 'activation' of gene expression is mediated through an enhancer element in the gene. In support of this, Menon *et al.* (1995) have identified a sequence 3–3.6 kb upstream of the transcriptional initiation site which enhances promoter activity in the growth hormone receptor gene in adult hepatocytes but has little effect in fetal cells. Further studies revealed that the 552 bp sequence bound a protein present only in adult hepatocytes, strongly suggesting that the enhancer element and the DNA–protein interactions in that region have an important role in the regulation of growth hormone receptor expression during development. However, the mechanism by which the protein-binding profile changes, from a fetal to a postnatal one, remains to be determined. A second example of sequences which determine cell-specific expression occurs in the production of pro-α1(I) collagen in osteoblasts. An enhancer sequence has been identified in the gene which binds an osteoblast-specific protein. This interaction probably confers tissue specificity for expression of this protein to osteoblasts (Rossert *et al.*, 1996).

Well-studied consensus regulatory elements include those which respond to steroid hormones (Chapter 4) and cAMP (Section 8.3.2). In addition to these, novel sequences through which transcription is modified have been described. One of these is the silencer element described in the parathyroid hormone gene. In this case, a sequence termed the calcium response element (CARE) has been located 3–4 kb upstream of the initiation site (Okazaki *et al.*, 1991). Through interactions with nuclear proteins this element reduces gene transcription in response to increased extracellular calcium concentrations.

1.5 Post-transcriptional regulation of gene expression

In early studies of gene regulation, changes in mRNA levels were often assumed to be due simply to transcriptional effects. Whilst this is true in many cases, there are other examples in which paradoxical results occur. In these instances, changes in the level of protein synthesized do not correlate with changes in transcription or mRNA levels. Close inspection reveals that there is regulation at post-transcriptional events. Such regulatory mechanisms provide a means whereby protein synthesis can be altered rapidly in response to acute changes in stimuli whereas responses to long-term changes in the cellular environment may be mediated through changes in gene transcription.

There are multiple points during protein synthesis at which post-transcriptional regulation occurs ranging from RNA processing to translation of mRNAs. For example, alterations in nuclear processing of RNA have been shown to modify expression of alkaline phosphatase (Kiledjian and Kadesch, 1991), prolactin (Billis *et al.*, 1992) and adenosine A1 receptors (Ren and Stiles, 1994) whilst changes in stability have been reported for many mRNAs including those for growth hormone (Paek and Axel, 1987; Murphy *et al.*, 1992), follicle-stimulating hormone (Attardi and Winters, 1993) and α1(I) collagen (Delany *et al.*, 1995). Regulation at the level of translation and mRNA–ribosomal interactions may involve stimulation of total (also known as global) protein synthesis, as is seen in insulin-induced protein synthesis in adipocytes (Pause *et al.*, 1994) or co-ordinate activation of dormant mRNA in germ cells (Bachvarova, 1992). However, there are also instances of gene-specific regulation through changes in the rates of initiation of translation, elongation of polypeptide chains or both to increase translational yield. Such changes in the efficiency of translation have been demonstrated during synthesis of insulin (Welsh *et al.*, 1986), ornithine aminotransferase (Muekler *et al.*, 1983), ornithine decarboxylase (Manzella and Blackshear, 1992), β-glucuronidase (Bracey and Paigen, 1987) and gonadotrophin-releasing hormone receptors (Tsutsumi *et al.*, 1995). For the purposes of this chapter, post-transcriptional regulation is divided into that which affects (i) RNA processing, (ii) mRNA stability and (iii) translation.

1.5.1 RNA processing

One of the constituents of processing is splicing of the RNA. While splicing of many transcripts is simple, that is the introns are removed and the exons joined, alternate splicing provides an opportunity to increase the diversity of the transcripts produced.

Constitutive splicing. To understand the mechanisms which regulate alternate splicing, a knowledge of the fundamentals of the process is essential (see reviews by Padgett *et al.*, 1986; Sharp, 1994). RNA contains a number of elements which are central to the splicing reaction. These sequences are conserved to greater or lesser extents, and include the 5′ and 3′ terminal dinucleotides in introns which are GU and AG, respectively (see *Figure 1.3*), and appear to be 100% conserved since a mutation in these bases results in abnormal splicing. In addition, the dinucleotide at the 3′ end of the exon is usually AG giving an exonic/intronic boundary sequence of AG/GU. Other important consensus sequences are the first six nucleotides of the intron (the 5′ donor site) which are GUAA/GGU, and the bases in the branch site towards the 3′ end of the intron (3′ acceptor site) which have the general formula of PyNPyPyPuAPy.

During splicing, cleavage occurs at the 5′ donor site of the intron and the G nucleotide forms a 5′–2′ bond with the A in the branch site in the intron to form a lariat structure. The intron is subsequently cleaved at the 3′ end and the lariat excised and degraded whilst the exonic sequences are ligated. This whole process is dependent on the assembly of a complex termed a spliceosome (for review see Sharp, 1994). This consists principally of the mRNA and small nuclear ribonucleoproteins (snRNPs) which are RNA molecules associated with proteins and include a number of species including U1, U2 and U4–6. In addition, some non-snRNP proteins are involved in the complex assembly. The U2 snRNP contains RNA sequences which are complementary to the branch site sequences and,

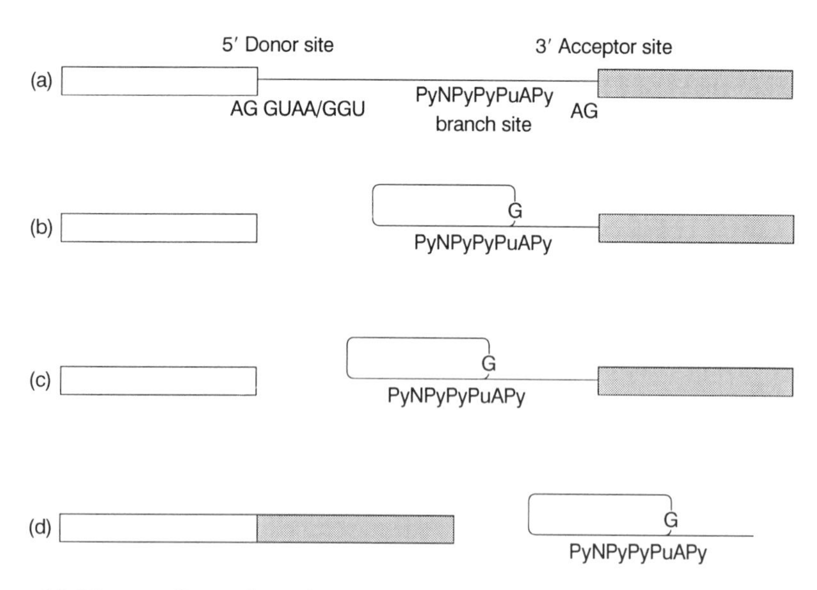

Figure 1.3. The exon/intron boundary sequences and the branch site consensus sequences in RNA are shown in (a). During splicing, the intron is cleaved at the 5′ donor site and the terminal G nucleotide forms a 5′–2′ bond with the A nucleotide in the branch site to form a lariat structure (b). The intron is cleaved at the 3′ end (c), the lariat excised and the exons ligated (d). Adapted from Lewin (1994).

through RNA–RNA interactions, contribute to the initial stages of spliceosome assembly. This process requires the presence of the non-snRNP factor, U2 auxiliary factor (Ruskin *et al.*, 1988). In general, it has been thought that the U1 snRNP interacts with the 5′ donor site. Although this is true, recent studies have demonstrated that, in the first instance, interactions between U1 snRNPs and the 3′ splice site are a necessary part of the process. Subsequently, there is interaction at the 5′ site. This sequence of events is described in the exon definition model (Robberson *et al.*, 1990), which predicts that the unit which is recognized during the splicing process is the exon and not the intron. In this model, the first step in the spliceosome assembly is the recognition of sequences within the 3′ acceptor site by snRNPs. This is followed by 'scanning' through the exon until a 5′ donor site is recognized. Thus, the exons are defined and removal of the intron proceeds (*Figure 1.4*). An intriguing question is how the appropriate 5′ and 3′ splice sites are matched to ligate adjacent exons and random exon skipping is avoided. One clue is provided by the model which indicates that a 5′ splice site must lie within a limited distance from the previously recognized 3′ splice site. Given that most internal exons are typically 200–300 bases long, if a 5′ donor site is not recognized within this limit, the exon will not be recognized as such. Thus, naturally occurring mutations within the acceptor or donor sites usually lead to exon skipping (Hawa *et al.*, 1996; Marvit *et al.*, 1987; Mottes *et al.*, 1994).

The choice of splice sites within RNA is dependent on a number of features. These include (i) the fidelity of the splice site sequences, (ii) the secondary structure around the splice sites and (iii) RNA–protein interactions. As regards (i), given that there is some flexibility within the consensus sequences, it is clear that within any

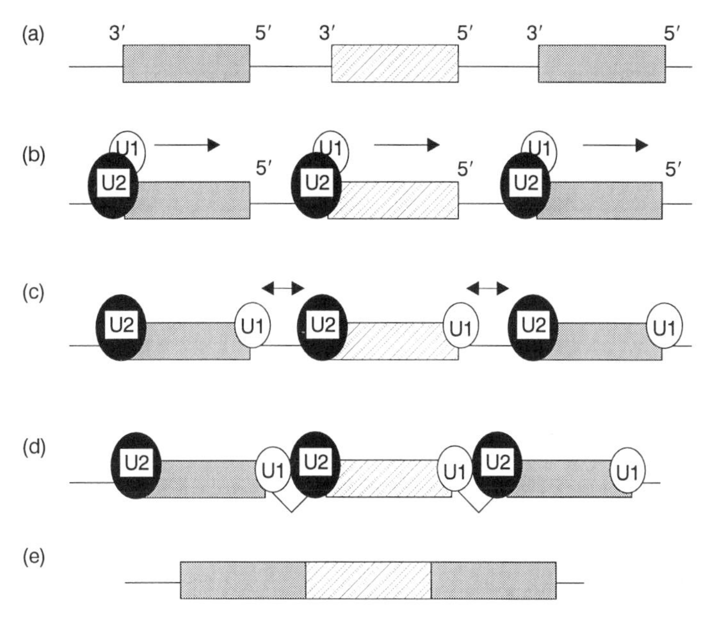

Figure 1.4. The exon definition model of splicing (adapted from Robberson *et al.*, 1990). Exons and the intronic 5′ and 3′ splice sites are shown (a). Initially, U1 and U2 snRNPs (and associated factors not shown) interact with the intronic 3′ acceptor sites (b) followed by the U1 assembly searching downstream for a recognizable 5′ donor site; that is, defining the exon (c). The adjacent snRNPs interact to cause looping of the introns (d) with subsequent removal of introns and ligation of exons (e).

RNA there are likely to be multiple elements which closely resemble splice sites. Those splice sites which are not perfect consensus sequences are referred to as 'weak' splice sites, and it is the balance between the strong and weak splice sites which is one of the determinants of the pattern of splicing. The relative 'strength' of a sequence can be altered by the surrounding nucleotides since the secondary structures may change the accessibility of particular splice sites to snRNPs. The importance of secondary structures can explain, in part, how sequences at a great distance from the consensus sequence can affect splice site selection. For example, sequences which influence alternate splicing of the fibronectin gene have been localized to more than 500 nucleotides away from the splice site (Huh and Hynes, 1993).

Alternate splicing. This is increasingly being reported as a mechanism of gene regulation. It provides a means whereby a single gene gives rise to multiple mRNA species with different properties and functions. There are several potential products of alternate splicing: (i) different proteins, (ii) isoforms of a single protein and (iii) mRNA species which differ in the 5′- and/or 3′-UTR. Probably the best known example of (i) is the synthesis of calcitonin or calcitonin gene-related peptide (CGRP) from the same gene. A more subtle consequence, whereby the splicing pattern yields a number of protein isoforms which differ to a small extent in their structure, is exemplified by the fibronectin gene. Processing of this gene leads to the generation of about 20 isoforms which differ slightly in their structural elements and function. Another example of this is PTHrP in which exon skipping, combined with the use of alternate promoters, leads to the production of three proteins which differ in their amino acid sequence (Mangin *et al.*, 1988) (see Chapter 6). Although a different function for these proteins has not been determined, the isoforms appear to be expressed in a tissue-specific fashion and may also differ between tumour and normal tissues (Campos *et al.*, 1994). In the third case, alternate splicing may generate multiple, heterogeneous mRNAs which are processed differently or exhibit different stabilities, etc. For example, differences in the 5′-UTR of mRNA affects the efficiency with which that variant is translated (see Section 1.6.2), whereas variation in the 3′-UTR may alter the half-life of the mRNA (see Section 1.5.2).

Thus, alternate splicing is a significant contributor to gene regulation. However, although the mechanisms of general splicing are becoming clearer, much remains to be discovered about the processes involved in the control of alternate splicing. Clearly, the process will be dependent on sequences within the RNAs and sensitive to RNA-binding proteins, which may be cell specific.

Some of the complexity of the processes involved in the regulation of alternate splicing can be illustrated by reference to the calcitonin/CGRP gene, the most studied gene in this respect. The gene contains six exons (*Figure 1.5*). Skipping of exon 4 during splicing yields CGRP, a process which occurs in neuronal cells. Conversely, the inclusion of exon 4 and the use of the polyadenylation signal contained therein produces an mRNA species coding for calcitonin restricted to the thyroid C cells. Several important factors have arisen from these studies which have shown that interaction between multiple sequences (often both exonic and intronic) and RNA-binding proteins determine the splicing patterns. For example, the branch site in intron 3 deviates from the consensus sequence in the human and rat genes (Adema *et al.*, 1990). This difference 'weakens' the splice site which may aid exon skipping in neuronal cells. However, in thyroid C cells, it is likely that other factors compensate for the suboptimal sequence and facilitate recognition of this site, and thus lead to inclusion of exon 4 and calcitonin production. Through the years, conflicting data has arisen as to which sequences are important in splicing of this gene. However, the current consensus is that *cis*-acting elements within exon 4 (Cote *et al.*, 1992; Emeson *et al.*, 1989; Lou *et al.*,

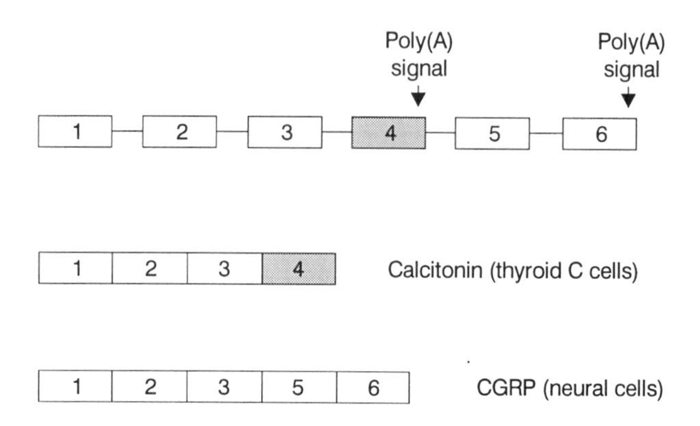

Figure 1.5. Products of alternate splicing of the calcitonin/CGRP gene.

1994) together with sequences within intron 4 (Zandberg *et al.*, 1995) are regulatory, whilst sequences in exons 5 and 6 have no role in splice site selection.

It is likely, that some, if not all, of these elements are important sites of protein interaction. Part of the mechanism whereby RNA-binding proteins alter splice site selection is via an effect on the secondary structures of RNA which are usually stabilized by bound proteins. Alternatively, proteins may bind close to a putative splice site to physically mask the site and thus effectively prevent recognition by splicing factors. In other tissues, this protein may be absent and the splice site becomes available for selection and processing. Binding of an uncharacterized 66 kDa protein to exon 4 sequences of the calcitonin gene has been described (Cote *et al.*, 1992) and this may be a factor which enhances recognition of the calcitonin-specific acceptor site. Further insight into the mechanism by which recognition of the suboptimal branch site is avoided in neuronal cells has been provided by the work of Roesser *et al.* (1993). These workers reported the partial purification of two brain-specific polypeptides which appear to inhibit selection of the calcitonin 3′ acceptor site in neuronal cells. The effect of these putative regulatory proteins may be to bind to the RNA and inhibit formation of the spliceosome around this region. Thus, the absence of these proteins in C cells effectively leads to an 'unmasking' of the acceptor site and inclusion of exon 4.

Finally, other general splicing factors and tissue-specific proteins have an important role in alternate splicing. One splicing factor in particular, SF2, discriminates between authentic and cryptic splice sites. Moreover, the effects of SF2 on splice site selection are dependent on its local concentration and thus it may have a role in alternate splicing (Krainer *et al.*, 1990).

1.5.2 Stability of mRNA

Change in the rate of turnover of a specific mRNA is an important mechanism of post-transcriptional regulation. By this process, the abundance of a particular mRNA may alter dramatically in the absence of transcriptional changes. There are numerous examples of genes which are controlled through changes in mRNA half-life. Not only might the half-life differ in response to exogenous stimuli, but cell-specific differences in stability can lead to differential expression of a protein. For example, the greater number of insulin receptors in HepG2 liver cells compared to fibroblasts (Hatada *et al.*, 1989; Reddy *et al.*, 1988) can be attributed partly to a longer half-life of insulin receptor mRNA in liver cells (Tewari *et al.*, 1991).

The 3′-UTR is well documented with regard to its influence on mRNA stability and several mechanisms have been defined by which variations within this region, or in its interactions with *trans*-acting factors, may alter the rate of mRNA degradation. These include: (i) the length of the poly(A) tail, (ii) the AU-rich *cis*-acting elements characteristic of the 3′-UTR of mRNAs with particularly short half-lives such as c-myc and (iii) other regulatory sequences. An introduction to these groups follows. However, for a detailed description of the measurement of mRNA stability and its determinants, the reader is referred to a comprehensive review by Ross (1995).

The poly(A) tract. The stability of a particular mRNA can be altered through events at the processing level such as the choice of polyadenylation signal or the length of the poly(A) tail. An example of the former is provided by the insulin-like growth factor I gene in which the use of multiple alternate polyadenylation signals generates several mRNA species differing in the sequence and length of the 3′-UTR and exhibiting different half-lives (Lund, 1994). Similarly, the poly(A) tail is known to affect the stability of transcripts and, in some instances, the length of the tail itself varies in response to exogenous stimuli. For example, glucocorticoids and thyroid hormone depletion stabilize growth hormone mRNA by increasing the length of the poly(A) tract (Murphy *et al.*, 1992; Paek and Axel, 1987). Conversely, the fall in rat thyrotropin β-subunit mRNA half-life induced by thyroid hormone is accompanied by a shortening of the poly(A) tail (Krane *et al.*, 1991). The mechanisms by which the degradation of mRNA is modulated by downstream sequences is likely to be mediated by *cis*- and/or *trans*-acting factors preventing or facilitating access of RNases to the 3′-UTR. This may occur either through protection of mRNA by proteins bound in the vicinity of RNase-sensitive sequences or through such interactions altering the conformation of the mRNA. One such RNA-binding protein is poly(A)-binding protein (PABP) which binds to the poly(A) tract. It has been suggested that, when bound, this protein restricts access of RNases to the mRNA; however, binding is less efficient or absent when the tail is short, and thus degradation is enhanced.

AU-rich sequences. A similar mechanism of RNA–protein interactions has been proposed for the destabilizing AU-rich motifs, AUUUA, present in mRNAs such as cytokines (Shaw and Kamen, 1986). mRNAs containing such multiple repeats are labile, and it has been suggested that binding of proteins to these elements 'targets' these mRNAs for rapid degradation by some unknown mechanism (Brewer, 1991; Malter, 1989). Alternatively, it has been proposed that the AU-rich sequences interact (through base pairing) with the poly(A) tract so preventing binding of PABP, and thus increasing degradation (Wilson and Treisman, 1988).

Other regulatory sequences. Novel sequences within the 3′-UTR which influence stability have also been recently identified. For example, 12 bases in the 3′-UTR of thyroid-stimulating hormone β (TSHβ) mRNA (TTAAATGT-GTTT) bind a protein with an apparent molecular weight of 80–85 kDa, the binding of which is regulated by T_3 (Leedman *et al.*, 1995). Similar RNA–protein interactions are implicated in the angiotensin II-induced rise in angiotensinogen synthesis which is mediated by an increase in mRNA stability. Klett *et al.* (1995) have demonstrated that binding of a 12 kDa protein to the 3′-UTR stabilizes angiotensinogen mRNA. Moreover, this rapid stabilization (less than 30 min) appears not to involve *de novo* synthesis of the protein but, instead, interaction with the 3′-UTR is activated by protein phosphorylation. This mechanism is consistent with the effects *in vivo* in which angiotensin II induces transient decreases in cAMP (Klett *et al.*, 1993). Steric hindrance of RNases is also likely to be the means by which RNA–protein interactions of the iron-responsive elements prolong the half-life of transferrin receptor mRNA (Casey *et al.*, 1988). This particular interaction is also important in regulating translation of ferritin mRNA and is discussed in more detail below (Section 1.6.1).

1.6 Translational regulation

The general processes involved in translation of RNA into proteins in eukaryotes are well described in many textbooks and, in more detail, in reviews by Hershey (1991), Merrick (1992) and Rhoads (1993). For the purposes of this chapter, a brief overview of the process follows. Initiation of translation requires the attachment of the 40S ribosomal subunit to the ^{7}methyl guanosine cap of mRNA, an event which is facilitated by the presence of initiation complex 4F (eIF-4F) which consists of the proteins eIF-4E, eIF-4A and p220 (Sonenberg, 1993). The initiation factor 4A exhibits helicase activity and 'melts' any secondary structures in the RNA in an ATP-dependent process. This event is accompanied by the binding of other initiation factors and movement of the ribosomal complex through the 5′-UTR until an initiation codon, AUG, is recognized. At this stage, the 60S ribo-somal unit attaches, forming a functional 80S ribosome, and production of the polypeptide chain is initiated. Addition of the appropriate amino acids and elongation of the polypeptide chain continue until a termination codon is recognized, after which the polypeptide chain is released. In most cases of protein synthesis, several ribosomes interact concomitantly with the mRNA forming polyribosomes (or polysomes) as a result of this re-initiation. During translation, the leader sequence of the polypeptide chain interacts with the ribosomal signal recognition particle (SRP) followed by interaction of this complex with the SRP receptor on the membrane of the endoplasmic reticulum. This is followed by a number of events including the initiation of transport across the membrane, release of the SRP and resumption of translation. Within the lumen of the endoplasmic reticulum, the signal peptide is removed by the actions of an endoplasmic reticulum-specific signal peptidase.

All of these processes are potential sites of regulation. However, to date, the most commonly reported stage of modulation during translation, is the initiation/re-initiation step. This may be the only change, or the rate of elongation of polypeptide chains may also increase concomitantly, in which case, the overall efficiency of translation is said to rise. Translational regulation may be be a global phenomenon whereby most, if not all, cellular proteins are co-ordinately increased, or it may occur in a gene-specific manner.

1.6.1 Regulation of global protein synthesis

In general, most of the changes in total protein synthesis are mediated through modifications of the general initiation factors, for example, phosphorylation of eIF-2α in response to heat shock (Hershey, 1989) or reduced phosphorylation of eIF-4E following serum starvation (Kaspar *et al.*, 1990). However, a novel mechanism whereby global protein synthesis in adipocytes is increased in response to insulin stimulation has recently been reported. A human gene has been cloned which encodes a protein regulator of 5′-cap function (Pause *et al.*, 1994). This protein interacts with eIF-4E to inhibit initiation, and thus translation, in the basal state. On exposure to insulin, the eIF-4E-binding protein becomes hyperphosphorylated and dissociates from eIF-4E, so relieving translational inhibition and leading to an increase in total cellular protein synthesis.

A more unexpected mechanism of co-ordinated regulation of protein synthesis is present in oocytes. In these cells, the transcriptional apparatus is inactive and the maternal mRNAs are dormant; that is, translation is inhibited (reviewed by Richter, 1991; Wickens, 1990). One means by which mRNAs are maintained in a dormant state is through an interaction of 'masking proteins' with the 3'-UTR. When bound these proteins inhibit initiation of translation, possibly by causing a conformational change which prevents access to the cap site of initiation factors, ribosomal units, etc. However, on dissociation, initiation occurs and protein synthesis proceeds. Some of these proteins have been identified in *Xenopus* oocytes and are termed Y-box proteins (Wolffe, 1994) and analogous proteins have been identified in mammalian germ cells (Kwon and Hecht, 1991).

1.6.2 Gene-specific regulation of translation

In recent years, the UTRs of specific mRNAs have been shown to have important regulatory functions at the translational level. This is not an unexpected role for the 5'-UTR given its association with the ribosomes. However, such a function for the 3'-UTR is more surprising. Nevertheless, the expression of an ever increasing number of genes has been shown to be regulated through both UTRs, a process which is often dependent on RNA–protein interactions.

The 5'-untranslated region. Translational regulation through this region, and the binding of proteins therein, could take many forms. For example, there may be a complete inhibition of translation through protein binding and the formation of messenger ribonucleoproteins (mRNPs), as has been described for ferritin mRNA (Rouault *et al.*, 1988). Alternatively, there may be modulation of the rates of initiation and/or elongation and translational efficiencies. A combination of such mechanisms has been implicated in translational regulation of insulin synthesis (Welsh *et al.*, 1986). These authors demonstrated that the glucose-induced increase in insulin production is mediated through increased rates of initiation and elongation, combined with increased recruitment of insulin mRNA from mRNP and free ribosomes on to polysomes. In a similar manner, stimulation of ornithine aminotransferase synthesis in liver is achieved predominantly through increased rates of initiation and elongation; that is, increased translational efficiency (Muekler *et al.*, 1983).

To understand the rationale by which the sequence of the 5'-UTR can affect translation of an mRNA requires a familiarity with the 'scanning model' of translation as proposed by Kozak (1991). This predicts that the initiation complex (see Section 1.6) moves along the mRNA until an initiation codon, AUG, is recognized (*Figure 1.6*). Recognition is facilitated if the codon is in a 'good context' with regard to adjacent nucleotides. The optimal sequence for recognition of the initiation codon is (-6) GCCA/GCC**AUG**G/A (+4) with positions -3 and +4 being most critical. However, there are many examples of initiation codons being in suboptimal contexts and, in these instances, secondary structures in the RNA may increase the probability of the codon being recognized. For example, a stem–loop structure downstream of the initiation codon will slow scanning of the ribosomal complex thus increasing the temporal interaction with the suboptimal complex. Moreover, a protein interaction with the secondary structure is likely to

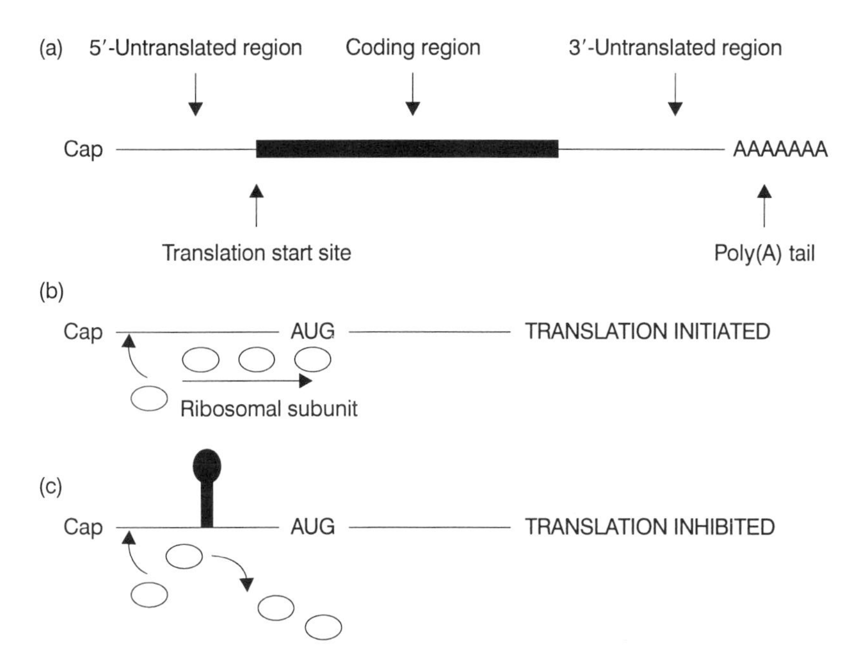

Figure 1.6. The general structure of mRNA is shown in (a). The Kozak model (Kozak, 1991) of translational initiation predicts that ribosomal subunits interact with mRNA at the cap site and scan through the mRNA until an initiation codon is recognized and translation occurs (b). The presence of secondary structures such as stem–loop structures upstream of the initiation codon prevents ribosomal scanning and access to the AUG codon and so inhibits translational initiation (c).

stabilize the structure and reduce scanning to a greater extent, since the energy required to 'melt' the sequence is increased. This mechanism, by which secondary structures compensate for suboptimal contexts, may be particularly important in those genes which have multiple AUG codons from which translation may be initiated. In this case, the secondary structures flanking the 'in-frame' AUG may augment recognition of the correct initiation codon. Conversely, a stem–loop structure between the cap site and the initiation codon may serve to inhibit translation since the 'scanning' ribosomal complex may not be able to 'melt' the structure (particularly if there is an interaction with protein) and thus not come into proximity with the AUG codon to initiate translation (*Figure 1.6c*). This mechanism has been shown to occur as a regulatory process of iron-dependent ferritin synthesis and is described in more detail as follows.

Ferritin mRNA is maintained in the cytoplasm in the form of mRNPs in which the mRNA is untranslatable. This control of ferritin mRNA translation is mediated through binding of a protein (the iron regulatory factor, IRF) in the 5′-UTR (Brown *et al.*, 1989; Rouault *et al.*, 1990). The particular sequences which interact with the protein, termed the iron-responsive element (IRE), form a stem–loop structure which lies between the cap site and the initiation codon (Hentze *et al.*, 1987). The effect of the bound protein is to stabilize the secondary structure so increasing the energy required to melt the RNA by the ribosomal complexes during scanning. Access to the initiation codon is therefore reduced and translation is prevented. However, exposure to a suitable stimulus causes the protein to dissociate from the mRNA and translation can proceed. A similar translational block caused by protein interactions within the 5′-UTR is an important component of regulation of synthesis of thymidylate synthase. However, in this case, the protein is the enzyme itself binding in the region of the start codon of its own mRNA to inhibit translation, an example of autoregulation of protein synthesis (Johnson, 1994).

There are several suggested mechanisms by which the IRF interaction with ferritin mRNA is controlled. These include the possibility that the RNA–protein interaction is regulated through availability of the IRF (Goessling *et al.*, 1992). This necessitates a regulated turnover of protein probably in response to iron concentrations whereas an alternative theory is that the interaction is determined by the conformation of the IRF, changes in which alter the affinity for the IRE (Tang *et al.*, 1992). This is supported by reports that a reduction in iron availability increases the IRE binding affinity. The IRE is not confined to the 5′-UTR of ferritin mRNA; multiple copies of the IRE are also present in the 3′-UTR of the transferrin receptor mRNA (Casey *et al.*, 1988). In this case, an interaction with the IRF is thought to hinder access of RNases to the UTR and thus prolong the half-life of the mRNA.

As described above, the nature of the 5′-UTR can affect the translatability of a transcript and this region may be a particularly important site of regulation of genes which are characterized by multiple, heterogeneous mRNA species. For example, switching production from a transcript which is poorly translated to one which is efficiently processed would lead to increased protein synthesis with no change in transcription. Differing translatabilities of transcripts have been described for the genes encoding insulin-like growth factor (IGF)-I (Yang *et al.*, 1995), IGF-II (De Moor *et al.*, 1995) and complement factor B (Garnier *et al.*, 1995).

In the case of IGF-I, at least five major transcripts differing in the length and sequence of the 5′-UTR are produced (Adamo *et al.*, 1991; Kim *et al.*, 1991). Importantly, the production and relative proportions of these transcripts differ in a cell-specific manner (Shemer *et al.*, 1992; West *et al.*, 1996). Some indication of the functional relevance of these multiple transcripts was originally provided by studies which revealed that transcripts with a short 5′-UTR are preferentially associated with polysomes, suggesting different translatabilities (Foyt *et al.*, 1992). Subsequently, Yang *et al.* (1995) confirmed that the translational efficiency of IGF-I mRNA is inversely proportional to the length of the 5′-UTR with a 10-fold difference in translatability between some transcripts. This difference in translation appears to be due, in part, to AUGUGA motifs which are present in the longer transcripts and which inhibit efficient translation. The presence of upstream AUG codons also inhibits translation of one of the two transcripts encoding murine complement factor B leading to a twofold difference in protein synthesis (Garnier *et al.*, 1995).

The presence of multiple AUG codons is not the only determinant of translatability of mRNA variants. In particular, sequences which may be involved in RNA–protein interactions and factors which regulate expression of such *trans*-acting RNA-binding proteins are important. For example, Teerink *et al.* (1994)

have demonstrated inhibitory sequences within the 5′-UTR of a poorly translated IGF-II mRNA variant and have shown that translation can be stimulated by the addition of extracts from undifferentiated, but not from differentiated, P19 embryonal carcinoma cells. De Moor *et al.* (1995) subsequently reported that proteins binding to the 5′-UTR of IGF-II transcripts influence translation. These data indicate that, in this case, translatability is dependent not only on RNA sequences but also on tissue-specific factors.

The signal leader sequence. An important, but not well studied, aspect of variation in the 5′-UTR of IGF-I mRNAs is that initiation of translation can occur at a number of AUG codons. Yang *et al.* (1995) report that peptides with signal sequences of different lengths are synthesized but that all variants are processed to the same mature IGF-I. However, variation in the signal peptide may affect the efficiency of processing and translocation of the protein at the endoplasmic reticulum. An instance of post-transcriptional regulation of gene expression at the endoplasmic reticulum has been reported during glucose-induced insulin synthesis. Welsh *et al.* (1986) have shown that glucose stimulation enhances the interaction of the signal sequence of insulin with the SRP. However, to date, there are few other examples of regulation at this level.

The 3′-untranslated region. RNA–protein interactions within the 3′-UTR also modify the initiation of translation or its efficiency (for review see Tanguay and Gallie, 1996). Global effects on protein synthesis in germ cells which are mediated through this region have been described above (Section 1.6.1); however, there are also examples of gene-specific regulation. For example, alternate use of polyadenylation sites in the amyloid precursor protein gene yields two mRNA species which differ only in the 3′-UTR. However, the transcript with the longer 3′-UTR has enhanced translational efficiency (de Sauvage *et al.*, 1992). Conversely, sequences within the 3′-UTR of human β-interferon (IFN-β) mRNA have been shown to reduce translation. This mRNA is characterized by the presence of multiple repeats of an octanucleotide sequence (UUAUUUAU) in the 3′-UTR, the removal of which increases translation 10-fold. Although this type of sequence is usually associated with effects on mRNA stability (Kruys *et al.*, 1987), in this case these motifs form a physical association with the poly(A) tail and reduce translational efficiency (Grafi *et al.*, 1993).

Similar motifs are present in other cytokine mRNAs including that encoding tumour necrosis factor (TNF). This mRNA is translationally repressed in quiescent macrophages but translation increases greatly in response to lipopolysaccharide (LPS), an effect abolished by removal of AU-rich motifs within the 3′-UTR. The mechanisms by which such effects are realized are currently unknown; however, there is evidence for the interaction of regulatory RNA-binding proteins with these AU-rich sequences. For example, a protein of approximately 38 kDA has been identified in macrophages which specifically binds to the AU-rich sequences in TNF mRNA and, since exposure to LPS reduces binding, this protein may act as a translational repressor (Kruys and Huez, 1994).

It seems likely that the effects of AU-rich sequences on mRNA stability and translation are linked such that the half-life of an mRNA alters after association with ribosomes. This type of mechanism may be important in post-transcriptional regulation of parathyroid hormone (PTH) synthesis (Hawa *et al.*, 1993). In this case, the low calcium-induced rise in PTH production occurs at the translational level and appears to alter the half-life of ribosomally associated PTH mRNA. This effect is mediated through the 3′-UTR and probably involves the interaction of proteins of which two with apparent molecular weights of approximately 48 and 70 kDa, have been shown to bind to this region (Vadher *et al.*, 1996).

The examples described above involve a subtle modulation of the rate of translation. Regulation of synthesis of other proteins may involve blocking translation. For example, the gene for erythroid 15-lipoxygenase in red blood cells is transcribed in the bone marrow during red blood cell development but the mRNA is not translated until the cells reach the peripheral circulation. This translational block is due to binding of a 48 kDa protein to pyrimidine-rich, 19 nucleotide repeat sequences within the 3′-UTR of the mRNA (Ostareck-Lederer *et al.*, 1994). To date, the 'signal' which activates translation is unknown but may involve modification of the protein by phosphorylation/dephosphorylation, etc., or changes in the rate of degradation of the repressor protein in the peripheral circulation.

Despite these examples, little is known about the mechanisms by which protein interactions at the 3′-UTR can influence initiation of translation at the 5′-UTR, although a number of possible explanations exist. For instance, proteins bound to the 3′-UTR may sterically hinder access of RNases to the region or cause conformational changes in the mRNA which reduce ribosomal access to the cap site.

1.7 Concluding remarks

Early studies in endocrinology were restricted to the measurement of hormone secretion. The subsequent rapid expansion of molecular biology has led to the identification and characterization of many of the genes involved. It has also provided an insight into the mechanisms by which these genes are regulated. The primary goal of this chapter has been to give an overview of some of the mechanisms which are important in the regulation of protein synthesis. Some of the processes involved are only beginning to be understood, but it is certain that the continuing development of new techniques and approaches to research will ensure rapid progress.

References

Adamo ML, Ben-Hur H, Roberts CT Jr, LeRoith D. (1991) Regulation of start site usage in the leader exons of the rat insulin-like growth factor-I gene by development, fasting, and diabetes. *Mol. Endocrinol.* 5: 1677–1686.

Adema GJ, van Hulst RAL, Baas PD. (1990) Uridine branch acceptor is a *cis*-acting element involved in regulation of the alternative processing of calcitonin/CGRP-I pre-mRNA. *Nucl. Acid Res.* **18:** 5365–5372.

Attardi B, Winters SJ. (1993) Decay of follicle-stimulating hormone-β messenger RNA in the presence of transcriptional inhibitors and/or inhibin, activin or follistatin. *Mol. Endocrinol.* **7:** 668–680.

Bachvarova RF. (1992) A maternal tail of poly(A): the long and the short of it. *Cell* **69**: 895–897.
Bennett VD, Adams SL. (1990) Identification of a cartilage-specific promoter within intron 2 of the chick α2(I) collagen gene. *J. Biol. Chem.* **265**: 2223–2230.
Bennett VD, Weiss IM, Adams SL. (1989) Cartilage-specific 5′ end of chick alpha2(I) collagen mRNAs. *J. Biol. Chem.* **264**: 8402–8409.
Billis WM, Delidow BC, White BA. (1992) Posttranscriptional regulation of prolactin (PRL) gene expression in PRL-deficient pituitary tumor cells. *Mol. Endocrinol.* **6**: 1277–1284.
Bracey LT, Paigen K. (1987) Changes in translational yield regulate tissue-specific expression of β-glucuronidase. *Proc. Natl Acad. Sci. USA* **84**: 9020–9024.
Brewer G. (1991) An A+U-rich element RNA-binding factor regulates c-myc mRNA stability in vitro. *Mol. Cell. Biol.* **11**: 2460–2466.
Brown PH, Daniels-McQueen S, Walden WE, Patino MM, Gaffield L, Bielser D, Thach RE. (1989) Requirements for the translational repression of ferritin transcripts in wheatgerm extracts by a 90 kDa protein from rabbit liver. *J. Biol. Chem.* **264**: 13383–13386.
Campos RV, Zhang L, Drucker DJ. (1994) Differential expression of RNA transcripts encoding unique carboxy-terminal sequences of human parathyroid hormone-related peptide. *Mol. Endocrinol.* **8**: 1656–1666.
Casey JL, Hentze MW, Koeller DM, Caughman SW, Rouault TA, Clausner RD, Harford JD. (1988) Iron-responsive elements: regulatory sequences that control mRNA levels and translation. *Science* **240**: 924–928.
Cote GJ, Stolow DT, Peleg S, Berget SM, Gagel RF. (1992) Identification of exon sequences and an exon binding protein involved in alternative RNA splicing of calcitonin/CGRP. *Nucl. Acid Res.* **20**: 2361–2366.
Delany AM, Gabbitas BY, Canalis E. (1995) Cortisol downregulates osteoblast α1(I) procollagen mRNA by transcriptional and posttranscriptional mechanisms. *J. Cell. Biochem.* **57**: 488–494.
De Moor CH, Jansen M, Bonte EJ, Thomas AAM, Sussenbach JS, Van den Brande JL. (1995) Proteins binding to the leader of the 6.0 kb mRNA of human insulin-like growth factor 2 influence translation. *Biochem. J.* **307**: 225–231.
Dynan WS, Tjian R. (1985) Control of eukaryotic messenger RNA synthesis by sequence-specific DNA-binding proteins. *Nature* **316**: 774–778.
Emeson RB, Hedjran F, Yeakley JM, Guise JW, Rosenfeld MG. (1989) Alternative production of calcitonin and CGRP mRNA is regulated at the calcitonin-specific splice acceptor. *Nature* **341**: 76–80.
Foyt HL, Lanau F, Woloschak M, LeRoith D, Roberts CT Jr. (1992) Effects of growth hormone on levels of differentially processed insulin-like growth factor-I mRNA in total and polysomal mRNA populations. *Mol. Endocrinol.* **6**: 1881–1888.
Garnier G, Circolo A, Colten HR. (1995) Translational regulation of murine complement factor B alternative transcripts by upstream AUG codons. *J. Immunol.* **154**: 3275–3282.
Goessling LS, Daniels-McQueen S, Bhattacharyya-Pakrasi M, Lin J-J, Thach RE. (1992) Enhanced degradation of the ferritin repressor protein during induction of ferritin messenger RNA translation. *Science* **256**: 670–673.
Grafi G, Sela I, Galili G. (1993) Translational regulation of human beta interferon mRNA: association of the 3′ AU-rich sequence with the poly(A) tail reduces translation efficiency *in vitro*. *Mol. Cell. Biol.* **13**: 3487–3493.
Hatada EN, McClain DA, Poter E, Ullrich A, Olefsky JM. (1989) Effects of growth and insulin treatment on the levels of insulin receptors and their mRNA in HepG2 cells. *J. Biol. Chem.* **264**: 6741–6747.
Hawa NS, O'Riordan JLH, Farrow SM. (1993) Post-transcriptional regulation of bovine parathyroid hormone synthesis. *J. Mol. Endocrinol.* **10**: 43–49.
Hawa NS, Cockerill FJ, Vadher S, Hewison M, Rut AR, Pike JW, O'Riordan JLH, Farrow SM. (1996) Identification of a novel mutation in hereditary vitamin D resistant rickets causing exon skipping. *Clin. Endocrinol.* **45**: 85–92.
Hentze MW, Caughman SW, Rouault TA, Barriocanal JG, Dancis A, Harford JB, Klausner RD. (1987) Identification of the iron-responsive element for the translational regulation of human ferritin mRNA. *Science* **238**: 1570–1573.
Hershey JWB. (1989) Protein phosphorylation controls translation rates. *J. Biol. Chem.* **264**: 20823–20826.
Hershey JWB. (1991) Translational control in mammalian cells. *Ann. Rev. Biochem.* **60**: 717–755.

Huh GS, Hynes RO. (1993) Elements regulating an alternatively spliced exon of the rat fibronectin gene. *Mol. Cell. Biol.* **9:** 5301–5314.

Johnson LF. (1994) Post-transcriptional regulation of thymidylate synthase gene expression. *J. Cell. Biochem.* **54:** 387–392.

Juang HH, Costello LC, Franklin RB. (1995) Androgen modulation of multiple transcription start sites of the mitochondrial aspartate aminotransferase. *J. Biol. Chem.* **270:** 12629–12634

Kaspar RL, Rychlik W, White MW, Rhoads RE, Morris DR. (1990) Simultaneous cytoplasmic redistribution of ribosomal protein L32 nmRNA and phosphorylation of eukaryotic initiation factor 4E after mitogenic stimulation of swiss 3T3 cells. *J. Biol. Chem.* **265:** 3619–3622.

Kiledjian M, Kadesch T. (1991) Post-transcriptional regulation of the human liver/bone/kidney alkaline phosphatase gene. *J. Biol. Chem.* **266:** 4207–4213.

Kim S-W, Lajara R, Rotwein P. (1991) Structure and function of a human insulin-like growth factor-I gene promoter. *Mol. Endocrinol.* **5:** 1964–1972.

Klett C, Nobling R, Gierschik P, Hackenthal E. (1993) Angiotensin II stimulates the synthesis of angiotensinogen in hepatocytes by inhibiting adenylyl cyclase activity and stabilising angiotensinogen mRNA. *J. Biol. Chem.* **268:** 25095–25107.

Klett C, Bader M, Schwemmle M, Ganten D, Hackenthal E. (1995) Contribution of a 12 kDa protein to the angiotensin II-induced stabilization of angiotensinogen mRNA: interaction with the 3′untranslated mRNA. *J. Mol. Endocrinol.* **14:** 209–226.

Kozak M. (1991) Structural features in eukaryotic mRNAs that modulate the initiation of translation. *J. Biol. Chem.* **266:** 19867–19870.

Krainer AR, Conway GC, Kozak D. (1990) The essential pre-mRNA splicing factor SF2 influences 5′ splice site selection by activating proximal sites. *Cell* **62:** 35–42.

Krane IM, Spindel ER, Chin WW. (1991) Thyroid hormone decreases the stability and the poly(A) tract length of rat thyrotropin β-subunit messenger RNA. *Mol. Endocrinol.* **5:** 469–475.

Kruys V, Huez G. (1994) Translational control of cytokine expression by 3′ UA-rich sequences. *Biochimie* **76:** 862–866.

Kruys V, Wathelet M, Poupart P, Contreras R, Fiers W, Content J, Huez G. (1987) The 3′ untranslated region of the human interferon-β mRNA has an inhibitory effect on translation. *Proc. Natl Acad. Sci. USA* **84:** 6030–6034.

Kwon YK, Hecht NB. (1991) Cytoplasmic binding to highly conserved sequences in the 3′ untranslated region of mouse protamine 2 mRNA, a translationally regulated transcript of male germ cells. *Proc. Natl Acad. Sci. USA* **88:** 3584–3588.

Latchman DS. (1990) *Gene Regulation: a Eukaryotic Perspective.* Unwin Hyman, London.

Leedman PJ, Stein AR, Chin WW. (1995) Regulated specific protein binding to a conserved region of the 3′-untranslated region of thyrotropin β-subunit mRNA. *Mol. Endocrinol.* **9:** 375–387.

Lewin B. (1994) *Genes V.* Oxford University and Cell Press, Oxford.

Lou H, Cote GJ, Gagel R. (1994) The calcitonin exon and its flanking intronic sequences are sufficient for the regulation of human calcitonin/calcitonin gene-related peptide alternative RNA splicing. *Mol. Endocrinol.* **8:** 1618–1626.

Lund PK. (1994) Insulin-like growth factor I: Molecular biology and relevance to tissue-specific expression and action. *Recent Prog. Horm. Res.* **49:** 125–148.

Malter J. (1989) Identification of an AUUUA-specific messenger RNA binding protein. *Science* **246:** 664–666.

Mangin M, Webb AC, Dreyer BE, Posillico JT, Ikeda K, Weir EC, Stewart AF, Bander NH, Milstone L, Barton DE, Francke U, Broadus AE. (1988) Identification of a cDNA encoding a parathyroid hormone-like peptide from a human tumor associated with humoral hypercalcemia of malignancy. *Proc. Natl Acad. Sci. USA* **85:** 597–601.

Manzella JM, Blackshear PJ. (1992) Specific protein binding to a conserved region of the ornithine decarboxylase mRNA 5′-untranslated region. *J. Biol. Chem.* **267:** 7077–7082.

Marvit J, DiLella AG, Brayton K, Ledley FD, Robson KJH, Woo SLC. (1987) GT to AT transition at a splice donor site causes skipping of the preceding exon in phenylketonuria. *Nucl. Acid Res.* **15:** 5613–5628.

Mathews LS, Hammer RE, Brinster RL, Palmiter RD. (1988) Expression of insulin-like growth factor I in transgenic mice with elevated levels of growth hormone is correlated with growth. *Endocrinology* **123:** 433–437.

McCuaig KA, Clarke JC, White JH. (1994) Molecular cloning of the gene encoding the mouse parathyroid hormone/parathyroid hormone-related peptide receptor. *Proc. Natl Acad. Sci.USA* **91:** 5051–5055.

McCuaig KA, Lee HS, Clarke JC, Assar H, Horsford J, White JH. (1995) Parathyroid hormone/parathyroid hormone related peptide receptor gene transcripts are expressed from tissue-specific and ubiquitous promoters. *Nucl. Acid Res.* **23:** 1948–1955.

Menon RK, Stephan DA, Singh M, Morris Jr SM, Zou L. (1995) Cloning of the promoter-regulatory region of the murine growth hormone receptor gene. Identification of a developmentally regulated enhancer element. *J. Biol. Chem.* **270:** 8851–8859.

Merrick WC. (1992) Mechanisms and regulation of eukaryotic protein synthesis. *Microbiol. Rev.* **56:** 291–315.

Mottes M, Sangalli A, Valli M, Forlino A, Gomez-Lira M, Antoniazzi F, Constantinou-Deltas CD, Cetta G, Pignatti PF. (1994) A base substitution as IVS-19 3′-end splice junction causes exon 20 skipping in proα2(I) collagen mRNA and produces mild osteogenesis imperfecta. *Hum. Genet.* **93:** 681–687.

Muekler MM, Merrill MJ, Pitot HC. (1983) Translational and pretranslational control of ornithine aminotransferase synthesis in rat liver. *J. Biol. Chem.* **258:** 6109–6114.

Murphy D, Pardy K, Seah V, Carter D. (1992) Posttranscriptional regulation of rat growth hormone gene expression: increased message stability and nuclear polyadenylation accompany thyroid hormone depletion. *Mol. Cell. Biol.* **12:** 2624–2632.

Okazaki T, Zajac JD, Igarishi T, Ogata E, Kronenberg HM. (1991) Negative regulatory elements in the human parathyroid gene. *J. Biol. Chem.* **266:** 21903–21910.

Ostareck-Lederer A, Ostareck DH, Standart N, Thiele BJ. (1994) Translation of 15-lipoxygenase mRNA is inhibited by a protein that binds to a repeated sequence in the 3′untranslated region. *EMBO J.* **13:** 1476–1481.

Padgett RA, Grabowski PJ, Konarska MM, Seiler S, Sharp P. (1986) Splicing of messenger RNA precursors. *Ann. Rev. Biochem.* **55:** 1119–1150.

Paek I, Axel R. (1987) Glucocorticoids enhance stability of human growth hormone mRNA. *Mol. Cell. Biol.* **7:** 1496–1507.

Pause A, Belsham GJ, Gingras A-C, Donze O, Lin T-A, Lawrence Jr JC, Sonenberg N. (1994) Insulin-dependent stimulation of protein synthesis by phosphorylation of a regulator of 5′-cap function. *Nature* **371:** 762–767.

Reddy SS-K, Lauris V, Kahn CR. (1988) Insulin receptor function in fibroblasts from patients with leprechaunism. *J. Clin. Invest.* **82:** 1359–1365.

Ren H, Stiles GL. (1994) Posttranscriptional mRNA processing as a mechanism for regulation of human A1 adenosine receptor expression. *Proc. Natl Acad. Sci. USA* **91:** 4864–4866.

Rhoads RE. (1993) Regulation of eukaryotic protein synthesis by initiation factors. *J. Biol. Chem.* **268:** 3017–3020.

Richter JD. (1991) Translational control during early development. *Bioessays* **13:** 179–183.

Robberson BL, Cote GJ, Berget SM. (1990) Exon definition may facilitate splice site selection in RNAs with multiple exons. *Mol. Cell. Biol.* **10:** 84–94.

Roesser JR, Liittschwager K, Leff SE. (1993) Regulation of tissue-specific splicing of the calcitonin/calcitonin gene-related peptide gene by RNA-binding proteins. *J. Biol. Chem.* **11:** 8366–8375.

Ross J. (1995) mRNA stability in mammalian cells. *Microbiol. Rev.* **59:** 423–450.

Rossert JA, Chen SS, Eberspaecher H, Smith CN, de Crombrugghe B. (1996) Identification of a minimal sequence of the mouse pro-alpha 1(I) collagen promoter that confers high-level ostoblast expression in transgenic mice and that binds a protein selectively present in osteoblasts. *Proc. Natl Acad. Sci. USA* **93:** 1027–1031.

Rouault TA, Hentze MW, Caughman SW, Harford JB, Klausner RD. (1988) Binding of a cytosolic protein to the iron-responsive element of human ferritin messenger RNA. *Science* **241:** 1207–1210.

Rouault T, Tang CK, Kaptain S, Burgess WH, Haile DJ, Samaniego F, McBride OW, Harford JB, Klausner RD. (1990) Cloning of the cDNA encoding an RNA regulatory protein – the human iron-responsive element-binding protein. *Biochemistry* **87:** 7958–7962.

Ruskin B, Zamore PD, Green MR. (1988) A factor, U2AF, is required for U2 snRNP binding and complex assembly. *Cell* **52:** 207–219.

de Sauvage F, Kruys V, Marinx O, Huez G, Octave JN. (1992) Alternative polyadenylation of the amyloid protein precursor mRNA regulates translation. *EMBO J.* **11:** 3099–3103.

Sharp PA. (1994) Split genes and RNA splicing. *Cell* **77:** 805–815.

Shaw G, Kamen R. (1986) A conserved AU sequence from the 3′ untranslated region of GM-CSF mRNA mediates selective mRNA degradation. *Cell* **46:** 659–667.

Shemer J, Adamo ML, Roberts Jr CT, LeRoith D. (1992) Tissue-specific transcription start site usage in the leader exons of the rat insulin-like growth factor-I gene: evidence for differential regulation in the developing kidney. *Endocrinology* **131:** 2793–2799.

Sonenberg N. (1993) Remarks on the mechanism of ribosome binding to eukaryotic mRNAs. *Gene Express.* **3:** 317–323.

Strachan T, Read AP. (1996) *Human Molecular Genetics*. BIOS Scientific Publishers, Oxford.

Tang CK, Chin J, Harford JB, Klausner RD, Rouault TA. (1992) Iron regulates the activity of the iron-responsive element binding protein without changing its rate of synthesis or degradation. *J. Biol. Chem.* **267:** 24466–24470.

Tanguay RL, Gallie DR. (1996) Translational efficiency is regulated by the length of the 3′ untranslated region. *Mol. Cell. Biol.* **16:** 146–156.

Teerink H, Kasperaitis MAM, De Moor CH, Voorma HO, Thomas AAM. (1994) Translation initiation on the insulin-like growth factor II leader 1 is developmentally regulated. *Biochem. J.* **303:** 547–553.

Tewari M, Tewari DS, Taub R. (1991) Posttranscriptional mechanisms account for differences in steady state levels of insulin receptor messenger RNA in different cells. *Mol. Endocrinol.* 5: 653–660.

Tsutsumi M, Laws SC, Rodic V, Sealfon SC. (1995) Translational regulation of the gonadotropin-releasing receptor in αT3–1 cells. *Endocrinology* **1306:** 1128–1136.

Vadher S, Hawa NS, O'Riordan JLH, Farrow SM. (1996) Translational regulation of parathyroid hormone gene expression and RNA: protein interactions. *J. Bone Min. Res.* **11:** 746–753.

Welsh M, Scherberg N, Gilmore R, Steiner DF. (1986) Translational control of insulin biosynthesis. Evidence for regulation of elongation, initiation and signal-recognition-particle-mediated translational arrest by glucose. *Biochem. J.* **235:** 459–467.

West CA, Arnett TR, Farrow SM. (1996) Expression of insulin-like growth factor I (IGF-I) mRNA variants in rat bone. *Bone* **19:** 41–46.

Wickens M. (1990) In the beginning is the end: regulation of poly(A) addition and removal during early development. *Trends Biochem. Sci.* **15:** 320–324.

Wilson T, Treisman R. (1988) Removal of poly(A) and consequent degradation of c-fos mRNA facilitated by 3′ AU-rich sequences. *Nature* **336:** 396–399.

Wolffe A. (1994) Structural and functional properties of the evolutionary ancient Y-box family of nucleic acid binding proteins. *Bioessays* **16:** 245–251.

Yang H, Adamo ML, Koval AP, McGuinness MC, Ben-Hur H, Yang Y, LeRoith D, Roberts Jr CT. (1995) Alternative leader sequences in insulin-like growth factor I. *Mol. Endocrinol.* **9:** 1380–1395.

Zandberg H, Moen TC, Baas PD. (1995) Cooperation of 5′ and 3′ processing sites as well as intron and exon sequence in calcitonin exon recognition. *Nucl. Acid Res.* **23:** 248–255.

2

Molecular analysis of gene structure and function

Paul M. Brickell

2.1 Introduction

It has become commonplace to reflect upon the enormous contribution made by recombinant DNA technology to the understanding and practice of medicine over the last 20 years. The basic techniques of gene cloning, blot hybridization and the polymerase chain reaction (PCR) are now taught to teenagers as part of the UK national curriculum, and there are many discussions of these methods in standard undergraduate and postgraduate text books (Brown, 1995). Similarly, much has been written about positional cloning, or reverse genetic, approaches to the identification of the genes that are mutated in inherited diseases (Watson *et al.*, 1992). It is to be hoped, particularly by those of us who have spent the last decade teaching molecular genetics to undergraduate medical students, that these basic approaches are now firmly embedded in the minds of the medical community. The aim of this chapter is therefore not to retread this old ground, but rather to introduce some of the more advanced technical developments that are being applied to the study of gene structure, expression and function in human disease. The fruits of these experimental approaches will be demonstrated over and over again in the chapters that follow.

2.2 Detecting mutations in genes

It is important to identify the precise mutations affecting a defective gene in a large number of patients. In many inherited diseases almost every patient has a unique mutation and so methods that allow rapid screening for mutations are required. Mutations involving large insertions or deletions of DNA are readily identifiable using Southern blotting in conjunction with agarose gel electrophoresis or pulsed field gel electrophoresis (Darling and Brickell, 1994), but the resolution of these techniques is too low to permit detection of point

Molecular Endocrinology, edited by G. Rumsby and S.M. Farrow.

mutations or small insertions and deletions. Such mutations can of course be detected by nucleotide sequencing (Brown, 1994), but this is too expensive and time-consuming to be useful as a screening technique. The genesis of PCR spawned a number of methods for rapidly screening genomic DNA samples for subtle mutations. Two of the most popular of these are single-stranded conformational polymorphism (SSCP) and denaturing gradient gel electrophoresis (DGGE) analysis, and these will now be described.

2.2.1 Single-stranded conformational polymorphism (SSCP) analysis

SSCP is a relatively straightforward and robust technique for screening DNA samples for subtle mutations, and it has been used in a large number of studies. One example was its use to analyse mutations of the *RET* gene in multiple endocrine neoplasia type 2A (Oishi *et al.*, 1995). Details of protocols for performing SSCP have been discussed recently by Bailey (1995) and the basis of this technique is shown schematically in *Figure 2.1*.

Techniques such as Southern blotting employ conventional non-denaturing gel electrophoresis to separate mixtures of double-stranded DNA molecules. The distance migrated by a double-stranded DNA molecule in these circumstances depends only on its length; the shorter the molecule, the further it migrates. In contrast, a single-stranded DNA molecule can form intramolecular hydrogen bonds under non-denaturing conditions, causing it to fold into a stable secondary structure. Upon electrophoresis, such molecules will migrate according to their overall size and shape, rather than their absolute length. The secondary structure adopted by a single-stranded DNA molecule is very sensitive to changes in nucleotide sequence, and a single nucleotide change can result in a different secondary structure and thus alter the rate of migration during electrophoresis (*Figure 2.1*). Wild-type and mutant DNA molecules can therefore be distinguished, although the success of SSCP in detecting mutations is a matter of dispute, with hit rates of between 35% and 100% being reported by different laboratories (Bailey, 1995).

In practice, when faced with a gene in which to search for mutations, it is important first to decide which regions to target. SSCP is best performed on fragments of just a few hundred nucleotides, and so it is usually impractical to screen the entire gene, which may be tens of thousands of nucleotides long. It is usual to begin by examining the coding exons, followed by other important regions of the gene such as the 5′ and 3′ untranslated regions, the promoter region and other sequences involved in regulating transcription. In some cases, previous work will have given clues as to which regions of the gene are most likely to be mutated. Having decided which region to examine, oligonucleotide primers are designed to allow its amplification by PCR. These are either unlabelled, or labelled at one end with a radioactive or fluorescent tag. PCR is then performed using DNA isolated directly from patients and healthy controls. The double-stranded DNA products of PCR are denatured by heating to 95°C, and are then rapidly cooled on ice. This allows secondary structure to form as a result of intramolecular hydrogen bonding, but there is not enough time for intermolecular hydrogen bonding to result in the formation of double-stranded DNA. The single-stranded DNA molecules are subjected to

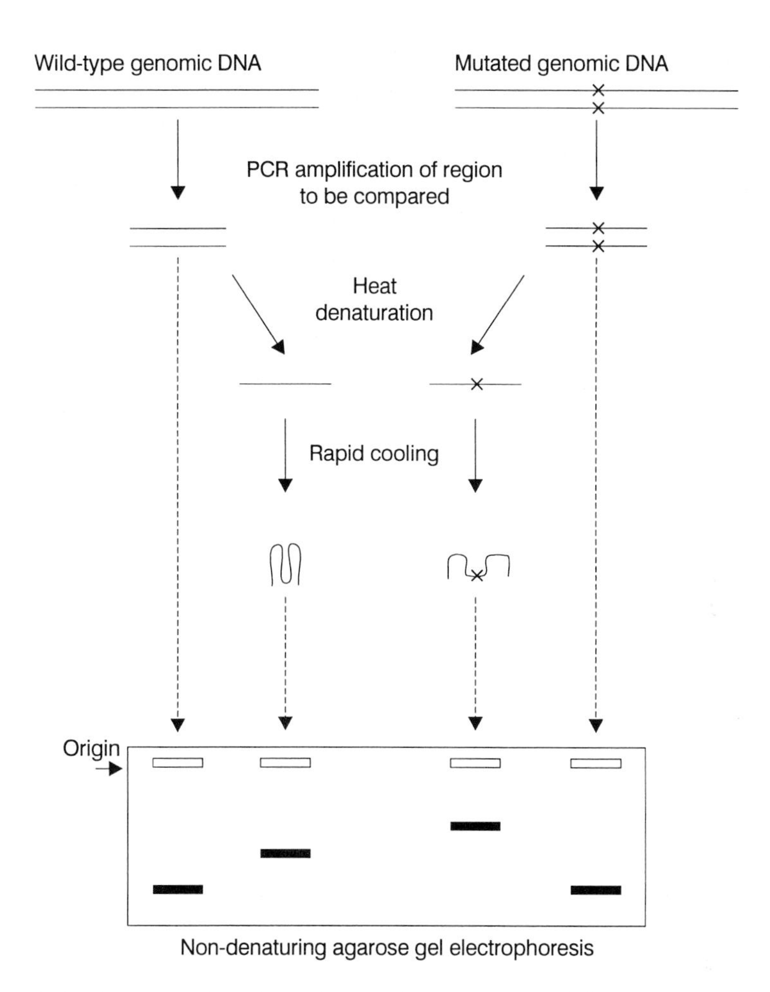

Figure 2.1. Single-stranded conformational polymorphism (SSCP) analysis.

polyacrylamide gel electrophoresis, using a gel made up to a recipe optimized for SSCP such as the 'mutation detection enhancement' (MDE) gel matrix (Bailey, 1995). Electrophoresed DNA is then detected by silver-staining (if unlabelled primers were used), autoradiography (for radiolabelled primers) or by performing electrophoresis in an automated DNA sequencer (for fluorescent-tagged primers).

Having used SSCP to scan through a gene to find a small region that contains a mutation, nucleotide sequencing of that region can be used to define the precise nature of the mutation.

2.2.2 Denaturing gradient gel electrophoresis (DGGE) analysis

Subtle mutations can also be detected by DGGE, in which double-stranded DNA is subjected to electrophoresis in a gel containing a gradient of increasing concentration of denaturing agents, such as formamide and urea. In a related technique, named temperature gradient gel electrophoresis (TGGE), the DNA is electrophoresed through a gradient of increasing temperature. A recent example of the use of DGGE to analyse mutations in the *RET* gene has been reported by Blank *et al.* (1996), and protocols for DGGE and TGGE have been discussed by Rolfs *et al.* (1993) and Henco *et al.* (1994), respectively. DGGE and TGGE are more difficult to perform than SSCP, but it is claimed that they are more successful at detecting mutations (Henco *et al.*, 1994).

In DGGE and TGGE, which are illustrated in *Figure 2.2*, the region of interest is first amplified by PCR from the genomic DNA of both a healthy individual with the wild-type DNA sequence and a patient carrying a mutation. As with SSCP, the PCR primers can be unlabelled, radiolabelled or labelled with a fluorescent tag. The double-stranded DNA PCR products are denatured by heating and then mixed together. The mixture is left to cool slowly to room temperature, allowing complementary DNA strands to form double-stranded DNA molecules. As shown in *Figure 2.2*, four types of DNA duplex will form. These will differ very slightly from each other in their stability when they are heated or treated with

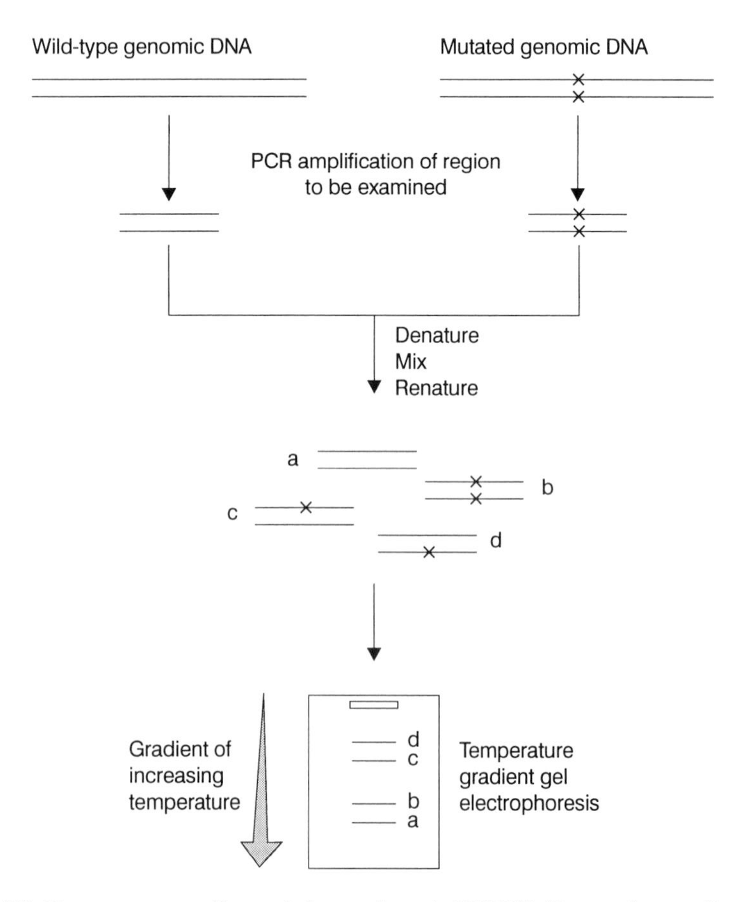

Figure 2.2. Temperature gradient gel electrophoresis (TGGE). Denaturing gradient gel electrophoresis (DGGE) is performed in the same way except that the gel contains a gradient of increasing concentration of formamide and urea, rather than being maintained with a temperature gradient.

denaturing chemicals. The DNA duplexes are loaded on to a polyacrylamide gel containing a gradient of formamide and urea, or down which a temperature gradient is maintained. As each DNA duplex migrates through the gel it will reach a position where the temperature (or concentration of urea/formamide) is high enough to cause it to begin to denature. The consequent increase in the bulk of the molecule causes its rate of migration to slow. Since the four species of duplex have different stabilities, this transition will occur at a different position on the gel for each duplex and their mobilities can therefore be distinguished. As with SSCP, once DGGE/TGGE has indicated that a small region of a gene contains a mutation, the nature of the mutation can be defined by nucleotide sequencing.

2.3 Analysing the regulation of gene expression

The great majority of point mutations found to be responsible for human disease to date result in aberrant splicing of primary gene transcripts or alteration of the protein coding region of the mRNA (Weatherall, 1991). However, there is growing interest in the idea that mutations in the transcriptional control regions of genes can cause disease, and that polymorphisms in these regions are important in normal variation between individuals (Semenza, 1994). For example, there is a considerable number of reports of cases of thalassaemia resulting from mutations in the transcriptional promoters or enhancers of globin genes (Balta *et al.*, 1994). The techniques of SSCP and DGGE/TGGE, followed by nucleotide sequencing, can be used to identify mutations in transcriptional control regions. The aim of this section is to describe techniques for testing whether such mutations do in fact have any effect on gene transcription.

2.3.1 Consensus sequences for transcription factor binding sites

Regulatory elements in the 5′-flanking region of genes (see section 1.3) are potential sites of mutation. An indication of the effect of the mutation can be gleaned by comparison with a database of consensus sequences for transcription factor binding sites, such as the TRANSFAC Database (Wingender *et al.*, 1996; accessible on the Internet at http://transfac.gbf-braunschweig.de/). This will reveal whether the mutation affects a sequence with the potential to bind a transcription factor. However, even if it does, the functional consequences of the mutation must be assayed directly.

2.3.2 Reporter gene assay

Reporter gene assays allow DNA fragments to be tested for their ability to regulate transcription. In the context of this chapter, they can be used to compare the activity of a DNA fragment containing a mutation with that of the wild-type DNA fragment (Balta *et al.*, 1994; Sun *et al.*, 1995).

Reporter gene assays make use of plasmid vectors such as that shown in *Figure 2.3*. To test a DNA fragment for promoter activity, it is cloned upstream of a reporter gene (Figure 2.3a). To test a DNA fragment for enhancer or silencer activity, it is cloned either upstream or downstream of a reporter gene that is

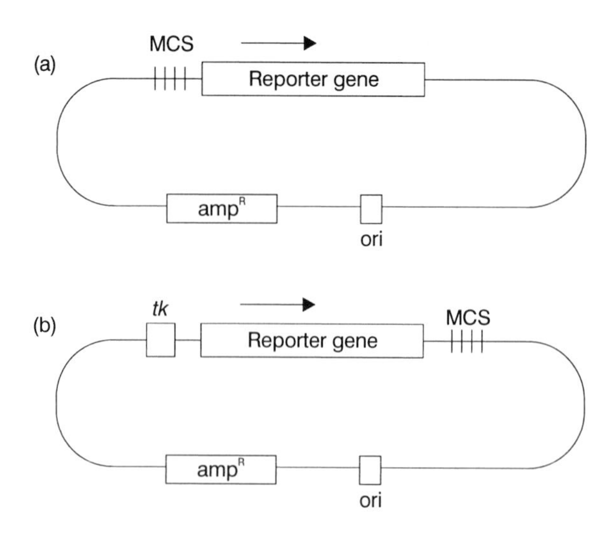

Figure 2.3. Typical structure of reporter plasmids used to test DNA fragments for promoter (a) or enhancer (b) activity. MCS, multiple cloning site; *tk*, herpes simplex virus thymidine kinase gene promoter; ori and amp^R, origin of replication and ampicillin resistance gene for propagating plasmid in *E. coli.* Direction of transcription shown by →.

already under the control of a promoter directing low level transcription, such as the herpes simplex virus thymidine kinase (*tk*) gene promoter (Figure 2.3b). The reporter plasmid, containing the fragment of interest, is placed inside cultured cells by a procedure called transfection, and the level of reporter gene expression is measured, usually after 2 or 3 days. Although reporter gene activity is usually measured in transient transfection assays of this type, it can also be assayed using transgenic animals. The elements of a reporter gene assay will now be described in more detail.

Reporter genes. A reporter gene should encode a protein that is not already expressed in the cell type to be transfected and for which there is a simple, quick and quantitative assay. A considerable number of reporter genes have been developed, but the three most popular are the *Escherichia coli* chloramphenicol acetyltransferase (*cat*) gene (Rosenthal, 1987), the luciferase (*luc*) gene from the firefly *Photinus pyralis* (Brasier and Ron, 1992) and the *E. coli* β-galactosidase (*lacZ*) gene (Vernet *et al.*, 1993). CAT activity can be measured by incubating lysates of transfected cells with radiolabelled chloramphenicol, separating the chloramphenicol from its acetylation products by thin layer chromatography or solvent extraction, and counting the amount of radioactivity in the substrate and products. Alternatively, levels of CAT protein can be measured directly by enzyme-linked immunoassay. To measure luciferase activity, lysates of transfected cells are incubated with the substrate luciferin, and the luminescent product is assayed in a luminometer. This system has been considerably refined in recent years, and the ease of the luciferase assay, coupled with its high signal-to-noise ratio, make it the current method of choice for reporter gene assays in cultured cells. β-Galactosidase activity can be measured by incubating lysates of transfected cells with a substrate such as X-Gal (5-bromo-4-chloro-3-indolyl-β-D-galactopyranoside) and assaying the blue-coloured product in a spectrophotometer. However, the *lacZ* reporter gene is most commonly used in transgenic animals, where its expression can be observed qualitatively in intact tissues by incubating them in X-Gal and examining the appearance of the blue reaction product under the microscope (Yu *et al.*, 1995). A fourth reporter gene system, which is likely to gain in popularity in the next few years, is based on green fluorescent protein (GFP) from the jellyfish *Aequorea victoria*. The hope is that it will be possible to measure GFP quantitatively in living cells (Prasher, 1995).

Host cells for reporter gene assays. As important as the choice of reporter gene is the choice of cell type in which to perform the assay. The properties of the cells to be transfected should be as close as possible to those of the cells in which the promoter is normally active, and some studies employ primary cultured cells. However, for a variety of reasons it is also very common to use established cell lines. For example, it may not be possible to establish primary cultures of the appropriate cell type, it may be difficult to obtain sufficient numbers of primary cells or it may be difficult to transfect primary cells efficiently.

Many established cell lines support correct promoter activity perfectly adequately. However, there are some situations that have proved very difficult to replicate in the tissue culture dish, either in primary or established cell cultures. For example, it has been difficult to reproduce correct, hormonally responsive expression of milk protein genes in cultured mammary gland epithelial cells, and studies of the promoters of these genes have relied heavily upon the use of reporter transgenes in transgenic mice (Barash *et al.*, 1996).

Transfection. There are a number of methods for transfecting cultured cells with DNA, but those most commonly used for reporter gene assays are calcium phosphate precipitation, electroporation and lipofection.

In the former, a solution of DNA in calcium chloride is gently mixed with a solution of sodium phosphate, creating a fine precipitate of calcium phosphate that traps the DNA molecules. The precipitate is then layered on to cells in a culture dish and left overnight, during which time the cells take up the DNA-containing particles by endocytosis. This method works best for cells that are attached to the culture dish. It is very widely used, but is not very effective for some cell types and is toxic for others. There are many different protocols, optimized for different cell types, and Hitt *et al.* (1994) provide one recent description. A related technique involves the formation of complexes of DNA and diethylaminoethyl (DEAE)-dextran, which cells also take up by endocytosis.

Electroporation involves the application of a very brief, high-voltage electric pulse to cells in suspension in medium containing DNA. This causes reversible electrical breakdown (REB) of cell membranes, followed by transient pore formation. DNA molecules pass through the pores by diffusion, electrical drift and electro-osmosis. The technique works for a great many cell types, but is nevertheless unsuitable for some. Once again, there are many variations on the basic protocol, and these have been discussed recently by Herr *et al.* (1994).

Finally, there are a growing range of commercially available products for transfecting cells by lipofection. A number of these involve adding DNA to a mixture of a polycationic lipid such as DOSPA (2,3-dioleyloxy-*N*-[2(spermine-carboxamido)ethyl]-*N*,*N*-dimethyl-1-propanaminiumtrifluoroacetate) and a neutral lipid such as DOPE (dioleoyl phosphatidylethanolamine). The resultant DNA–lipid complexes fuse with cell membranes and the DNA is transferred into the cells. As with the other methods for transfection, lipofection works extremely well with some cell types and very poorly with others. A recent description of its use to transfect primary epidermal keratinocytes has been given by Dlugosz *et al.* (1995).

Judicious use of one or other of these methods means that most cell types can now be transfected with DNA. Transfection efficiencies vary greatly between cell types, but it is reasonable to expect 10–20% of cells in a culture dish to take up DNA in a typical experiment.

Once the reporter plasmid DNA has entered the cell it is transported to the nucleus and, if the cloned DNA fragment has promoter activity, the reporter gene is transcribed. Plasmids are typically maintained in the nuclei of eukaryotic cells for at least 72 h, after which time they are lost. In a very small proportion of transfected cells, the plasmid DNA becomes stably integrated into chromosomal DNA in linear form, allowing the isolation of stably transfected cells. However, reporter gene assays are usually performed on transiently transfected cells harvested 48 to 72 h after transfection, allowing time for expression of the reporter gene to occur without significant loss of reporter plasmids from the cells.

2.3.3 Gel mobility shift assay

The gel mobility shift, or gel retardation, assay identifies protein-binding sites on DNA molecules and so can be used to determine whether a mutation within a putative or proven transcription factor-binding site affects the binding of the factor. For example, Sun *et al.* (1995) used it to show that a mutation in the promoter of the low-density lipoprotein gene from a patient with familial hypercholesterolaemia resulted in reduced binding affinity for the Sp1 transcription factor. Reporter gene assays performed in parallel showed that the mutation resulted in reduced levels of transcription from the promoter. The principle of the assay is illustrated in *Figure 2.4*. A small, linear, double-stranded DNA fragment containing the protein-binding site is labelled at its ends with a radioactive tag, and is incubated with an extract of nuclear proteins from the appropriate cell or tissue, under conditions that allow proteins to bind specifically to their recognition sites in the DNA. The DNA–protein complexes are then subjected to non-denaturing polyacrylamide gel electrophoresis, alongside a sample of the labelled DNA fragment that was not incubated with nuclear protein extract. Protein–DNA complexes will be larger than the DNA fragment alone and so will migrate more slowly, as shown in *Figure 2.4*. Abolition of such a mobility shift by a mutation in the DNA fragment would indicate that the protein could no longer bind to the mutated sequence, whilst a decrease in the intensity of the shifted band might indicate a decreased affinity of the protein for its mutated binding site.

This technique can also be used as a basis for a more versatile assay based on competitive binding. For example, a short double-stranded oligonucleotide containing

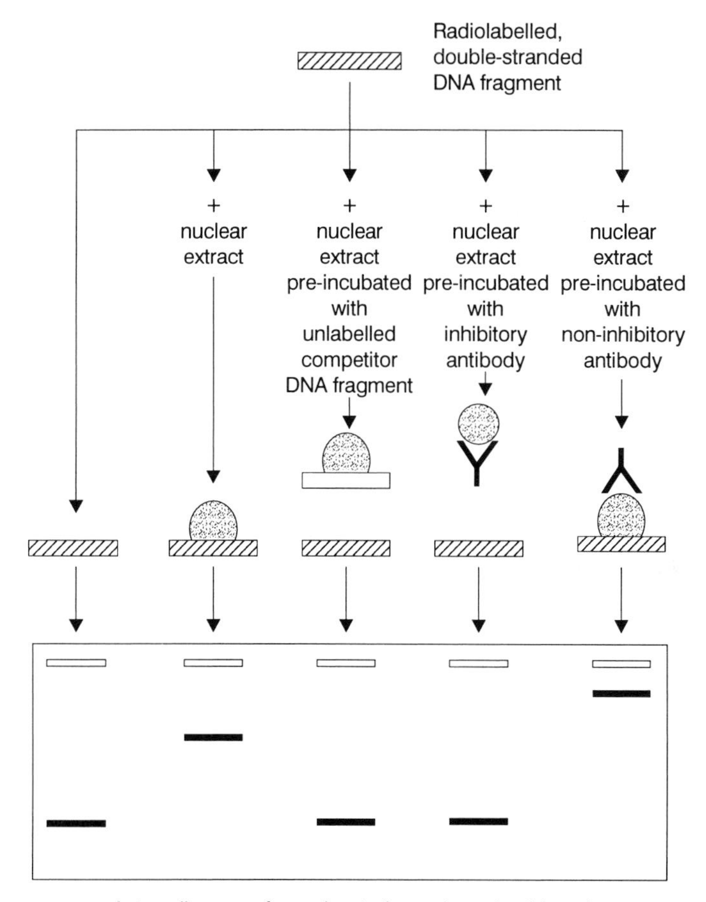

Figure 2.4. Gel mobility shift assay.

the wild-type protein-binding site will compete effectively with the labelled wild-type DNA fragment for binding to the protein, and will abolish the gel mobility shift if it is present at a sufficiently high concentration (*Figure 2.4*). An oligonucleotide containing the mutated protein-binding site can then be tested for its ability to compete with the labelled wild-type fragment. Comparison of the extent of inhibition of protein binding by the wild-type and mutant oligonucleotides will indicate the extent to which the mutation affects the affinity of protein binding.

The identity of the protein factor binding to the shifted DNA fragment can be confirmed if an antibody against it is available. Incubation of the nuclear protein extract with the antibody may abolish binding of the protein factor to its DNA target. In this case, no mobility shift will be seen upon electrophoresis. Alternatively, the antibody, protein factor and DNA target may form a single large complex which will migrate even more slowly than the protein–DNA complex. Such an effect is termed a 'supershift'.

Of course, this technique can also be used to examine the affinity of a mutated transcription factor for its binding site. For example, Lobaccaro *et al.* (1996) used it to show that, in affected individuals from two families with inherited androgen insensitivity, the androgen receptors were unable to bind to their target sequences. Again, they complemented these studies with reporter gene assays showing that the mutated receptors lacked the ability to transactivate transcription.

2.3.4 Deoxyribonuclease I (DNase I) protection assay

DNase I protection, often called DNase I footprinting, provides another means of detecting protein–DNA interactions. The principle of the technique is that the binding of a protein factor to DNA protects the DNA against digestion by DNase I. As shown in *Figure 2.5*, the DNA fragment to be tested is labelled at one end with a radioactive tag and incubated with nuclear protein extract. The mixture is then treated with a tiny quantity of DNase I, such that each DNA molecule is cut, on average, at just one position. If a protein has bound to the DNA molecule then the region occupied by the protein will be protected from cutting by the DNase I. Finally, the mixture is subjected to denaturing polyacrylamide gel electrophoresis followed by autoradiography. If no protein has bound to the DNA molecule, a ladder of bands, each differing in size from the next by one nucleotide, will appear on the autoradiograph. If, on the other hand, a protein has bound, then the DNA fragments that would have been generated by cutting within the protein-binding site will be absent, and a clear region, or 'footprint' will be found on the autoradiograph. This technique is more difficult to perform than the gel mobility shift assay, but gives a much more precise indication of the location of the protein-binding site in a DNA fragment. It is frequently used to complement the gel mobility shift assay.

As with the gel mobility shift assay, this technique can be used to compare the ability of wild-type and mutant DNA fragments to bind protein factors. For example, Balta *et al.* (1994) used it to compare the binding of a transcription factor to the wild-type and mutant enhancer of the human $^{A}\gamma$-globin gene, integrating this with a study of the activity of the two DNA sequences in a reporter gene assay.

2.4 Molecular 'fingerprinting' of DNA and RNA populations

Developments in PCR technology have made it possible to obtain 'fingerprints' of specific DNA or RNA populations by a family of techniques known as 'arbitrarily primed PCR' or AP-PCR (McClelland *et al.*, 1995). For example, it is possible to compare the electrophoretic profile of PCR-generated fragments from wild-type and mutant genomes in order to identify mutated DNA sequences (Ionov *et al.*, 1993), to analyse changes in chromosome number in cancer (Peinado *et al.*, 1992), and to compare the genomes of organisms for purposes of phylogenetic classification (Welsh and McClelland, 1993). When applied to the comparison of DNA genomes in this way, AP-PCR is often called random amplified polymorphic DNA (RAPD) fingerprinting. A similar approach can be used to compare populations of RNA molecules in two different cell or tissue types, or in a single cell type

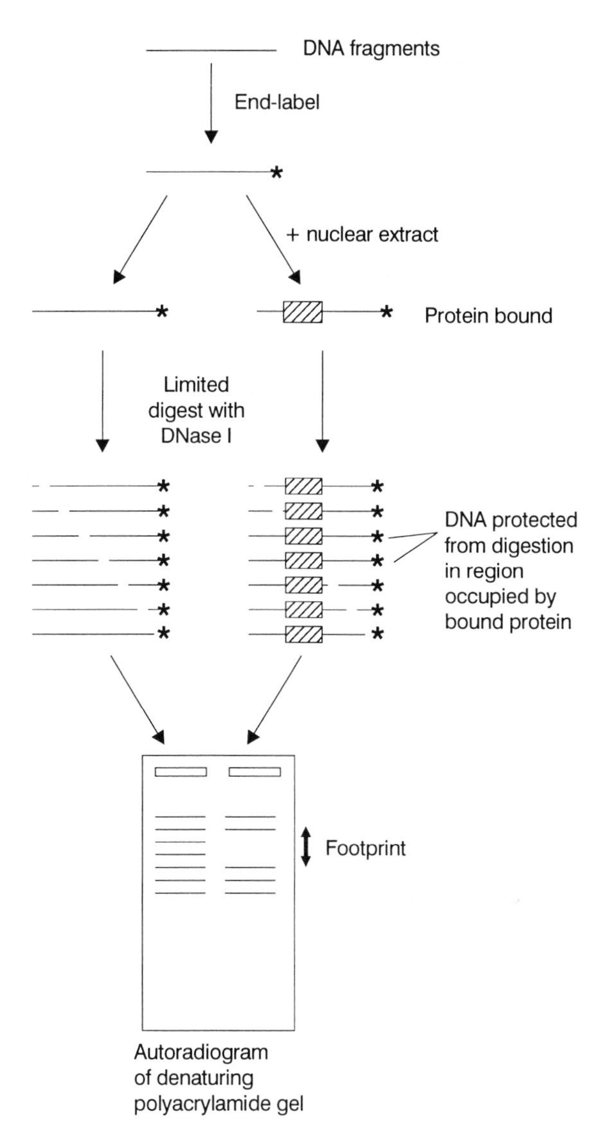

Figure 2.5. Deoxyribonuclease I (DNase I) protection assay.

before and after application of an external stimulus. When used in this way, to analyse differential gene expression, the technique is usually called 'differential display' (McClelland *et al.*, 1995). To choose just two examples from many, differential display has led to the identification of genes that are differentially expressed in different brain tumours (Uchiyama *et al.*, 1995), and in aortic smooth muscle cells before and after treatment with glucose (Nishio *et al.*, 1994).

To illustrate the approach, a typical protocol for differential display is shown in *Figure 2.6*. As might be expected, there are many variations on this theme (McClelland *et al.*, 1995). In the protocol shown, polyadenylated RNA is first isolated from the cells or tissues to be compared. Clearly, there is no theoretical limit

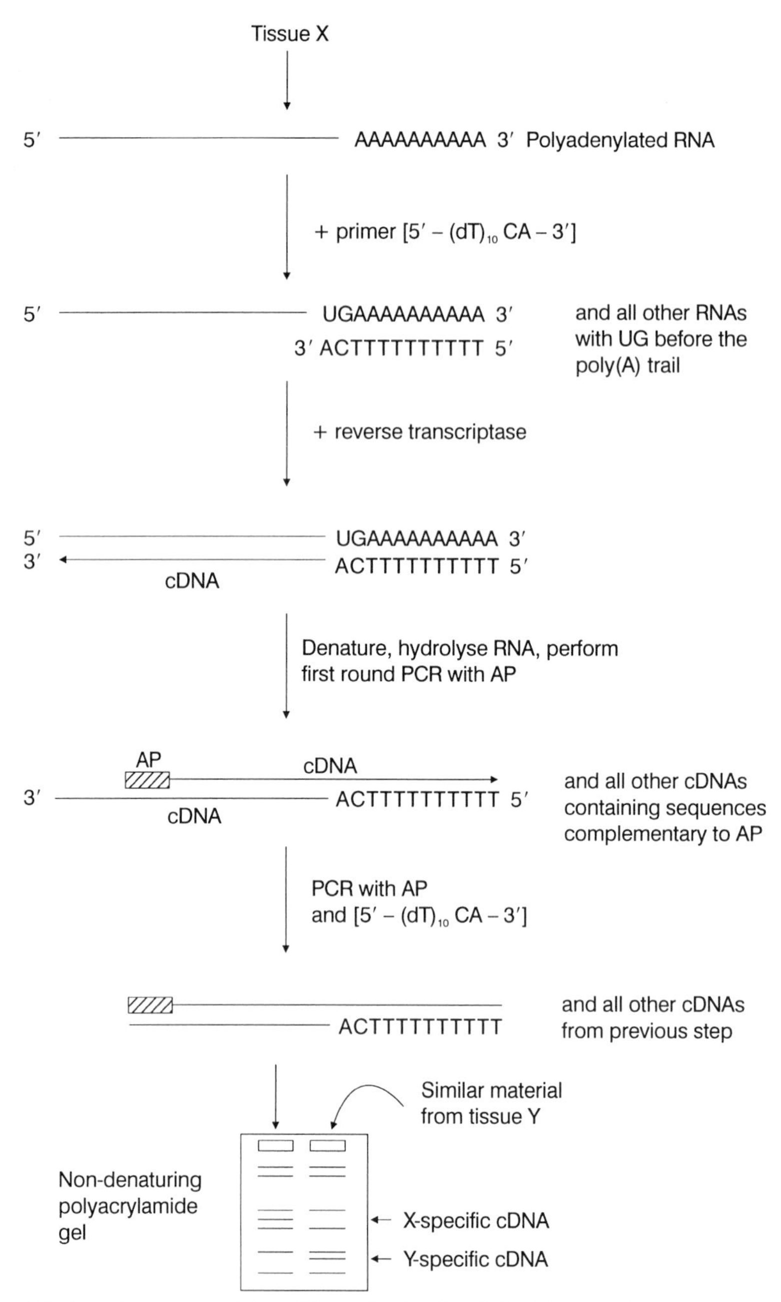

Figure 2.6. A typical protocol for differential display. AP, arbitrary primer.

to the number of samples that can be compared at the same time. Synthesis of a single-stranded cDNA copy of the RNA template by reverse transcriptase is then performed using an oligo (dT) primer with an anchor of two bases at its 5′ end. For example, the primer used in *Figure 2.6* has the sequence 5′–$(dT)_{10}$CA–3′. This will only prime cDNA synthesis from RNA molecules with the sequence UG immediately in front of the poly(A) tail, which represent one twelfth of all RNA molecules. After cDNA synthesis, the DNA–RNA hybrid is entered into a standard PCR protocol, using the 5′–$(dT)_{10}$CA–3′ primer and a so-called 'arbitrary primer'. This is typically an oligonucleotide of 10 nucleotides with an arbitrarily chosen sequence. For example, a recent study used an oligonucleotide with the sequence 5′–CTGATCCATG–3′. This sequence will appear in a subset of the RNAs that were copied into cDNA in the reverse transcriptase step, and these species will be amplified in the PCR step. The PCR products are analysed by denaturing polyacrylamide gel electrophoresis and, if the 'arbitrary primer' is radiolabelled, are detected by autoradiography. PCR fragments that appear in one cell type and not another are of potential interest and can be cut out of the gel and cloned into a plasmid vector for further study.

A typical differential display experiment might result in the appearance of around 100 bands on the autoradiograph, and therefore represents a comparison between a relatively small subset of the RNA molecules in the different cell types used; namely, those RNA molecules that can be copied by the two primers. Use of different primer pairs allows comparison of different subsets of RNA molecules, but it would be quite an undertaking to perform enough different reactions to survey the whole of the RNA populations. Differential display is therefore not normally used to provide an exhaustive comparison between the RNA populations in different cell types, but rather to identify a few of the many possible interesting differences between them. There are methods for performing exhaustive comparisons between RNA populations, but these are based on the subtractive hybridization protocols discussed below.

RAPD fingerprinting is performed in an essentially similar fashion to differential display, without the need for the initial conversion of RNA to cDNA by reverse transcriptase (Welsh *et al.*, 1995).

2.5 Isolating differentially expressed genes by subtractive hybridization

A simple method for identifying cDNA clones corresponding to mRNAs present in tissue X but not in tissue Y would be to construct a cDNA library from X mRNA and to screen duplicate filters with radiolabelled cDNA probes synthesized using X mRNA or Y mRNA as templates (*Figure 2.7a*). Clones hybridizing with the X probe but not the Y probe would correspond to X-specific mRNAs. The problem with such an approach is that hybridization with X-specific clones, which are likely to be rare, is very difficult to detect amongst the vast number of clones that hybridize with both probes. A better approach is to construct and screen a cDNA library that has already been highly enriched for clones corresponding to X-specific mRNAs. This can be achieved by using subtractive hybridization techniques. These provide a means of removing all the sequences that are held in common by two RNA populations and of cloning only those that are specific to one of the RNA populations. For example, such techniques have been used to clone RNAs that appear in cultured cells after treatment with fibroblast growth factor (Li *et al.*, 1994) or that appear in amphibian tadpole tail tissue following treatment with thyroid hormone (Wang and

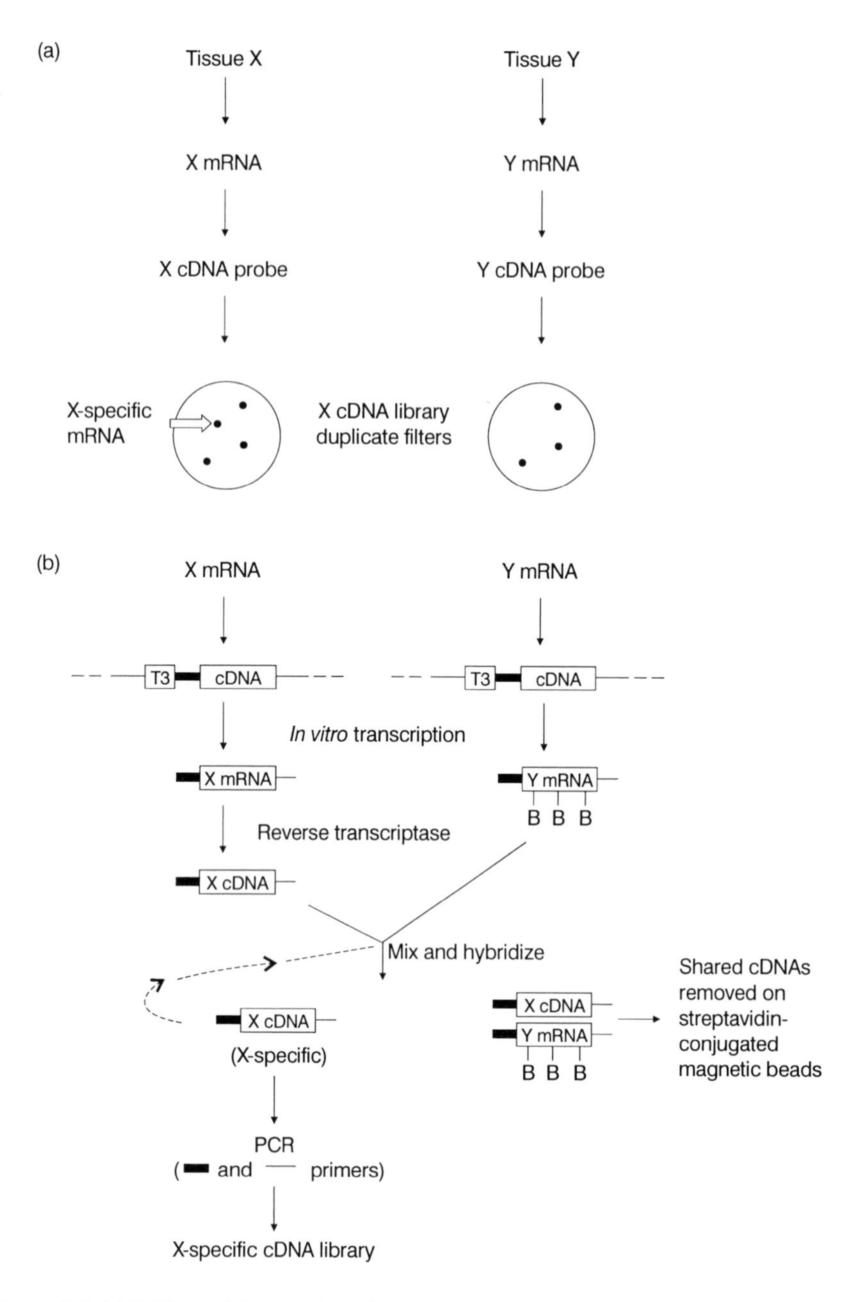

Figure 2.7. (a) Differential screening of a cDNA library. (b) Construction of a sequence-enriched cDNA library using subtractive hybridization. B, biotin; T3, T3 RNA polymerase promoter.

Brown, 1991). Subtractive hybridization has had a long and chequered history, and it is arguable that the considerable energy expended on it by many groups has not been justified by its successes. Recent developments in PCR technology and in the ease of handling tiny amounts of nucleic acid have certainly improved matters, but it is still not an enterprise to be undertaken lightly. Nevertheless, subtractive hybridization offers the theoretical possibility of being able to isolate all of the RNA molecules that are expressed differentially in two RNA populations, including both quantitative and qualitative differences (Wang and Brown, 1991). This represents a considerable theoretical advantage over the differential display techniques discussed above. There are, naturally, myriad versions of subtractive hybridization protocols. The one to be discussed here is similar to that used by Klar *et al.* (1992) to isolate cDNA clones of RNAs expressed specifically in either the ventral or dorsal parts of the developing spinal cord.

As shown in *Figure 2.7b*, the first step is the isolation of polyadenylated RNA from the two tissues to be compared (X and Y) and the construction of two cDNA libraries in a vector such as bacteriophage λZAP (Short and Sorge, 1992). The libraries are constructed in such a way that all of the cDNA inserts are in the sense orientation with respect to the T3 RNA polymerase promoter located in the vector. This means that DNA can be prepared from the two libraries and transcribed with T3 RNA polymerase to yield large quantities of mRNA corresponding to each tissue. In the example shown, the Y cDNA library is transcribed in the presence of biotinylated UTP, yielding biotin-labelled mRNA. The X mRNA is then copied into single-stranded cDNA using reverse transcriptase, and the RNA removed by hydrolysis. The cDNA is synthesized in such a way that short stretches of vector sequence are attached to each end. The X cDNA is next hybridized with a 30-fold molar excess of the biotin-labelled Y mRNA. X cDNA molecules that correspond to mRNAs present in both X and Y tissues will form hybrids with the biotin-labelled Y mRNA, whilst cDNAs that are unique to X tissue will remain single stranded. The mixture is incubated with magnetic beads coupled to streptavidin, which binds with high affinity to the biotin-labelled hybrids. These can then be removed from the mixture using a magnet. The solution is now enriched for cDNAs corresponding to X-specific mRNAs. To be sure that as many as possible of the sequences held in common are removed, it is usual to perform further rounds of hybridization as shown in *Figure 2.7b*. Finally, the highly enriched X cDNA is amplified by PCR, using primers corresponding to the short stretches of vector sequence at each end, and cloned. Despite being highly enriched for X-specific clones, the library is likely still to contain clones of RNAs found in both X and Y. The final step is therefore to plate the library out, transfer the clones to duplicate nylon membranes and to hybridize them with radiolabelled cDNA prepared from polyadenylated X or Y RNA. Clones that hybridize with the X probe but not with the Y probe are X specific and can be picked for further study.

2.6 Analysing gene function in transgenic mice

There are essentially two methods available for generating transgenic mice: direct microinjection of fertilized eggs and manipulation of embryonic stem (ES) cells

(Watson *et al.*, 1992). The former is much easier to perform than the latter and is practised by far more laboratories. Microinjection can be used to express a gene at higher levels than normal, or in cell types that do not normally express it at all. It can also be used to express mutant genes either to investigate the effect of a mutation *in vivo* or to abolish gene function (knockout or null). If the protein product of such a mutant gene is able to inhibit activity of the endogenous wild-type protein in a dominant fashion (dominant negative), then microinjection can be used to create mutant phenotypes. For example, Moller *et al.* (1996) made transgenic mice that expressed a dominant negative insulin receptor in their striated muscle and showed that they developed increased adiposity, impaired glucose tolerance and dyslipidaemia. Whilst the microinjection approach is therefore extremely valuable, the great advantage of the ES cell route to the creation of transgenic mice is that it can be used to introduce specific mutations into endogenous genes. This has proved to be a tremendously powerful method for exploring gene function (Capecchi, 1994; Watson *et al.*, 1992).

2.6.1 Producing transgenic mice by microinjection

To generate transgenic mice by microinjection (Gordon, 1994), female mice are first induced to superovulate and then mated. Fertilized eggs are collected by washing out the oviduct, and a tiny quantity of DNA is injected directly into the male pronucleus of the fertilized egg using microinjection apparatus. In a proportion of cases, the microinjected DNA will subsequently become inserted into the host cell chromosome. It is usual for many copies of the DNA to be inserted in a tandem array at a single chromosomal site although, since integration occurs at random, this site will differ in every case. The microinjected eggs are transferred to the oviduct of a foster mother and allowed to develop. The percentage of eggs that develop to term varies, but is usually between 10% and 30%. Most of the newborn mice will not contain any foreign DNA, but some, perhaps as many as 40%, will have developed from eggs in which the microinjected DNA became stably integrated, and will carry this DNA in every cell of their body. These are termed transgenic mice. To determine whether a newborn mouse is transgenic, the tip of the tail is cut off, DNA is extracted and Southern hybridization or PCR is used to look for the foreign DNA.

This approach has subsequently been developed to allow the production of transgenic animals of many different mammalian species, including rats, rabbits, sheep, pigs and cows (Hammer *et al.*, 1985; Ward and Nancarrow, 1995).

2.6.2 Producing transgenic mice using ES cells

ES cell lines are derived by culturing the inner cell mass of mouse blastocysts (Abbondanzo *et al.*, 1993). If grown in tissue culture medium alone, ES cells will differentiate and give rise to cultures containing a range of differentiated cell types, such as muscle cells, endothelial cells, cartilage cells, haematopoietic cells and neurons. However, if they are grown on a layer of fibroblasts, or in medium containing a protein named leukaemia inhibitory factor (LIF), they will remain undifferentiated (Abbondanzo *et al.*, 1993). Remarkably, such undifferentiated ES cells can be injected into the inner cell mass of a normal mouse blastocyst and will contribute to all tissues of the mouse that develops from it (Stewart, 1993). Such mice contain a mixture of cells, some derived from the cells of the host blastocyst and others derived from the ES cells, and are termed chimaeric mice. For example, if ES cells derived from a black mouse are put into blastocysts from white mice, the chimaeric offspring will have black and white coats. To make transgenic mice by this route, the cultured ES cells are transfected with foreign DNA by one of the methods described previously (Section 2.3.2). In addition to the gene of interest, the foreign DNA includes a selectable marker such as the *neo* gene, which confers resistance to the antibiotic G418. When G418 is added to the transfected cells, only those in which the foreign DNA has become stably integrated will survive. These cells are injected into host blastocysts, which give rise to chimaeric mice comprising a mixture of wild-type cells and ES-derived cells containing the foreign DNA. The eggs and sperm of these chimaeric mice are a similar mixture, such that when they are mated with wild-type mice, their offspring are either wild-type or heterozygotes in which every cell contains one set of wild-type chromosomes and one set of chromosomes carrying the foreign DNA. Interbreeding between these heterozygotes will give rise to homozygous transgenic mice.

2.6.3 Producing transgenic mice carrying specific mutations by homologous recombination in ES cells

There are a few examples in the literature of microinjected DNA integrating into an endogenous gene and disrupting its function. For example, during studies of the activity of the c-*myc* proto-oncogene, a line of transgenic mice was derived which had deformed limbs. It transpired that the c-*myc* transgene had integrated into an endogenous mouse gene named *ld*, for 'limb deformity', which is required for correct development of a number of structures, including the limbs (Woychik *et al.*, 1985). Such events are very rare, however, and there is no way of controlling which gene is disrupted. In contrast, specific mutations targeted to a gene of choice can be generated by the process of homologous recombination in ES cells (Ramírez-Solis *et al.*, 1993; Watson *et al.*, 1992).

When a dish of ES cells is transfected with foreign DNA containing a mouse gene, most of the cells that take up DNA subsequently lose it over a period of a few days. In a very small proportion of the cells, however, the foreign DNA becomes stably integrated into the host chromosomes. In the great majority of such cells, integration occurs at random sites in the genome by a process called heterologous recombination, where the foreign DNA recombines with unrelated host DNA sequences. However, in approximately one in every 1000 stably transfected cells, integration occurs by homologous recombination, in which the foreign DNA recombines with the identical sequence in the host genome. This makes it possible to derive an ES cell line in which a specific mutation has been introduced into a specific gene, and hence to derive transgenic mice that are homozygous for the mutation. There are a number of different procedures for achieving this, and one of the simplest is illustrated in *Figure 2.8*. This method, which is called 'positive–negative selection', employs a 'targeting vector' in which the *neo* gene is

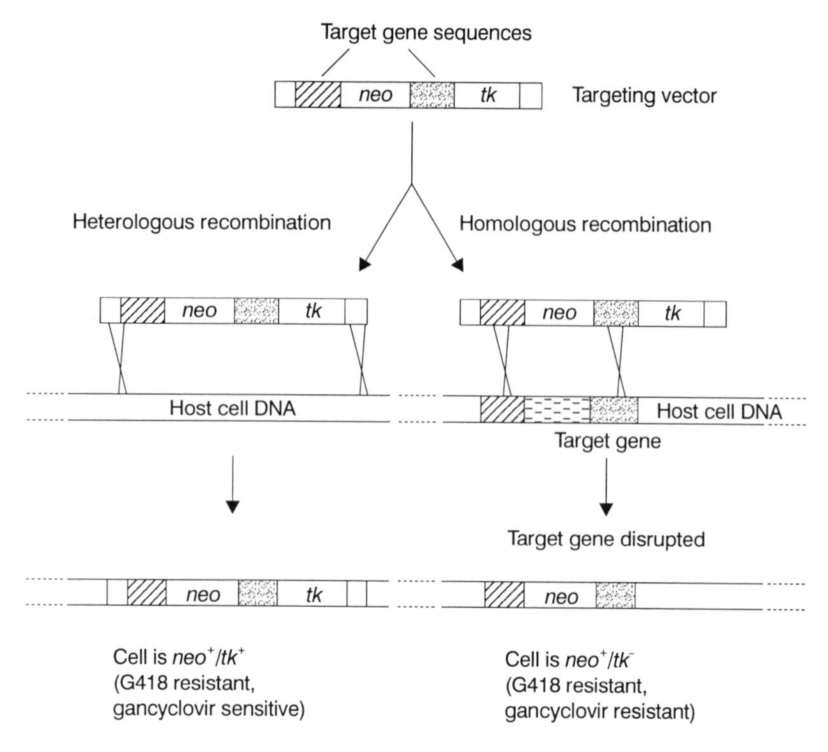

Figure 2.8. Positive–negative selection method for isolating ES cells in which a specific gene has been disrupted by homologous recombination. *neo*, neomycin resistance gene; *tk*, herpes simplex virus thymidine kinase gene.

inserted into a cloned piece of the mouse gene to be targeted. Downstream of the mouse gene sequence is another selectable marker gene, encoding the herpes simplex virus thymidine kinase (*tk*) gene. The targeting vector is transfected into ES cells in culture, and grown in the presence of G418 and gancyclovir. As discussed above, G418 will kill any cells that do not contain the *neo* gene. In contrast, gancyclovir is a drug that is not itself toxic to cells but is converted into a toxic product by thymidine kinase. Hence, it will kill any cells that do express the *tk* gene. As illustrated in *Figure 2.8*, cells that do not become stably transfected will be neo^-/tk^- and will be killed by the G418, whilst cells that become stably transfected by heterologous recombination at random sites will be neo^+/tk^+ and will die in the presence of gancyclovir. Only cells that have become stably transfected by homologous recombination between the mouse gene sequences in the targeting vector and the identical sequences in the mouse genome will be neo^+/tk^- and will survive the double selection. These cells can be grown, injected into mouse blastocysts and so used to generate lines of mutant mice. In the vernacular of molecular biologists, mice of this sort are often called 'knockout' or 'null' mice, but it is important, when reading this term, to find out exactly what kind of mutation has been introduced. In the procedure described above, the targeted gene is disrupted by insertion of a large piece of DNA, the *neo* gene. Variations of the method can be used to delete regions of DNA from the targeted gene or to create small mutations, including point mutations. Similarly, the technique has been used to correct mutations in mice by replacing the mutated region of the endogenous gene with the wild-type sequence. For example, Thompson *et al.* (1989) used it to correct a defect in the hypoxanthine phosphoribosyl transferase (HPRT) gene, mutations in which are responsible for Lesch–Nyhan syndrome in humans.

Once a homozygous mutant strain has been derived, both copies of the target gene will be disrupted in every cell of the mice. This may sometimes be inconvenient. For example, if disruption of a target gene results in early embryonic lethality, it is clearly not possible to examine the specific effects of mutating that gene on structures that arise later in development or on adult physiology. To overcome this problem, a number of groups have developed a system for deleting a gene in specific cell types, or in response to a specific exogenous signal. This is based on the Cre/*lox* system of bacteriophage P1 (Sauer, 1993). Cre (causes recombination) is a protein that mediates efficient recombination of specific DNA sequences called *lox* (locus of x-ing-over) and can therefore excise regions of DNA that are flanked by *lox* sites. Homologous recombination in ES cells can be used to produce transgenic mice carrying the *Cre* gene, and with *lox* sites inserted at each end of the target gene. If the *Cre* gene is placed under the control of a tissue-specific promoter, then the target gene will be deleted only in those tissues that express Cre. Sauer (1993) has discussed the wider applications of the Cre/*lox* system, and Li *et al.* (1996) provide one example of its recent use, to create a large deletion in the amyloid precursor protein gene.

2.6.4 Transgenic mice as models of human disease

One application of targeted gene disruption is the creation of mouse strains carrying mutations responsible for human disease. For example, targeted disruption of the *cftr* (cystic fibrosis transmembrane conductance regulator) gene in transgenic mice resulted in symptoms very similar to those seen in humans with cystic fibrosis (Wilson and Collins, 1992), and such mice have been used subsequently to test protocols for somatic gene therapy of cystic fibrosis (Dorin, 1995; Hyde *et al.*, 1993).

2.7 Identifying protein–protein interactions

A great number of biological processes are mediated by pathways of interacting proteins, and the development of methods for identifying specific protein–protein interactions has been central to the growth in our understanding of such pathways. In particular, our current view of intracellular signal transduction pathways owes much to these experimental approaches. There are a number of techniques in current use, but the most commonly used are the 'pull-down' and 'yeast two-hybrid' methods, and these will now be discussed.

2.7.1 'Pull-down' techniques

This approach can be used to identify cellular proteins that bind to a known protein or subdomain, for which a cDNA clone is available. The principle, illustrated in

Figure 2.9, is to express the protein or subdomain in *E. coli* in such a way that it can be readily purified by binding to Sepharose beads. This usually involves expressing it with six histidine residues (a 'His tag'; Kroll *et al.*,1993) or the *Schistosoma japonicum* glutathione *S*-transferase (GST) protein (Smith and Johnson, 1988) fused to its amino terminus. His tags bind with high affinity to nickel-charged Sepharose, whilst GST fusion binding proteins bind with high affinity to glutathione-coupled sepharose. Sepharose beads carrying fusion protein are then incubated with cellular protein extracts under conditions that allow protein–protein interactions to take place. After washing the beads, the bound cellular proteins are eluted and analysed by sodium dodecyl sulphate (SOS) –polyacrylamide gel electrophoresis. These proteins can then be identified by repeating the experiment on a preparative scale, cutting the protein bands out of the gel and applying proteolytic microsequencing techniques to determine the amino acid sequences of peptides derived from them. This

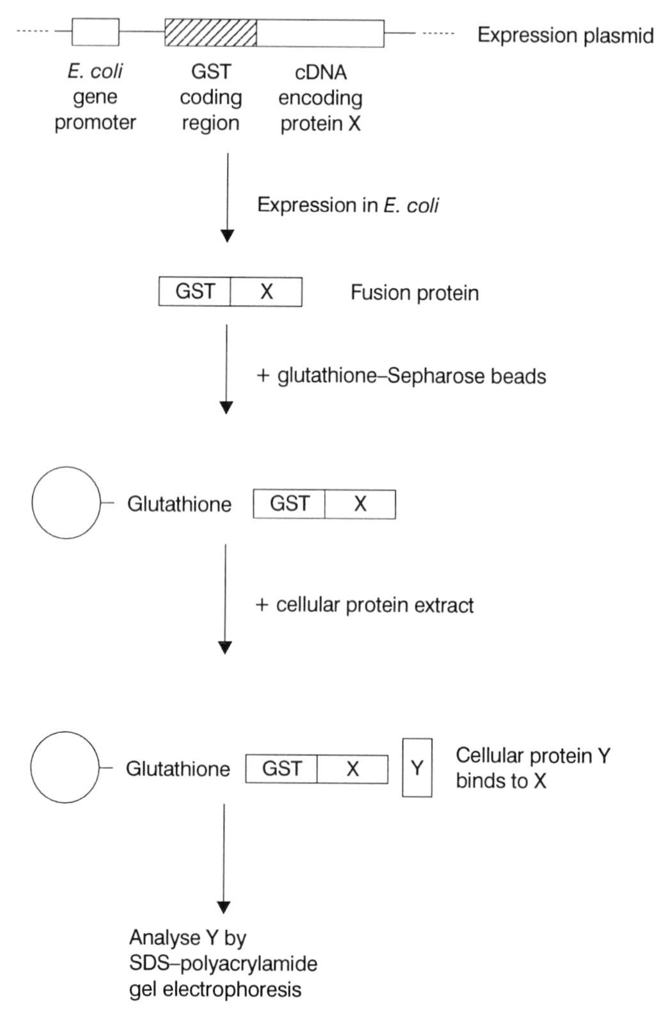

Figure 2.9. Glutathione *S*-transferase (GST) 'pull-down' technique.

procedure has been applied successfully on numerous occasions. For example, we recently used it to identify an interaction between the Fyn cytoplasmic protein-tyrosine kinase and WASp, the protein that is mutated in Wiskott–Aldrich syndrome (Banin *et al.*, 1996). However, it suffers from the drawback that protein–protein interactions defined thus *in vitro* may prove not to occur *in vivo*. It is therefore essential to determine whether the interaction does occur *in vivo*, for example by co-immunoprecipitation of the interacting proteins from cell lysates (Banin *et al.*, 1996).

2.7.2 The yeast two-hybrid system

The yeast two-hybrid system (*Figure 2.10*) is a method for identifying protein–protein interactions *in vivo*, albeit in yeast cells rather than mammalian cells. The most commonly used version of this technique is based on the yeast Gal4 protein (Fields and Sternglanz, 1994). In common with other transcription factors, this protein consists of a DNA-binding domain (DBD), which binds to a specific nucleotide sequence upstream of the promoter of target genes, and an activation

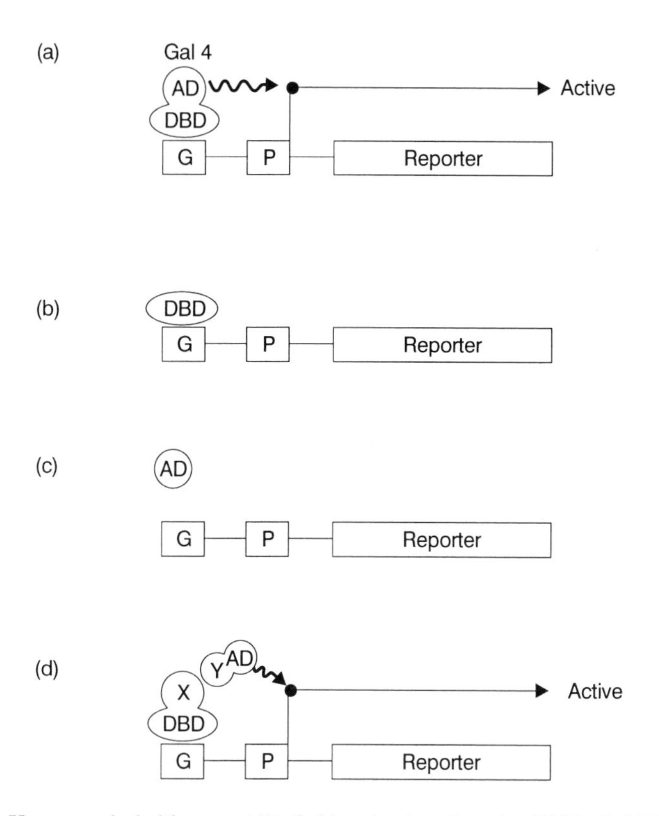

Figure 2.10. Yeast two-hybrid assay. AD, Gal4 activation domain; DBD, Gal4 DNA-binding domain; G, Gal4-binding site; P, promoter; X, protein X; Y, protein Y. Intact Gal4 (a) activates reporter gene transcription. Isolated Gal4 DBD (b) or AD (c) does not activate reporter gene transcription. Gal4 AD and GBD non-covalently linked by interaction between fusion partners, proteins X and Y (d), activate reporter gene transcription.

domain (AD), both of which are required for normal function (*Figure 2.10*). Surprisingly, AD activates transcription not only when covalently linked to DBD, but also when bound to it non-covalently. Thus, if two interacting proteins (X and Y; *Figure 2.10*) are produced in yeast cells as fusion proteins with DBD and AD, respectively, they will activate expression of a reporter gene controlled by a promoter containing Gal4-binding sites. Two commonly used reporter genes are the *E. coli lacZ* and yeast *HIS3* genes. Activation of *lacZ* is assayed by the blue coloration of yeast colonies grown in the presence of X-Gal, while activation of *HIS3* is assayed by the ability of colonies to grow on medium lacking histidine. A typical experiment to identify cellular proteins that bind to a known target protein involves construction of a plasmid in which the sequence encoding the target protein is fused in frame downstream of the Gal4 DBD coding sequence, and of a cDNA library in which the cDNAs are fused in frame downstream of the Gal4 AD coding sequence. These constructs are co-transfected into a yeast strain carrying *lacZ* and *HIS3* reporter plasmids, and the yeast cells plated out at high density on medium lacking histidine. The only yeast cells that should form colonies are those in which the Gal4 DBD becomes associated with the Gal4 AD as a result of a specific interaction between the target protein and proteins encoded by the cDNAs. As a second test, to reduce the frequency of false positives, yeast colonies surviving this selection should be blue when plated on medium containing X-Gal.

Although the yeast two-hybrid system allows identification of protein–protein interactions *in vivo*, it is still necessary to demonstrate that the interactions detected occur in the authentic cell type. One of the many successful applications of this technique is the recent finding of the direct interaction of vitamin D receptor with transcription factor IIB (MacDonald *et al.*, 1995).

2.8 Conclusion

It should be clear from the above that there is no slowing in the pace of technological innovation as gene cloning enters its third decade. The contribution made by these new techniques to the understanding, diagnosis and treatment of endocrine disorders will become abundantly apparent in the chapters to come.

References

Abbondanzo SJ, Gadi I, Stewart CL. (1993) Derivation of embryonic stem cell lines. *Meth. Enzymol.* 225: 803–823.

Bailey AL. (1995) Single-stranded conformational polymorphisms. In: *PCR Strategies* (eds MA Innis, DH Gelfand, JJ Sninsky). Academic Press, San Diego, CA, pp. 121–129.

Balta G, Brickner HE, Takegawa S, Kazazian HH, Papayannopoulou T, Forget BG, Atweh GF. (1994) Increased expression of the $^{G}\gamma$ and $^{A}\gamma$ globin genes associated with a mutation in the $^{A}\gamma$ enhancer. *Blood* **83:** 3727–3737.

Banin S, Truong O, Katz D, Waterfield M, Brickell PM, Gout I. (1996) Wiskott-Aldrich syndrome protein (WASp) is a binding partner for c-Src family protein-tyrosine kinases. *Curr. Biol.* **6:** 981–988.

Barash I, Nathan M, Kari R, Ilan N, Shani M, Hurwitz DR. (1996) Elements within the β-lactoglobulin gene inhibit expression of human serum albumin cDNA and minigenes in transfected cells but rescue their expression in the mammary gland of transgenic mice. *Nucl. Acids Res.* **24:** 602–610.

Blank RD, Sklar CA, Martin ML. (1996) Denaturing gradient gel electrophoresis to diagnose multiple endocrine neoplasia type 2. *Clin. Chem.***42**: 598–603.

Brasier AR, Ron D. (1992) Luciferase reporter gene assay in mammalian cells. *Meth. Enzymol.* **216**: 386–397.

Brown TA. (1994) *DNA Sequencing: the Basics.* IRL Press at Oxford University Press, Oxford.

Brown TA. (1995) *Gene Cloning: an Introduction*, 3rd Edn. Chapman and Hall, London.

Capecchi MR. (1994) Targeted gene replacement. *Sci. Am.* **270**: 52–59.

Darling DC, Brickell PM. (1994) *Nucleic Acid Blotting: the Basics.* IRL Press at Oxford University Press, Oxford.

Dlugosz AA, Glick AB, Tennenbaum T, Weinberg W, Yuspa SH. (1995) Isolation and utilization of epidermal keratinocytes for oncogene research. *Meth. Enzymol.* **254**: 3–20.

Dorin JR. (1995) Development of mouse models for cystic fibrosis. *J. Inherit. Metab. Dis.* **18**: 495–500.

Fields S, Sternglanz R. (1994) The two-hybrid system: an assay for protein:protein interactions. *Trends Genet.* **10**: 286–292.

Gordon JW. (1994) Production of transgenic mice by pronuclear injection. In: *Cell Biology: a Laboratory Handbook*, Vol. 3 (ed. JE Celis). Academic Press, San Diego, CA, pp. 106–111.

Hammer RE, Pursel VG, Rexroad CE, Wall RJ, Bolt DJ, Ebert KM, Palmiter RD, Brinster RL. (1985) Production of transgenic rabbits, sheep and pigs. *Nature* **315**: 680–683.

Henco K, Harders J, Wiese U, Riesner D. (1994) Temperature gradient gel electrophoresis (TGGE) for the detection of polymorphic DNA and RNA. In: *Methods in Molecular Biology*, Vol. 31, *Protocols for Gene Analysis* (ed. AJ Harwood). Humana Press, Totowa, NJ, pp. 211–228.

Herr S, Pepperkok R, Saffrich R, Wiemann S, Ansorge W. (1994) Electroporation of cells. In: *Cell Biology: a Laboratory Handbook*, Vol. 3 (ed. JE Celis). Academic Press, San Diego, CA, pp. 37–43.

Hitt M, Bett AJ, Prevec L, Graham FL. (1994) Construction and propagation of human adenovirus vectors. In: *Cell Biology: a Laboratory Handbook*, Vol. 1 (ed. JE Celis). Academic Press, San Diego, CA, pp. 479–490.

Hyde SC, Gill DR, Higgins CF, Trezise AEO, MacVinish LJ, Cuthbert AW, Ratcliff R, Evans MJ, Colledge WH. (1993) Correction of the ion transport defect in cystic fibrosis transgenic mice by gene therapy. *Nature* **362**: 250–255.

Ionov Y, Peinado MA, Malkhosyan S, Shibata D, Perucho M. (1993) Ubiquitous somatic mutations in simple repeated sequences reveal a new mechanism for colonic carcinogenesis. *Nature* **363**: 558–561.

Klar A, Baldassare M, Jessell TM. (1992) F-spondin: a gene expressed at high levels in the floor plate encodes a secreted protein that promotes neural cell adhesion and neurite extension. *Cell* **69**: 95–110.

Kroll DJ, Abdel-Malik-Abdel-Hafiz H, Marcell T, Simpson S, Chen CY, Gutierrez-Hartmann A, Lustbader JW, Hoeffler JP. (1993) A multifunctional prokaryotic protein expression system: overproduction, affinity purification and selective detection. *DNA Cell Biol.* **12**: 441–453.

Li W-B, Gruber CE, Lin J-J, Lim R, D'Alessio JM, Jessee JA. (1994) The isolation of differentially expressed genes in fibroblast growth factor stimulated BC3H1 cells by subtractive hybridization. *BioTechniques* **16**: 722–729.

Li ZW, Stark G, Gotz J, Rulicke T, Muller U, Weissmann C. (1996) Generation of transgenic mice with a 200-kb amyloid precursor protein gene deletion by Cre recombinase-mediated site-specific recombination in embryonic stem cells. *Proc. Natl. Acad. Sci. USA* **93**: 6158–6162.

Lobaccaro JM, Poujol N, Chiche L, Lumbroso S, Brown TR, Sultan C. (1996) Molecular modelling and in vitro investigations of the human androgen receptor DNA-binding domain: application for the study of two mutations. *Mol. Cell. Endocrinol.* **116**: 137–147.

MacDonald PN, Sherman DR, Dowd DR, Jefcoat SC Jr, DeLisle RK. (1995) The vitamin D receptor interacts with general transcription factor IIB. *J. Biol. Chem.* **270**: 4748–4752.

McClelland M, Mathieu-Daude F, Welsh J. (1995) RNA fingerprinting and differential display using arbitrarily primed PCR. *Trends Genet.* **11**: 242–246.

Moller DE, Chang PY, Yaspelkis BB, Flier JS, Wallberg-Henriksson H, Ivy JL. (1996) Transgenic mice with muscle-specific insulin resistance develop increased adiposity, impaired glucose tolerance and dyslipidemia. *Endocrinology* **137**: 2397–2405.

Nishio Y, Aiello LP, King GL. (1994) Glucose induced genes in bovine aortic smooth muscle cells identified by mRNA differential display. *FASEB J.* **8**: 103–106.

Oishi S, Sato T, Takiguchi-Shirahama S, Nakamura Y. (1995) Mutations of the RET proto-oncogene in multiple endocrine neoplasia type 2A (Sipple's syndrome). *Endocr. J.* **42:** 527–536.

Peinado MA, Malkhosyan S, Velazquez A, Perucho M. (1992) Isolation and characterization of allelic losses and gains in colorectal tumors by arbitrarily primed polymerase chain reaction. *Proc. Natl. Acad. Sci. USA* **89:** 10065–10069.

Prasher DC. (1995) Using GFP to see the light. *Trends Genet.* **11:** 320–323.

Ramírez-Solis R, Davis AC, Bradley A. (1993) Gene targeting in embryonic stem cells. *Meth. Enzymol.* **225:** 855–878.

Rolfs A, Schuller I, Finckh U, Weber-Rolfs I. (1993) Denaturing gradient gel electrophoresis (DGGE). In: *PCR: Clinical Diagnostics and Research* (eds A Rolfs, I Schuller, U Finckh, I Weber-Rolfs). Springer, Heidelberg, pp. 159–163.

Rosenthal N. (1987) Identification of regulatory elements of cloned genes with functional assays. *Meth. Enzymol.* **152:** 704–720.

Sauer, B. (1993) Manipulation of transgenes by site-specific recombination: use of Cre recombinase. *Meth. Enzymol.* **225:** 890–900.

Semenza GL. (1994) Transcriptional regulation of gene expression: mechanisms and pathophysiology. *Hum. Mutat.* **3:** 180–199.

Short JM, Sorge JA. (1992) *In vivo* excision properties of bacteriophage λZAP expression vectors. *Meth. Enzymol.* **216:** 495–508.

Smith DB, Johnson KS. (1988) Single-step purification of polypeptides expressed in Escherichia coli as fusions with glutathione S-transferase. *Gene* **67:** 31–40.

Stewart CL. (1993) Production of chimeras between embryonic stem cells and blastocysts. *Meth. Enzymol.* **225:** 823–855.

Sun XM, Neuwirth C, Wade DP, Knight BL, Soutar AK. (1995) A mutation (T-45C) in the promoter region of the low-density-lipoprotein (LDL)-receptor gene in a Welsh patient with a clinical diagnosis of heterozygous familial hypercholesterolaemia (FH). *Hum. Mol. Genet.* **4:** 2125–2129.

Thompson S, Clarke AR, Pow AM, Hooper ML, Melton DW. (1989) Germ line transmission and expression of a corrected HPRT gene produced by gene targeting in embryonic stem cells. *Cell* **56:** 313–321.

Uchiyama CM, Zhu J, Carroll RS, Leon SP, Black PM. (1995) Differential display of messenger ribonucleic acid: a useful technique for analyzing differential gene expression in human brain tumors. *Neurosurgery* **37:** 464–469.

Vernet M, Bonnerot C, Briand P, Nicolas J-F. (1993) Application of *LacZ* gene fusions to preimplantation development. *Meth. Enzymol.* **225:** 434–451.

Wang Z, Brown DB. (1991) A gene expression screen. *Proc. Natl. Acad. Sci. USA* **88:** 11505–11509.

Ward KA, Nancarrow CD. (1995) The commercial and agricultural applications of animal transgenesis. *Mol. Biotechnol.* **4:** 167–178.

Watson JD, Gilma M, Witkowski J, Zoller M. (1992) *Recombinant DNA*, 2nd Edn. Scientific American Books, New York.

Weatherall DJ. (1991) *The New Genetics and Clinical Practice*, 3rd Edn. Oxford University Press, Oxford.

Welsh J, McClelland M. (1993) The characterization of pathogenic microorganisms by genomic fingerprinting using arbitrarily primed polymerase chain reaction (AP-PCR). In: *Diagnostic Molecular Microbiology* (eds DH Persing, TF Smith, FC Tenover, TJ White). ASM Press, Washington DC, pp. 595–602.

Welsh J, Ralph D, McClelland M. (1995) DNA and RNA fingerprinting using arbitrarily primed PCR. In: *PCR Strategies* (eds MA Innis, DH Gelfand, JJ Sninsky). Academic Press, San Diego, CA, pp. 249–276.

Wilson JM, Collins FS. (1992) More for the modellers. *Nature* **359:** 195–196.

Wingender E, Dietze P, Karas H, Knueppel R. (1996) TRANSFAC: a database of transcription factors and their DNA binding sites. *Nucl. Acids Res.* **24:** 238–241.

Woychik RP, Stewart TA, Davis LG, D'Eustachio P, Leder P. (1985) An inherited limb deformity created by insertional mutagenesis in a transgenic mouse. *Nature* **318:** 36–40.

Yu BD, Hess JL, Horning SE, Brown GAJ, Korsmeyer SJ. (1995) Altered *Hox* expression and segmental identity in *Mll*-mutant mice. *Nature* **378:** 505–508.

3

Mechanism of action of peptide hormones

Stewart Barker

3.1 Introduction

Over the last decade, or so, there has been a huge expansion in our knowledge concerning the mechanisms involved in the transduction of extracellular signals into cellular responses. This expansion has, to a large extent, been due to the development of improved and novel molecular biology techniques, including the polymerase chain reaction (PCR), homology cloning, *in vitro* mutagenesis, and the use of heterologous expression systems. In addition, the development of specific antibodies to numerous protein components of the signal transduction machinery, including antibodies raised against motifs containing phosphorylated tyrosine residues, and their use for immunoprecipitation, has played an important role in these advances. During this same period, our concept of the classical endocrine system, in which a hormone is secreted in one tissue and circulated in the blood to a second *target* tissue, has acquired a degree of flexibility as the techniques of immunocytochemistry and *in situ* hybridization have revealed local sites of synthesis and action (paracrine or autocrine) for molecules previously regarded as classical hormones.

This chapter focuses on both the common and divergent features of peptide hormone action and includes representative examples of the major signalling pathways. Detailed consideration is given to structural characteristics of the transmembrane receptor proteins and the downstream consequences of receptor activation. Particular emphasis is given to guanine nucleotide-binding protein (G-protein)-linked receptor-mediated events. During the course of this chapter examples are referred to from both non-hormonal and non-peptide systems. This is dictated by the greater progress of research in some fields and also reflects a conservation of receptor structure, and shared components of the intracellular signal transduction machinery, which extends beyond the nature of the ligand moiety itself.

3.2 Receptor protein structure–function relationships

Each peptide hormone interacts with a *specific* high affinity receptor which is an integral protein spanning the plasma membrane. Three main receptor superfamilies

Molecular Endocrinology, edited by G. Rumsby and S.M. Farrow.

have been identified which interact with peptide hormone ligands. The tyrosine kinase receptor family includes receptors for epidermal growth factor/transforming growth factor-α, platelet-derived growth factor and a number of proto-oncogenes, such as c-*erb* B-2 and c-*sis* (Yarden and Ullrich, 1988). The receptors for insulin and insulin-like growth factor-I (IGF-I) are also members of this family. Insulin is a classical *poly*-peptide hormone and yet its receptor has structural and functional characteristics in common with receptors of locally acting growth factors. The key feature of this receptor family is the presence of an intrinsic tyrosine kinase activity in the intracellular domain of the receptor. Signal transduction occurs through phosphorylation of specific tyrosine residues of substrate molecules and auto-phosphorylation sites within the intracellular domain of the receptor itself.

A second superfamily, the cytokine or haemopoietin receptor family, together with receptors for interleukins 2, 3, 4, 6, and 7, includes receptors for growth hormone (GH) and prolactin (PRL) (Cosman *et al.*, 1990). The sequence homology which underlies the structural similarities between these receptors occurs within the extracellular ligand-binding domain. GH receptor and PRL receptor were the first family members to be cloned and show surprising homology within this region, considering their diverse effects.

The third and largest family comprises the seven-transmembrane-domain receptors (7TMDRs), so named because of their characteristic membrane-spanning heptahelical structure. This family consists almost entirely of receptors which couple directly to G-protein heterotrimeric complexes, and which, therefore, will be subsequently referred to as G-protein-linked receptors (GPLRs). To date, more than 140 members of this family (excluding the odorant receptors) have been identified, many through the use of homology cloning, together with a number of orphan receptors for which the cognate ligand has yet to be defined (Parmentier *et al.*, 1995). Ligands whose effects are mediated through binding to members of this receptor family include angiotensin II (Ang II), vasopressin (AVP), parathyroid hormone (PTH), and glycoprotein hormones such as luteinizing hormone (LH), follicle-stimulating hormone (FSH) and thyroid-stimulating hormone (TSH) (see *Table 3.1*). Although the seven transmembrane helices of GPLRs are relatively highly conserved, there is considerable diversity of structure in both the extracellular and intracellular domains (see Sections 3.2.3 and 3.3.2).

In addition to these receptor families, a further group exists known as the receptor guanylate cyclase family. To date, no direct evidence for a peptide *hormone*-activated receptor guanylate cyclase has been identified; however, the effects of atrial natriuretic peptides are known to be mediated by certain members of this receptor family (Garbers, 1994).

3.2.1 Insulin/IGF-I receptors

The insulin receptor is a tetrameric protein which consists of two, identical, extracellular α-subunits responsible for ligand binding. These subunits are linked by disulphide bridges to one another, and to two β-subunits which traverse the plasma membrane and contain tyrosine kinase activity. The IGF-I receptor is structurally very similar and IGF-I binding initiates a virtually identical series of signalling events (Cheatham and Kahn, 1995; LeRoith *et al.*, 1995). The multimeric structure of the insulin receptor is the basis of a subgroup within the family of tyrosine kinase receptors. The formation of this multimeric complex effectively eliminates the necessity for receptor dimerization, which is characteristic of the other members of this receptor family. The immediate consequence of ligand binding is that the receptor undergoes autophosphorylation at multiple phosphorylation sites on the intracellular domains of the β-subunits. This leads to an association between the insulin receptor and its primary substrate molecule, insulin receptor substrate-1 (IRS-1), which is in turn phosphorylated by the insulin receptor on multiple tyrosine residues. It is this rapid phosphorylation of IRS-1 which is the key step responsible for directing the divergent pathways associated with insulin action (*Figure 3.1*).

IRS-1 is a 180 kD protein which contains at least 10 known sites for tyrosine phosphorylation consisting of the amino acid sequence, YMXM, or similar motifs (Dey *et al.*, 1996). These motifs represent specific 'docking sites' for up to 10 different signalling molecules which contain *src* homology 2 (SH2) domains, including the regulatory subunit of phosphatidylinositol-3-kinase (PI3-kinase), and the adaptor molecule Grb2 (Skolnik *et al.*, 1993). The two main links to physiological endpoints are provided, firstly, by insulin receptor–IRS-1–Grb2 input into the p21*ras*-mediated mitogen-activated protein (MAP) kinase pathway (Robbins *et al.*, 1994) (leading to the effects of insulin on gene expression and mitogenesis) and, secondly, by insulin receptor–IRS-1-mediated activation of PI3-kinase, which is involved in the regulation of glucose transporter function and modulation of S6 kinase during protein synthesis (Cheatham and Kahn, 1995). As will be seen with other signalling pathways described below, there are further levels of complexity due to the existence of multiple isoforms of these signalling molecules which arise from alternative splicing. These may be differentially expressed, differentially regulated and differentially located within a given cell type, to allow the specificity of insulin action under a given set of physiological conditions. For example, functional investigation of IRS-1 knockout mice has led to the discovery of a second insulin receptor substrate, known as IRS-2 (Patti *et al.*, 1995). Studies in which IRS-1, IRS-2 or both were transfected into cells which do not express these molecules endogenously have permitted a limited classification of their involvement in insulin receptor-mediated effects (*Figure 3.1*). There is also evidence for an alternative pathway for insulin/IGF-I-driven *ras* activation which can occur by direct interaction with another adaptor molecule known as Shc (Dey *et al.*, 1996) which, when phosphorylated, is able to interact with Grb2 thus eliminating the requirement for IRS-1. Although it might seem that multiple inputs impinge upon essentially identical signalling cascades, this apparent redundancy of signalling pathways is likely to underlie a far more complex and highly regulated system of co-ordinated receptor-initiated cellular responses.

3.2.2 GH/PRL receptors

As a consequence of their sequence homology, the mechanisms of action of GH and PRL receptors are thought to be similar, although the GH receptor is the best characterized (Postel-Vinay and Finidori, 1995). Although members of different

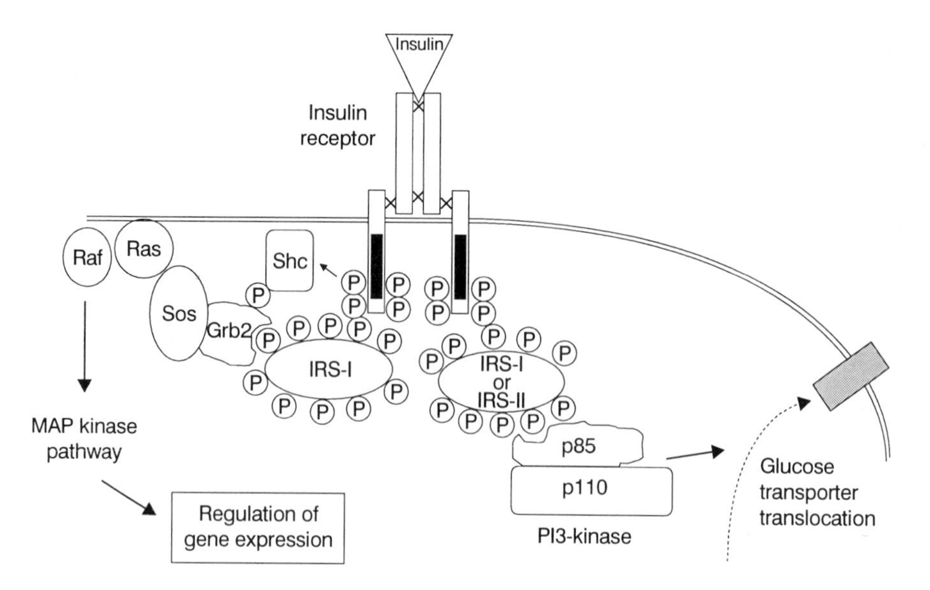

Figure 3.1. Alternative signalling pathways initiated by binding of insulin to its transmembrane receptor protein. Central to this scheme are the insulin receptor substrates, IRS-1 and IRS-2. Insulin receptor-dependent phosphorylation of these proteins at multiple tyrosine residues (denoted by encircled 'P') enables them to direct distinct and diverse intracellular signalling routes. Interaction of IRS-1 with the adaptor molecule Grb2 leads to activation of the MAP (mitogen-activated protein) kinase pathway, while both IRS-1 and IRS-2 are able to activate phosphatidylinositol-3-kinase (PI3-kinase) leading to effects on glucose transport. The mitogenic effect of insulin can also be exerted without the involvement of IRS-1, by direct interaction between the insulin receptor and the adaptor molecule Shc. Solid bars represent intrinsic tyrosine kinase activity in intracellular domains of the insulin receptor. p85 and p110 are regulatory and catalytic subunits of PI3-kinase, respectively.

families, GH and insulin receptors have certain features in common with regard to signalling, for example hormone-dependent receptor dimerization is required to initiate the signalling cascade (*Figure 3.2*). GH binds to two GH receptor extracellular domains in a sequential manner, with each receptor molecule binding to one of two distinct sites on the ligand, resulting in the characteristic bell-shaped response curve for GH (Wells *et al.*, 1993). A truncated form of the GH receptor also exists, known as GH-binding protein (GHBP), which is the entire extracellular domain of the GH receptor and is thought to arise from protease cleavage. GHBP prolongs the half-life of circulating GH and may have other functions which are yet to be defined.

The GH receptor becomes phosphorylated upon ligand activation in a process catalysed by a soluble tyrosine kinase of the Janus kinase family JAK2. This kinase associates with a proline-rich motif within a region known as Box 1 in the N-terminal portion of the cytoplasmic domain of the GH receptor (Postel-Vinay and Finidori, 1995). The association of JAK2 with the GH receptor provides a platform for several different GH-activated signalling pathways. Firstly, phosphorylation of the transcription factor STAT 1 (one of a family of signal transducers and activators of transcription; Schindler and Darnell, 1995), results in the homodimerization of this protein and its subsequent migration into the nucleus where it directly interacts with specific response elements (Meyer *et al.*, 1994). A second pathway involves the docking of adaptor molecules via their SH2-binding domains at specific tyrosine-containing motifs in the GH receptor intracellular domain. In this way, the GH receptor also has an input into the MAP kinase pathway through Shc–Grb2 interactions (VanderKuur *et al.*, 1995). A third pathway, through activated JAK2 phosphorylation of IRS-1, leads to effects on the activation of PI3-kinase and glucose transport which GH has in common with insulin (Argetsinger *et al.*, 1995; see also *Figure 3.1*). The PRL receptor has also been shown to interact with, and activate, JAK2, STAT 1 (David *et al.*, 1994) and STAT 5 (Wakao *et al.*, 1994).

It is clear from both the insulin receptor and GH/PRL receptor signalling cascades that there is considerable potential overlap between pathways. However, the fidelity of cellular responses is likely to depend on tissue distribution of receptors, multiple isoforms of kinases, adaptor molecules, transcription factors and associated molecules, and differences in the persistence of the signal elicited through a particular set of signalling pathway components.

3.2.3 G-protein-linked receptors (GPLRs)

Structural features. The common features amongst GPLRs form the basis of a structural model which has undergone considerable evolutionary divergence (Strader *et al.*, 1994). The seven transmembrane helices form a hydrophobic core and contain many of the conserved amino acids present in GPLRs (Probst *et al.*, 1992) (*Figure 3.3*). The positioning of these key residues suggests a structural role in maintaining receptor conformation for full activity. In most GPLRs, two cysteine residues in the second and third extracellular loops are essential for full function. Another feature which is common to many GPLRs is the amino acid motif DRY present at the junction between the third transmembrane helix and the N-terminal end of the second intracellular loop. Mutation of the initial aspartate of this motif in the β-adrenergic receptor (β-AR), for example, results in a receptor which binds ligand avidly but is unable to activate adenylate cyclase. Asparagine residues within the N-termini and extracellular loops represent putative sites for receptor glycosylation, which is extensive in many GPLRs. However, in most cases, mutation of these residues has little effect on receptor binding or signalling, although glycosylation may be important for normal expression and targeting of some receptors (Rands *et al.*, 1990). An overall perspective of the basic structure of these receptors suggests that the conserved transmembrane helices represent a structural scaffold which creates the hydrophobic pore. This structure is likely to exhibit some flexibility within the membrane which would allow for ligand-dependent conformational changes in the receptor to be transmitted across the plasma membrane and so alter the extent and/or nature of receptor interaction with G-protein complexes.

GPLRs range in size from the compact minimal structure of the gonadotrophin-releasing hormone (GnRH) receptor (327 amino acid residues) (Reinhart *et al.*, 1992) to the extensive calcium (Ca^{2+})-sensing receptor (1085 amino acid residues) (Brown *et al.*, 1993). The 'signature' of each GPLR receptor species is provided by the length

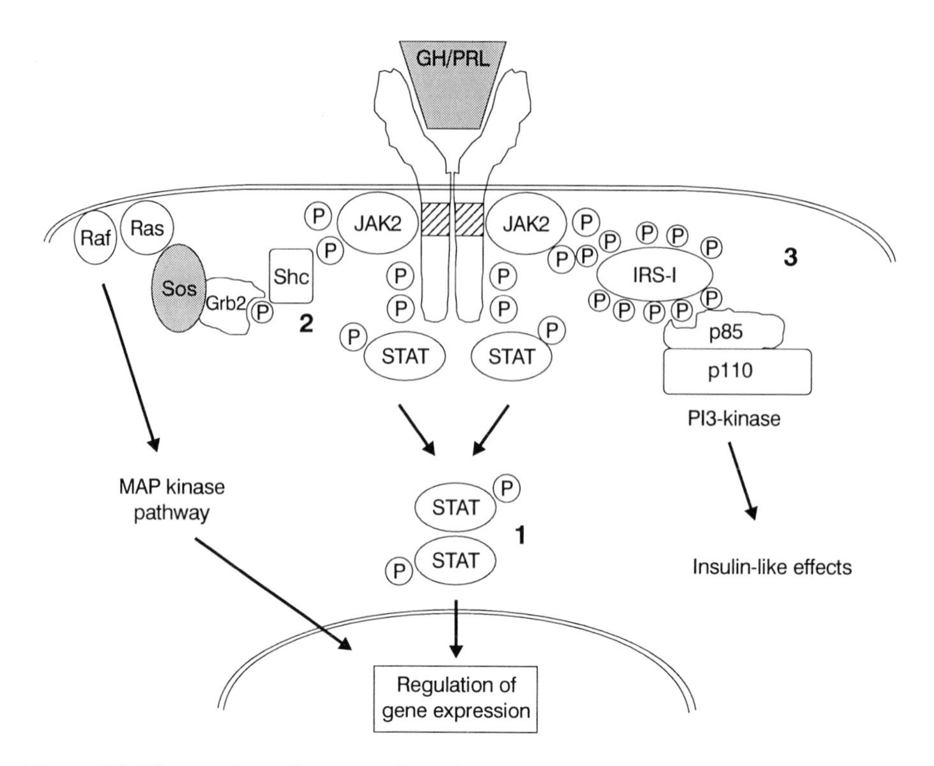

Figure 3.2. Three main pathways mediated by the growth hormone/prolactin (GH/PRL) receptor through association between JAK2 (Janus kinase family) kinase and the Box 1 regions (hatched areas) of the intracellular domains of the GH/PRL receptor dimer. (1) GH-dependent phosphorylation of STAT (signal transducers and activators of transcription) proteins, followed by migration of activated STAT dimers to the nucleus where they interact directly to regulate gene transcription; (2) activated JAK2 phosphorylation of the adaptor molecule Shc, leading to Ras-mediated progression through the MAP (mitogen-activated protein) kinase pathway; (3) JAK2 phosphorylation of the insulin receptor substrate, IRS-1, resulting in insulin-like effects through the activation of PI3-kinase. p85 and p110 are the regulatory and catalytic subunit of PI3-kinase, respectively.

and complexity of the extracellular loops, the N-terminus (which provide sites for ligand interaction) and the component intracellular loops which determine the specificity of interactions with different G-protein complexes.

There is a tremendous diversity of ligands which act via GPLRs ranging from Ca^{2+} and the relatively small bioactive amines (e.g. adrenaline and dopamine), through short peptides such as Ang II and GnRH, to glycoprotein hormones such as LH and TSH. Although these are not all peptide hormones, it is useful to compare the structures of different ligand–receptor systems to identify features which might be important for ligand binding (see *Table 3.1*).

It has been shown by mutagenesis studies that the bioactive amines interact directly within the hydrophobic pore of their receptors (Strader *et al.*, 1994) and removal of the extracellular loops from β-AR has no effect on ligand binding. Contact points for the non-peptide Ang II receptor antagonist, DuP753 (or

Losartan) have also been shown to lie within the hydrophobic pore of the Ang II (AT1 subtype) receptor, whereas the native ligand, Ang II, an octapeptide, interacts both within the pore and with specific amino acid residues within the extracellular loops of its receptor (Hunyady *et al.*, 1996). In contrast with the majority of GPLRs, the receptors for the glycoprotein hormones, LH (and human choriogonadotrophin, hCG), FSH and TSH, have relatively extensive N-terminal portions of 346 (Segaloff and Ascoli, 1993), 348 (Heckert *et al.*, 1992) and 398 (Vassart and Dumont, 1992) amino acids, respectively. These N-terminal extensions, which are encoded by multiple exons, have been shown to provide several sites for ligand–receptor interaction. The Ca^{2+}-sensing receptor, despite the comparatively small size of its ligand, also has an extended N-terminal domain (613 amino acids long), which is thought to bind calcium ions with a relatively low affinity, reflecting the relatively high level of available ligand (Brown *et al.*, 1993).The intracellular domains of GPLRs show an equal degree of variation, the importance of which is discussed in Section 3.3.2.

Receptor subtypes and receptor antagonists. The stoichiometry of binding for GPLR ligand–receptor interactions is 1:1, perhaps with the exception of the Ca^{2+}-sensing receptor. However, in some instances, more than one species of ligand is able to bind to a given receptor with equal affinity. Examples of this phenomenon include, the melanocortin-3 (MC3) receptor which binds all forms of melanocyte-stimulating hormone (α-, β- and γ-MSH) and adrenocorticotrophic

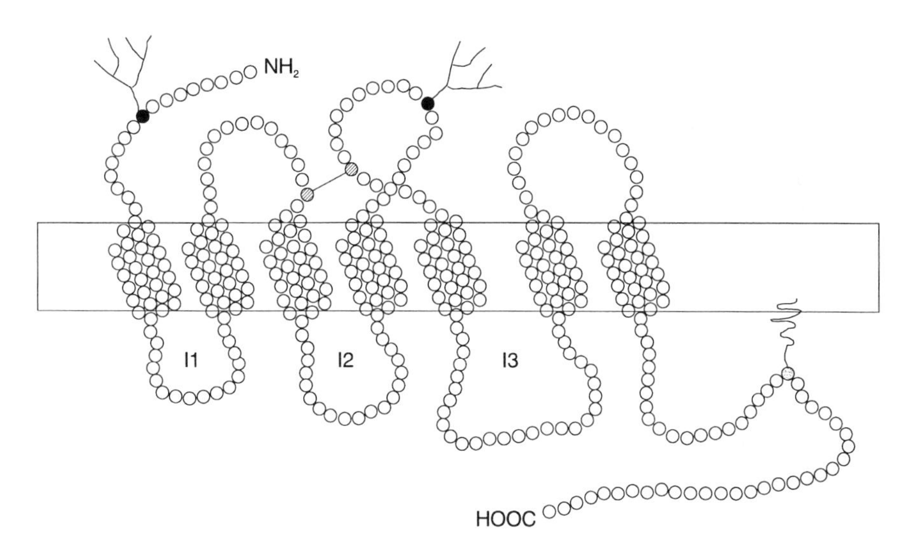

Figure 3.3. Schematic representation of G-protein-linked receptor (GPLR) structural features. Seven transmembrane helices, embedded in the plasma membrane, support extracellular domains involved in ligand binding interactions and intracellular domains (I1–I3) involved in signal transduction (circles represent individual amino acid residues). Two extracellular cysteine residues (hatched circles) form an intra-loop disulphide bridge. A cysteine residue in the C-terminal tail (stippled circle) may be palmitoylated in some GPLRs and provide a fourth intracellular loop. Asparagine residues in the extracellular domains (filled circles) provide potential sites for receptor glycosylation.

Table 3.1. Selectivity and promiscuity of some G-protein-linked receptors

Ligand	Receptor(s)	Gα-subunit	Effect
Catecholamines	β1-, β2-, β3-AR	G_s	AC (+)
Ang II	AT_1	G_q	PLC (+)
		G_i	AC (–)
		G_o?	VOC[a]
	AT_2	?	?
Bradykinin	BK_2	G_q	PLC (+)
Endothelin	ET_A/ET_B	G_q	PLC (+)
		G_s	AC (+)
		G_i	AC (–)
ACTH	MC2	G_s	AC (+)
MSH	MC1, MC3–5	G_s	AC (+)
AVP	V_{1A}	G_q	PLC (+)
	V_{1B}	G_q	PLC (+)
	V_2	G_s	AC (+)
Oxytocin	OT	G_q	PLC (+)
Calcium	Ca^{2+}-sensing	G_q	PLC (+)
		G_i	AC (–)
Calcitonin	CT1/CT2	G_s	AC (+)
		G_q	PLC (+)
		G_i	AC (–)
PTH/PTHrP[b]	PTH1	G_s/G_q	AC (+)/PLC (+)
PTH	PTH	G_q?	?
Glucagon	Glucagon	G_s	AC (+)
Somatostatin	SSTR1–SSTR5	G_s	AC (+)
	SSTR2	G_q	PLC (+)
		G_o	VOC
CRH[c]	CRH	G_s	AC (+)
TRH[d]	TRH	G_q	PLC (+)
GnRH	GnRH	G_q	PLC (+)
		G_s	AC (+)
FSH	FSH	G_s/G_q	AC (+)/PLC (+)
LH/hCG	LH/hCG	G_s/G_q	AC (+)/PLC (+)
TSH	TSH	G_s/G_q	AC (+)/PLC (+)

(+) Stimulated; (–) inhibited. [a] Voltage-gated Ca^{2+} channel; [b] PTH-related peptide; [c] corticotrophin-releasing hormone; [d] thyrotrophin-releasing hormone.
See text for other abbreviations.

hormone (ACTH) with similar affinity (Gantz *et al.*, 1993), and the vasopressin 1b and oxytocin receptors which each bind both AVP and oxytocin (Schlosser *et al.*, 1994). However, each of these ligand sets comprises highly related peptides and *in vivo* it would be expected that differential distribution and expression of both receptors and ligands would limit the extent to which cross-activation occurs. This is clearly demonstrated by the melanocortin receptor family (Cone *et al.*, 1993) where the MSH (MC1) receptor gene is expressed predominantly in melanocytes, whereas the ACTH (MC2) receptor gene is almost exclusively expressed in the adrenal gland and is probably under the influence of a tissue-specific promoter.

An equally important phenomenon, physiologically, is the existence of multiple receptors which respond to a particular ligand species. Some of these *receptor subtypes* may have arisen either as distinct receptor genes through gene duplication or through reverse transcription of a pre-existing mRNA and incorporation into the genome (Probst *et al.*, 1992). Examples include dopamine receptor subtypes D_2, D_3, and D_4, the subtypes of the rat type I Ang II receptor, namely AT_{1A} and AT_{1B}, and PTH receptors 1 and 2. Receptor subtypes may also occur as a result of alternative splicing of a single gene transcript yielding receptors with alternate N-termini (human AT_1 receptor; Curnow *et al.*, 1995) or C-termini (prostaglandin E receptors; Namba *et al.*, 1993). Alternative splicing may also lead to receptor subtypes with variable loops as exemplified by sequence differences in the first intracellular loop of two calcitonin receptor variants (Zolnierowicz *et al.*, 1994).

The existence of receptor subtypes provides a potentially useful means by which a given cell can respond to different prevailing 'environmental' conditions. Subtypes can be expressed differentially, and exert distinctly different effects. Furthermore, expression of the gene for each receptor subtype may be under the control of different regulatory mechanisms. An example of this diversity is provided by the receptors for Ang II. The two angiotensin receptor subtypes found in many mammalian species are denoted AT_1 and AT_2 and share the characteristic heptahelical configuration of a GPLR. However, there is limited overall sequence or functional homology despite being able to bind Ang II with equal affinity (Nahmias and Strosberg, 1995). Most of the classical actions of Ang II are in fact mediated through the AT_1 subtype, while the AT_2 subtype is thought to play a role in growth inhibition and apoptosis in certain angiotensin target tissues (Hayashida *et al.*, 1996). The development of selective inhibitors for these receptor subtypes represents an important consideration since angiotensin exerts its effects systemically, through the classical renal renin–angiotensin system, and also acts as a local factor produced from within target tissues. A number of non-peptide compounds which distinguish between AT_1 and AT_2 receptors have already been successfully developed. These have identified the different tissue distributions of these two receptor subtypes and provide the potential for selective therapeutic intervention (Timmermans *et al.*, 1991). An annual update of receptor subtype-selective agonists/antagonists for many other GPLRs can be found in the *Receptor and Ion Channel Supplement* published with *Trends in Pharmacological Sciences*.

The development of antagonists to receptor subtypes has been an extensively pursued goal. A number of novel approaches are now being taken to facilitate the design of a new generation of non-peptide peptide mimetics. The basis for this arises from studies to identify the precise residues involved in the interaction between a ligand and receptor. One approach has been to identify amino acid residues which are conserved between the same receptor in different species. Their functional significance can then be assessed using systematic site-directed mutagenesis (alanine scanning) in conjunction with determination of ligand binding kinetics and/or activation of a particular signalling pathway. This type of analysis can also be used to identify key residues in the peptide ligand which are essential for receptor activation. For example, in one study, a series of degenerate oligonucleotides was used for site-directed mutagenesis of the 25–34 region of PTH, previously determined by deletion analysis to be important for receptor binding (Gardella *et al.*, 1993). These experiments yielded a series of PTH (1–84) mutants the properties of which could be compared with wild-type peptide in functional assays. In this way an essential role for hydrophobic residues in ligand–receptor interactions was demonstrated. This finding is likely to have important implications with regard to the binding of PTH-related peptide to this receptor (see Section 6.6).

Other strategies have combined information obtained from biophysical approaches, such as X-ray crystallography of larger peptide ligands (e.g. >40 amino acid residues) or nuclear magnetic resonance studies of smaller peptides, with molecular biological techniques and functional assays. An example of this type of multidisciplinary approach is provided by work relating to the differences between the endothelin receptor subtypes ET_A and ET_B (Krystek *et al.*, 1994). In this study, target residues to be mutated were selected using a molecular model for the ET_A receptor, based on that of bacteriorhodopsin (the archetypal 7TMDR), together with sequence comparisons between ET_A and ET_B receptors. This led to the identification of a single residue, thought to be surface-exposed, within the second transmembrane domain which, when mutated, changed the ligand–binding profile of the ET_A receptor to that of the ET_B for a whole series of endothelin receptor ligands and antagonists.

These and other techniques are now being applied to the development of minimal forms of peptide ligands thereby yielding a peptide mimic containing the essential elements for receptor interaction presented in the appropriate spatial configuration. This has already been achieved for atrial natriuretic peptide (Li *et al.*, 1995).

3.3 Signal transduction pathways

As the signal transduction pathways initiated by insulin/IGF-I and GH/PRL receptors are described above (see Sections 3.2.1 and 3.2.2), this section concentrates on pathways mediated by GPLRs. In basic terms, binding of ligand to the GPLR is thought to lead to activation of a G-protein heterotrimeric complex through contacts between the intracellular loops and C-terminal tail of the receptor and specific regions of the G-protein complex. This process and the structural/sequence requirements for these contacts are the major focus of this section.

3.3.1 The G-protein heterotrimeric complex

There have been a number of excellent reviews dealing with the composition, function and diversity of G-protein heterotrimeric complexes (Gαβγ) (for example see Neer, 1995; Offermanns and Schultz, 1994). However, our understanding of the precise details of receptor–G-protein interactions has advanced in recent years through the application of X-ray crystallography and structural modelling techniques to G-protein structure combined with detailed mutational analysis of the individual subunits which comprise Gαβγ. There are now known to be at least 21 Gα-, five Gβ-, and 11 Gγ-subunit species. Since the association between Gβ and Gγ subunits is extremely tight and can only be disrupted under denaturing conditions (Mende *et al.*, 1995), Gα and Gβγ can be regarded as the only two functional entities. The large number and variation of Gβ and Gγ subunits would suggest considerable heterogeneity of Gβγ moieties; however, there is a certain degree of specificity in the interactions such that only certain combinations occur.

Figure 3.4 shows a schematic representation of the events which are thought to occur on agonist binding to a GPLR and is derived from a number of sources (Conklin and Bourne, 1993; Lefkowitz *et al.*, 1993; Wall *et al.*, 1995). It was originally perceived that GPLRs exist in two states, an inactive state with low affinity for agonist which is uncoupled from G-protein, and an activated G-protein-coupled state with high affinity for agonist (DeLean *et al.*, 1980). In this model, the binding of ligand favours formation of a ligand–receptor–G protein ternary complex. More recently, however, substantial evidence has accumulated to suggest that the unliganded receptor itself occurs in two states: an inactive uncoupled (R) state and a G-protein-coupled (R*) state which has an innate, and probably

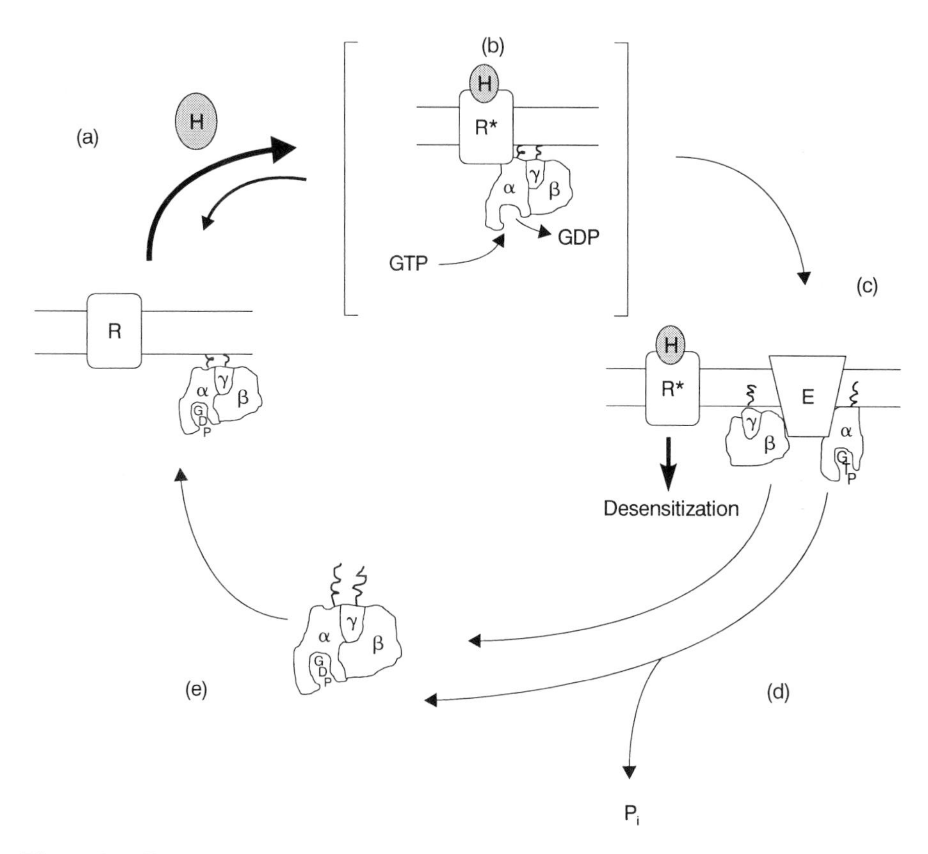

Figure 3.4. Hormone receptor-activated G-protein cycle. (a) Binding of hormone (H) induces a shift in the equilibrium from inactive uncoupled receptor (R) towards activated G-protein-coupled receptor (R*); (b) 'transition state' complex between agonist-bound receptor and Gαβγ. Guanine nucleotide exchange represents the main rate-limiting step of this cycle and binding of GTP to Gα results in the dissociation of the HR–Gα–Gβγ ternary complex; (c) $G\alpha_{GTP}$ and Gβγ modulate effector molecules (E) while the receptor undergoes desensitization; (d) hydrolysis of Gα-bound GTP by an intrinsic GTPase in Gα liberates inorganic phosphate (P_i), allowing reassociation of $G\alpha_{GDP}$ with Gβγ; (e) Gαβγ returns to the G-protein pool.

receptor-specific, ability to activate the G-protein. Thus, the level of activation in the basal state (that is, in the absence of agonist) reflects an equilibrium between these two conditions. In this new 'allosteric ternary complex model' (Lefkowitz *et al.*, 1993), agonist acts both to shift the equilibrium in the direction of R^* and to stabilize the ternary complex. Evidence for this model has arisen from two main sources. Firstly, some receptor antagonists have been found to cause a reduction in the basal level of G-protein activation thereby acting as 'inverse agonists' (Leeb-Lundberg *et al.*, 1994; Schütz and Freissmuth, 1992). Secondly, receptor mutants have been identified in *in vitro* studies which exhibit constitutive activity in the absence of agonist (Samama *et al.*, 1993). These findings may have important implications in terms of antagonist design and also pathophysiology.

G-protein activation is presumed to be the consequence of conformational changes in the receptor which result from the transition between R and R^* (*Figure 3.4a* and *b*). These, in turn, cause associated changes in the conformation of $G\alpha$ such that the rate of dissociation of GDP is increased to be replaced by GTP. Binding of GTP to $G\alpha$ reduces the affinity of R^* for $G\alpha\beta\gamma$ and causes the dissociation of $G\alpha$ from $G\alpha\beta\gamma$ (*Figure 3.4c*). $G\alpha_{GTP}$ is then able to activate downstream effector molecules such as adenylyl cyclase (AC) and phospholipase C (PLC). $G\beta\gamma$ also modulates the activity of a number of effector molecules, including some forms of AC, and is particularly important in some mechanisms of receptor desensitization (see Section 3.4.3). By virtue of an intrinsic GTPase, $G\alpha$ is returned to the inactive GDP-bound state which allows it to re-associate with an appropriate $G\beta\gamma$ (*Figure 3.4d* and *e*) thus returning it to the G-protein pool, available once more to associate with another receptor molecule.

3.3.2 Basis of diversity and specificity of the G-protein heterotrimeric complex

An awareness of certain structural features of the individual G-protein components is important to our understanding of the way in which $G\alpha\beta\gamma$ interacts with the receptor. According to the recently resolved crystal structures for the heterotrimer (Wall *et al.*, 1995) and individual $G\alpha$ subunits (Rens-Domiano and Hamm, 1995), both the C- and N-termini of $G\alpha$ lie in close proximity to each other on the side of the molecule which faces the plasma membrane (Conklin and Bourne, 1993). This face of $G\alpha$ also presents itself to receptors, $G\beta\gamma$ and effector molecules. Lipid modification of the C- and/or N-terminus of $G\alpha$, together with isoprenylation of a CAA*X* motif and processing of the C-terminus of $G\gamma$, provide a form of anchoring of $G\alpha\beta\gamma$ at the membrane (Wedegaertner *et al.*, 1995).

The $G\alpha$ subunit. In general terms, there are four major subfamilies or classes of $G\alpha$ subunit: $G_s\alpha$, $G_q\alpha$, $G_o\alpha$ and $G_i\alpha$. These are known to characteristically activate AC, PLC and ion channels, and to inhibit AC, respectively. Peptide hormone receptors which evoke these effects may interact with one or more of these $G\alpha$ subunits (see *Table 3.1*) and thus activate more than one effector. Promiscuity of receptor–G-protein interaction is a relatively common occurrence *in vitro*, as can be seen from the proportion of different receptors which appear to be able to stimulate both AC and PLC. For example, the AT_1 receptor stimulates PLC through activation of $G_q\alpha$ and inhibits AC by activation of $G_i\alpha$. In some cases, divergence of response can be attributed to the activation of different receptor subtypes. Indeed, two calcitonin receptor subtypes, expressed in a kidney proximal tubule cell line (LLC-PK1 cells), activate distinct signalling pathways and are differentially expressed at different stages of the cell cycle (Horne *et al.*, 1994). However, in certain cases a *single* cloned receptor can elicit dual or even multiple effects in a given cell type. In some systems, the signalling pathway initiated by a receptor may depend on the agonist concentration, as is the case for the TSH, calcitonin and LH/hCG receptors which primarily stimulate the activation of AC at low agonist concentrations, and at higher concentrations lead to the activation of PLC (Gudermann *et al.*, 1992). In contrast with this apparent promiscuity, many receptors activate only one Gα class in all circumstances, at least *in vitro*. The β1-, β2- and β3-ARs, for instance, activate only AC. Furthermore, several different receptor species may be expressed in a single cell type which act upon one particular signal cascade.

A degree of specificity between Gα subunits and receptors is due in part to the C-terminus as shown in co-transfection studies with the D_2 receptor and chimaeric Gα subunits. D_2 receptors normally activate $G_{i2}\alpha$ subunits exclusively to inhibit AC. However, replacement of the C-terminal 30 amino acids of the $G_q\alpha$ subunit by those from $G_{i2}\alpha$ leads to an uncharacteristic interaction between D_2 and $G_q\alpha$ causing activation of PLC (*Figure 3.5*) (Conklin *et al.*, 1993). Other lines of evidence including mutagenesis and/or deletion of residues of the C-terminus support the importance of these residues for receptor interaction (Conklin and Bourne, 1993; Rens-Domiano and Hamm, 1995). However, other regions of Gα and also regions of Gβ (Wall *et al.*, 1995) and Gγ (Kleuss *et al.*, 1993) are likely to confer additional specificity to receptor Gαβγ interactions.

G-protein-linked receptors (GPLR). There are certain structural features of the receptor–G-protein interface that determine the selectivity, or otherwise, of receptors for a particular set of G-proteins. In addition to the aforementioned regions of the Gαβγ complex which can be recognized by certain receptors, there are regions within the intracellular domains of GPLRs which are important for conferring G-protein selectivity. Much of the pioneering work in this area has centred upon studies of the β-AR. This receptor has come to be regarded as the classical GPLR model for relating structure–function relationships, although it is important not to assume that identical features determine the action of all other GPLRs. Mutational analysis of the β-AR has identified specific regions within the intracellular domains, in particular the N- and C-terminal portions of the third intracellular loop (I_3) and sections of the C-terminal tail, as being critical for G-protein coupling and activation of AC (Savarese and Fraser, 1992). In a similar manner, mutation of the AT_1 receptor has suggested that comparable regions within the intracellular domains of this receptor are important for activation of PLC (Sandberg, 1994). However, since β-AR and the AT_1 receptor initiate distinct, and in some situations opposing, G-protein-mediated effects, the nature of the component residues within these regions is most likely to determine the extent of G-protein selectivity for a given receptor.

Further evidence to support this hypothesis has been gained from experiments involving the construction of chimaeric receptors. Again, the β-AR has been

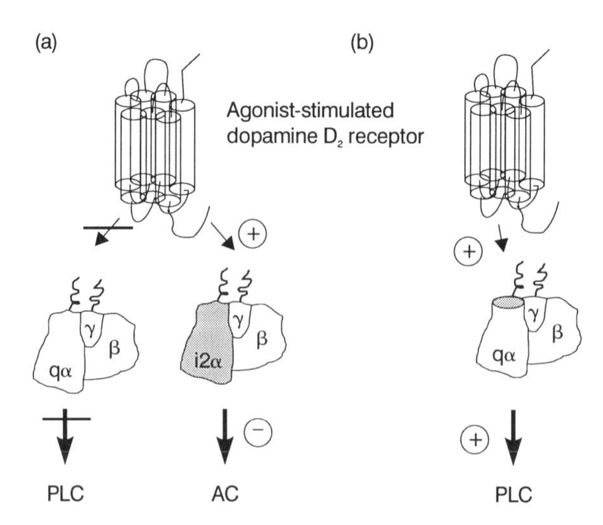

Figure 3.5. Diagrammatic representation of the use of chimaeric proteins to identify structural regions responsible for receptor–G-protein selectivity/specificity (Conklin *et al.*, 1993). (a) Wild-type interaction of dopamine D_2 receptor selectively activates $G_{i2}\alpha$, resulting in the inhibition of adenylyl cyclase (AC), but has no effect on phospholipase C (PLC) activity mediated through the $G_2\alpha$ complex; (b) switching of specificity using chimaeric $G_q\alpha$, containing C-terminal residues from $G_{i2}\alpha$ (shaded portion), leading to uncharacteristic D_2 receptor-mediated stimulation of PLC.

extensively used in this context. In one study, systematic substitution of limited stretches of sequence encoding the intracellular domains of the β-AR ($G_s\alpha$-activating) into the m1-acetylcholine receptor ($G_q\alpha$-activating) resulted in a switching of signalling pathways from the m1-acetylcholine wild-type stimulation of the inositol phosphate pathway to the generation of cAMP which is characteristic of β-AR (Wong and Ross, 1995).

There are also common structural elements within the intracellular domains of all GPLRs which are necessary for G-protein activation but which may not necessarily contribute to selectivity. Thus, the snake venom peptide mastoparan, which forms an amphipathic helix, is able to activate G-proteins in the absence of receptor (Wong and Ross, 1995). This implies that specific regions within the intracellular loops of GPLRs which can take up this conformation are responsible for the physical activation of the G-protein. Polypeptides corresponding to regions of the intracellular domains of receptors have also been used to study their effect on G-protein activation in the absence of intact receptor (Okamoto and Nishimoto, 1992). In addition, use of minigenes encoding similar polypeptides has shown that specific peptides can competitively and selectively inhibit the endogenous signal transduction pathways for a particular receptor *in vitro* (Hawes *et al.*, 1994).

The C-terminal tail of the GPLR (*Figure 3.3*) may also play a role in determining selectivity. Splice variants of the prostaglandin E receptor, which have different lengths of C-terminus, exhibit differing degrees of promiscuity, with the shortest form interacting exclusively with $G_s\alpha$ (Namba *et al.*, 1993). Conversely, analysis of mutant forms of PTH receptor with progressive truncations of the C-terminus exhibited an increase in promiscuity as the length of the C-terminus was reduced (Schneider *et al.*, 1994). In some GPLRs, a C-terminal motif containing a palmitoylated cysteine residue may allow the formation of a fourth intracellular loop by anchoring a portion of the C-terminal tail within the plasma membrane (Bouvier *et al.*, 1995). One of the effects of this modification may be to reveal specific structural epitopes to certain $G\alpha\beta\gamma$ complexes.

Another feature of the GPLR which influences G-protein coupling is the aspartate residue of the DRY motif (see Section 3.2.3). Point mutation of this residue in an α2A-AR mutant caused a loss of ability to increase potassium currents, while the ability to couple to $G_i\alpha$ (to inhibit AC) and inhibit voltage-gated calcium currents remained intact (Surprenant *et al.*, 1992). It is possible that this key aspartate residue in GPLRs is important in maintaining the conformation of the receptor to allow for appropriate G-protein coupling.

Apart from these aspects of receptor structure, there are many other factors in a cell which will determine which, and to what extent, different signalling pathways are activated. These may include receptor density, and the relative expression of the various G-protein and effector molecule species.

3.3.3 Effector molecules

AC and PLC are the main effector molecules which are the targets of peptide hormone receptor-initiated signal cascades.

Adenyl cyclase. AC is a membrane-bound enzyme which catalyses the conversion of ATP to cAMP and inorganic pyrophosphate inside the cell. There are now known to be at least 10 AC isoforms which share a basic structure consisting of two sets of six membrane-spanning domains and two cytoplasmic domains, each of which contain highly homologous catalytic domains. Eight of these isoforms have been well characterized and fall into three broad groups based on sequence homology and shared regulatory mechanisms. Although all AC isoforms are stimulated by $G_s\alpha$, the different isoforms are regulated in a complex manner by several different modulators (Sunahara *et al.*, 1996) (*Table 3.2*). The effects of calcium provide a typical example of the differential regulation of AC. ACI and ACVIII (expressed in specific regions of the brain) are both stimulated by Ca^{2+}/calmodulin so that increases in $[Ca^{2+}]_i$ could theoretically lead to increases in cAMP. In contrast, in peripheral tissues, elevated $[Ca^{2+}]_i$ can cause direct inhibition of ACs V and VI. The net outcome is dependent on, for example, the number and nature of receptor systems activated and the relative expression of AC isoforms. In addition, intracellular [cAMP] is also under the considerable influence of phosphodiesterases (see Section 3.4.2).

The primary effect of cAMP as a second messenger is to provoke the release of the catalytic domain of cAMP-dependent protein kinase (PKA) from regulatory subunits. This leads to the downstream effects of PKA such as phosphorylation of cAMP response element-binding protein (CREB), which has a direct effect on gene transcription (Hunter and Karin, 1992). PKA can also phosphorylate various signalling molecules, including GPLRs, in a heterologous manner, and may exert negative feedback effects by phosphorylation of specific AC isoforms.

Table 3.2. Features of the mammalian adenylyl cyclase (AC) isoforms

AC isoform	Positive modulators	Negative modulators	Tissue distribution
ACI	$G_s\alpha$, Ca^{2+}/CaM[a]	$G_i\alpha/G_o\alpha$, $G\beta\gamma$	Brain
ACIII	$G_s\alpha$, Ca^{2+}/CaM	$G_i\alpha/G_o\alpha$, $G\beta\gamma$	Brain, heart, lung, olf. ep.[b]
ACVIII	$G_s\alpha$, Ca^{2+}/CaM	$G_i\alpha/G_o\alpha$, $G\beta\gamma$?	Brain
ACII	$G_s\alpha$, $G\beta\gamma$, PKC		Brain>lung, olf. ep.
ACIV	$G_s\alpha$, $G\beta\gamma$		Widespread
ACVII	$G_s\alpha$, $G\beta\gamma$?		Widespread
ACV	$G_s\alpha$, PKC	$G_i\alpha$, PKA, Ca^{2+}	Widespread
ACVI	$G_s\alpha$	$G_i\alpha$, PKA, Ca^{2+}	Widespread

[a] calmodulin; [b] olfactory epithelium. See text for other abbreviations.

Phospholipase C (PLC). The other main G-protein-activated effector molecule is PLC which has three major forms PLC-β, -γ, and -δ. In addition, PLC-β occurs as four different isoforms, β_1–β_4, which range in size from 130 to 150 kDa, are activated by G-proteins from the $G_q\alpha$ subfamily and are, in some cases, inhibited by members of the $G_i\alpha$ subclass. PLC-γ is activated by receptor tyrosine kinases. Both PLC-β and PLC-γ catalyse the hydrolysis of phosphatidylinositol (4,5) bisphosphate (PIP_2) to generate the two second messengers, inositol (1,4,5) trisphosphate (IP_3) and diacylglycerol (DAG) (Exton, 1996) which cause release of calcium from intracellular stores, via the IP_3 receptor, and activation of Ca^{2+}-dependent isoforms of protein kinase C (PKC), respectively (for review see Berridge, 1993). Calcium release from stores also leads to an influx of calcium into the cell, via calcium release-activated calcium (CRAC) channels, by a mechanism referred to as capacitative calcium entry. The process of calcium mobilization is thought to be de-activated in a number of ways. For example, PLC-β can act as a GTPase-activating protein (GAP) thus increasing the rate of GTP hydrolysis from the activated Gα subunit (Berstein *et al.*, 1992) whilst PLC-β_2 is inhibited by direct phosphorylation by PKA (Liu and Simon, 1996). Also, calcium release from stores is transient due to the action of a Ca^{2+} ATPase which is responsible for actively replenishing the calcium stores. Finally, Ca^{2+}/DAG activation of PKC is thought to inhibit capacitative calcium entry, perhaps by phosphorylation of CRAC channels.

Ion channels. Ion channels can also be regarded as effector molecules which can be regulated by the binding of hormone to a GPLR (Hille, 1994). These fall into three main categories: calcium channels, non-selective ion channels (facilitating calcium or sodium flux) and potassium channels. Ion channels are multimeric proteins (consisting of α, β and γ subunits) which associate to form a limited pore whose opening and closing is tightly controlled (Catterall, 1993). In endocrine cells, most calcium channels are voltage-sensitive L-type channels for which opening is primarily directed by the α subunit in response to membrane depolarization. The effects of GPLRs on these channels can be exerted in a number of ways including stimulation as a result of direct phosphorylation by PKA or by interaction with $G_s\alpha$. Inhibition of calcium channel opening can be mediated by activation of GPLRs coupled to specific members of the $G_o\alpha$ subfamily (Kleuss *et al.*, 1993) and may be one general mechanism for inhibition of hormone secretion. Non-selective cation channels represent a separate but large and diverse family which includes members which are activated by cyclic nucleotides, such as cAMP (Kaupp, 1991) and also by inositol (1,4,5) trisphosphate (IP_3) and inositol (1,3,4,5) tetrakisphosphate (IP_4) (Berridge, 1993; Irvine, 1992).

Other effector molecules. In addition to the more established modulation of effector molecules by Gα subunits, Gβγ is an important regulator of these molecules in its own right (Müller and Lohse, 1995). Work in this field has so far centred upon the impact of Gβγ on signal transduction pathways mediated by GPLRs other than those which respond to peptide hormones. However, a number of important aspects of Gβγ action are equally applicable to many peptide hormone GPLRs. Two issues are of particular interest. Firstly, Gβγ has been shown to regulate the activity of all main classes of effector molecule, although these effects may be restricted or preferential towards certain effector molecule isoforms. Thus, ACII and ACIV are selectively stimulated by Gβγ while PLC-β isoforms show a graded response to Gβγ, with PLC-β_4 unresponsive. Some of the effects of Gβγ are synergistic with Gα, some are antagonistic and some are independent of Gα.

Secondly, specific combinations of Gα, Gβ and Gγ can preferentially transduce the signal from a particular receptor to a given effector molecule. This has been elegantly demonstrated by studying the effects of antisense oligonucleotides (directed against various G-protein subunits) on ion channel function (Kalkbrenner *et al.*, 1995; Kleuss *et al.*, 1993). In addition, when Gβγ is acting independently of Gα, as is the case with its ability to activate certain GPLR kinases (GRKs; see Section 3.4.3), this also appears to be Gβγ-combination dependent. This specificity of receptor and effector interaction is probably the result of specific micro-structural differences at the interface between these molecules. It is also therefore possible that different regions of these interacting proteins are responsible for contact with different effector molecule species. However, use of a peptide representing a short sequence from ACII has been shown to block Gβγ effects on several different effector molecules, suggesting a common recognition sequence in these molecules (Chen *et al.*, 1995).

3.4 Regulation of the response

3.4.1 General

In addition to the modulatory effects of individual signalling molecules on other components of the signal cascade, there are a number of other mechanisms which have profound and opposing effects on the progression and duration of a given signal transduction cascade. Potentially one of the most widespread and far-reaching of these involves the reversal of kinase-mediated events by removal of phosphate from phosphorylated proteins by protein phosphatases. There are several families of phosphatase which are categorized on the basis of their substrate preference (Hunter, 1995). Thus, phosphotyrosine phosphatases oppose the effects of tyrosine kinases including receptor tyrosine kinases (e.g. insulin receptor) and soluble tyrosine kinases such as JAK2; serine/threonine kinases can reverse the effects of kinases such as PKC and PKA; and there is also a family of dual-specificity phosphatases which are thought to act upon proteins phosphorylated by dual-specificity kinases including MAP kinase. However, despite the obvious importance of these proteins, this area of our knowledge is one of the least well developed.

Other mechanisms which are also not clearly defined at the present time, and which have an impact on G-protein-mediated events, include the GTPase-activating proteins, or GAPs (see Section 3.3.3), which increase the rate of GTP hydrolysis thus accelerating the deactivation of G-proteins (Ross, 1995). A further group of proteins, recently discovered by yeast two-hybrid screening approaches (Section 2.7.2), known as G-alpha-interacting proteins (GAIPs), appear to interact specifically with $G_i\alpha$ and therefore may have a role in regulating $G\alpha$ subunit activity (DeVries *et al.*, 1995). An additional way in which the magnitude and duration of G-protein-mediated signals can be controlled is by downregulation of the particular G-protein itself. This seems to be particularly pertinent for G_q- and $G_{11}\alpha$ species which have been shown to be effectively eliminated in an agonist-dependent manner, for instance in response to GnRH (Milligan *et al.*, 1995).

Finally, there are two other regulatory mechanisms of considerable potential importance and which are specifically relevant to GPLR-mediated signalling events, namely, the phosphodiesterase-catalysed degradation of cyclic nucleotides and the regulation of receptor expression and function. These are discussed in greater detail below (see Sections 3.4.2 and 3.4.3, respectively).

3.4.2 Phosphodiesterase (PDE)

PDEs catalyse the hydrolysis of the 3′ 5′phosphodiester bond of the second messengers cAMP and/or cGMP to produce the inactive 5′AMP or 5′GMP. There are now known to be seven PDE families (PDE1–PDE7) totalling at least 50 different members which are products of several different genes, although many are splice variants of a parent gene. Aside from a highly conserved catalytic core domain these enzyme families exhibit different substrate specificities and means of regulation, and highly specific tissue and cell type distributions (for a comprehensive review see Beavo, 1995).

The most obvious role for a PDE is to attenuate the response to GPLRs coupled to $G_s\alpha$ by terminating a cAMP-dependent effect. One example of this is provided by PDE4s expressed in Sertoli cells where FSH-induced cAMP-dependent activation of PKA leads to phosphorylation and thereby activation of PDE4 and a reduction in steady state cAMP levels. In contrast, certain members of the PDE1 family are Ca^{2+}/calmodulin-activated and also act as a substrate for PKA. In cells expressing PDE1A, for example, cAMP production is maintained in what is known as a 'feed-forward' cycle. In this case, activation of AC initiates the cycle. Phosphorylation of PDE1A by PKA inhibits the PDE1A–calmodulin interaction decreasing the level of PDE1A available to hydrolyse cAMP. Therefore cAMP concentrations continue to rise. This cycle can, however, be interrupted by increases in intracellular calcium which complexes with calmodulin to increase the amount of catalytically active PDE with a subsequent reduction in cAMP levels. Another example of the diversity of PDE regulation is found in glomerulosa cells of the adrenal cortex which express high levels of PDE2. This PDE is stimulated by cGMP leading to a reduction in cAMP levels and is central to the regulation of aldosterone production by atrial natriuretic peptide. These distinctive modes of action and high degree of PDE isozyme tissue distribution make PDEs potentially important targets for selective inhibitory therapies.

3.4.3 Receptor regulation and desensitization

There are a number of mechanisms by which receptors themselves can be regulated. These include receptor internalization followed by degradation or re-cycling of the receptor (as exemplified by members of the receptor tyrosine kinase family; Yarden and Ullrich, 1988), or downregulation of receptor expression (usually involving a reduction in steady-state mRNA levels) often resulting from prolonged exposure to agonist for a period of hours. However, in the case of GPLRs, a much more rapid process may also occur (in seconds to minutes) which is known as receptor desensitization. This generally involves phosphorylation of the receptor on serine or threonine residues present within the intracellular loops and C-terminal tail. The kinases responsible may include PKA, PKC, or one of a more recently identified family of GPLR kinases or GRKs (Premont *et al.*, 1995). The key characteristic of GRKs is that they preferentially phosphorylate the activated form of the receptor in question. This phosphorylation renders the receptor a suitable substrate for one of yet another family of proteins known as arrestins (Gurevich *et al.*, 1995). Arrestin binding to GRK-phosphorylated receptors has been proposed to mediate the sequestration of phosphorylated receptor in intracellular vesicles or endosomes (Ferguson *et al.*, 1996; Lohse *et al.*, 1990) which then allows the receptor to be de-phosphorylated/re-sensitized by a GPLR phosphatase (Pitcher *et al.*, 1995). Much of the progress in this area has been due to work carried out by Lefkowitz and colleagues which has generated a classical paradigm for this process of rapid homologous receptor desensitization based around their studies of the β-AR (for review see Lefkowitz, 1993), although this should not necessarily be regarded as the only possible mechanism.

Six mammalian GRKs have so far been cloned which can be classified according to the nature of their C-terminal regulatory domain (Premont *et al.*, 1995). β-AR kinase (β-ARK) 1 and -2 (GRK2 and -3, respectively) possess extended C-termini which interact with G-protein βγ subunits and are thereby directed to the receptor substrate, presumably by virtue of the lipid-modified C-terminus of the Gγ subunit. Other GRKs have lipid modifications of their own which facilitate their interaction with membrane lipids independently of Gβγ. The GRKs exhibit some degree of tissue distribution specificity. GRK4 expression, for example, is almost exclusive to the testes while GRK2 and -3, although widespread, are particularly abundant in peripheral blood leukocytes (see *Table 3.3*). Some substrate specificity has also been demonstrated for GRKs in terms of receptor selectivity. This information has been gained from reconstitution studies using purified receptors and kinases in lipid vesicles, and also from studies in which receptors and various GRKs are over expressed in a heterologous cell

Table 3.3. Characteristics of the family of G-protein-linked receptor kinases (GRKs)[a]

Family member	Common name	Mode of activation	Tissue distribution
GRK1	Rhodopsin kinase	Light/$[Na^+]\downarrow$	Retina>>pineal
GRK2	β-ARK1	Requires Gβγ	PBL[b]>brain>heart> lung>kidney
GRK3	β-ARK2	Requires Gβγ	Brain>spleen>heart, lung, kidney
GRK4	IT-11	Phospholipid-directed auto-phosphorylation	Testis>>>brain
GRK5	GRK5	Phospholipid-directed auto-phosphorylation	Heart, placenta, lung>sk[c]>brain, liver, pancreas >kidney
GRK6	GRK6	Phospholipid-directed auto-phosphorylation	Brain, sk>pancreas >heart, lung, kidney, placenta>liver

[a] For further information see Inglese *et al.* (1993).[b] Peripheral blood leukocytes; [c] skeletal muscle.

line. However, the results of these studies have not absolutely defined receptor–GRK preferences, suggesting that other factors apart from receptor-specific interactions are important *in vivo*. Nevertheless, a number of GPLRs have C-terminal domains with numerous serine/threonine residues which could provide sites for phosphorylation by a GRK. Furthermore, the first definitive identification of the GRK phosphorylation sites in a GPLR has indicated that it is indeed specific C-terminal serine residues which are phosphorylated by GRK2 in the β-AR (Fredericks *et al.*, 1996). Information regarding GRK phosphorylation of peptide hormone receptors has not been as forthcoming as that for the bioamine receptors but some evidence has recently emerged to suggest that the AT_1 receptor is a substrate for phosphorylation by some GRKs (Barker *et al.*, 1995a,b; Oppermann *et al.*, 1996). In contrast, the human GnRH receptor, which has virtually no C-terminus, does not undergo the process of rapid homologous desensitization. Instead this receptor is desensitized over a much longer time course which probably reflects the context in which this receptor functions (Davidson *et al.*, 1994).

An understanding of these desensitization mechanisms is not purely of esoteric importance. It has already been shown, for example, that an increase in β-ARK in the hearts of patients suffering from chronic heart failure results in a diminished response to β-agonists (Ungerer *et al.*, 1993). Furthermore, studies of cardiac function in transgenic mice in which β-ARK or an inhibitor of β-ARK are overexpressed have demonstrated the *in vivo* significance of such mechanisms in the functional regulation of physiological processes (Koch *et al.*, 1996). It is entirely possible, therefore, that abnormalities in the desensitization apparatus could have equally profound pathophysiological consequences in other systems in which peptide hormone receptors play a major role.

3.5 Future perspectives

The great expansion in our knowledge concerning the many different families of signalling molecules, and within them the wide diversity of individual members, has had two major impacts on our understanding of cellular function. The identification of multiple protein isoforms at each step of each signal transduction pathway has advanced an explanation for the diverse effects which can be elicited by a particular hormone in a given cell/tissue context. However, this diversity of signalling molecules presents a new challenge in trying to establish how these multiple pathways are selectively coordinated. This challenge is even more pronounced in the face of increasing evidence which suggests a potential for significant cross-talk between receptor tyrosine kinase- and GPLR-mediated pathways, which were previously regarded as quite independent (Post and Brown, 1996). In addition, certain key molecules contain common interaction motifs such as the SH2 and SH3 domains and the pleckstrin homology (PH) domains, the latter being present in PLC-β, -γ and -δ, β-ARK, and in Shc and IRS-1(Lemmon *et al.*, 1996), and may play a role in directing these molecules to interact with membrane phospholipid.

The new generation of techniques now becoming available, such as the yeast two-hybrid system (see Section 2.7.2), should allow a more systematic appraisal of protein–protein interactions and thereby make it possible to define which are potentially valid. However, it is likely that additional constraints are imposed upon such interactions in the physiological context by so far unresolved mechanisms. For example, selective compartmentalization of different sets of signalling molecules within distinct intracellular 'microzones' (Neubig, 1994) may maximize specific interactions. Other techniques such as differential display reverse transcriptase PCR (see Section 2.4) should provide us with direct information regarding stimulus-specific changes in gene expression in a given cell type. This, together with a more rigorous use of antisense oligonucleotide and associated technologies, will allow us to assess the relative contributions of particular molecules within signal transduction pathways which lead to a specific cellular response. Alongside these advances, two of the major goals of the investigator remain unchanged, namely to identify the specific loci of pathological disturbances and to pinpoint targets for selective therapeutic intervention.

References

Argetsinger LS, Hsu GW, Myers MG Jr, Billestrup N, White MF, Carter-Su C. (1995) Growth hormone, interferon-gamma, and leukemia inhibitory factor promoted tyrosyl phosphorylation of insulin receptor substrate-1. *J. Biol. Chem.* **270:** 14685–14692.

Barker S, Kapas S, Fluck RJ, Clark AJL. (1995a) Effects of the selective protein kinase C inhibitor Ro 31–7459 on human angiotensin II receptor desensitisation and intracellular calcium-release. *FEBS Lett.* **369:** 263–266.

Barker S, Müller S, Cammas FM, Straub A, Winstel R, Clark AJL, Lohse MJ. (1995b) Direct phosphorylation of the angiotensin (AT1 receptor) by protein kinase C alpha and βARK-1. *J. Endocrinol.* **146S:** P312.

Beavo JA. (1995) Cyclic nucleotide phosphodiesterases: functional implications of multiple isoforms. *Physiol. Rev.* **75:** 725–748.

Berridge MJ. (1993) Inositol trisphosphate and calcium signaling. *Nature* **361:** 315–325.

Berstein G, Blank JL, Jhon D-Y, Exton JH, Rhee SG, Ross EM. (1992) Phospholipase C-β1 is a GTPase-activating protein for $G_{q/11}$, its physiologic regulator. *Cell* **70:** 411–418.

Bouvier M, Moffett S, Loisel TP, Mouillac B, Hebert T, Chidiac P. (1995) Palmitoylation of G-protein-coupled receptors: a dynamic modification with functional consequences. *Biochem. Soc. Trans.* **23:** 116–120.

Brown EM, Gamba G, Riccardi D *et al*. (1993) Cloning and characterization of an extracellular Ca^{2+}-sensing receptor from bovine parathyroid. *Nature* **366:** 575–580.

Catterall WA. (1993) Structure and function of voltage-gated ion channels. *Trends Neurosci.* **16:** 500–506.

Cheatham B, Kahn CR. (1995) Insulin action and insulin signaling network. *Endocr. Rev.* **16:** 117–142.

Chen J, DeVivo M, Dingus J *et al.* (1995) A region of adenyl cyclase 2 critical for regulation by G protein βγ subunits. *Science* **268:** 1166–1169.

Cone RD, Mountjoy KG, Robbins LS, Nadeau JH, Johnson KR, Roselli-Rehfuss L, Mortrud MT. (1993) Cloning and functional characterization of a family of receptors for the melanotropic peptides. *Ann. NY. Acad. Sci.* **680:** 342–363.

Conklin BR, Bourne HR. (1993) Structural elements of Gα subunits that interact with Gβγ, receptors, and effectors. *Cell* **73:** 631–641.

Conklin BR, Farfel Z, Lustig KD, Julius D, Bourne HR. (1993) Substitution of three amino acids switches receptor specificity of $G_q\alpha$ to that of $G_i\alpha$. *Nature* **363:** 274–276.

Cosman D, Lyman SD, Idzerda RL, Beckmann MP, Park LS, Goodwin RG, March CJ. (1990) A new cytokine receptor superfamily. *Trends Biochem. Sci.* **15:** 265–269.

Curnow KM, Pascoe L, Davies E, White PC, Corvol P, Clauser E. (1995) Alternatively spliced human type 1 angiotensin II receptor mRNAs are translated at different efficiencies and encode two receptor isoforms. *Mol. Endocrinol.* **9:** 1250–1262.

David M, Petricoin III EF, Igarashi K, Feldman JM, Finbloom DS, Larner AC. (1994) Prolactin activates the interferon-regulated p91 transcription factor and the JAK2 kinase by tyrosine phosphorylation. *Proc. Natl Acad. Sci. USA* **91:** 7174–7178.

Davidson JS, Wakefield IK, Millar RP. (1994) Absence of rapid desensitization of the mouse gonadotropin-releasing hormone receptor. *Biochem. J.* **300:** 299–302.

DeLean A, Stadel JM, Lefkowitz RJ. (1980) A ternary complex model explains the agonist-specific binding properties of the adenylate cyclase-coupled β-adrenergic receptor. *J. Biol. Chem.* **255:** 71108–71117.

DeVries L, Mousli M, Wurmser A, Farquhar MG. (1995) GAIP, a protein that specifically interacts with the trimeric G protein $G\alpha_{i3}$, is a member of a protein family with a highly conserved core domain. *Proc. Natl Acad. Sci. USA* **92:** 11916–11920.

Dey BR, Frick K, Lopaczynski W, Nissley SP, Furlanetto RW. (1996) Evidence for the direct interaction of the insulin-like growth factor I receptor with IRS-1, Shc, and Grb10. *Mol. Endocrinol.* **10:** 631–641.

Exton JH. (1996) Regulation of phosphoinositide phospholipases by hormones, neurotransmitters, and other agonists linked to G proteins. *Ann. Rev. Pharmacol. Toxicol.* **36:** 481–509.

Ferguson SG, Downey III WE, Colapietro A-M, Barak LS, Menard L, Caron MG. (1996) Role of β-arrestin in mediating agonist-promoted G protein-coupled receptor internalization. *Science* **271:** 363–366.

Fredericks ZL, Pitcher JA, Lefkowitz RJ. (1996) Identification of the G protein-coupled receptor kinase phosphorylation sites in the human β_2-adrenergic receptor. *J. Biol. Chem.* **271:** 13796–13803.

Gantz I, Konda Y, Tashiro T, Shimoto Y, Miwa H, Munzert G, Watson SJ, DelValle J, Yamada T. (1993) Molecular cloning of a novel melanocortin receptor. *J. Biol. Chem.* **268:** 8246–8250.

Garbers DL. (1994) Guanylyl cyclase receptors. *J. Biol. Chem.* **269:** 30741–30744.

Gardella TJ, Wilson AK, Keutmann HT, Oberstein R, Potts JT, Kronenberg HM, Nussbaum SR. (1993) Analysis of parathyroid hormone's principal receptor-binding region by site-directed mutagenesis and analog design. *Endocrinology* **132:** 2024–2030.

Gudermann T, Birnbaumer M, Birnbaumer L. (1992) Evidence for dual coupling of the murine luteinizing hormone receptor to adenyl cyclase and phosphoinositide breakdown and Ca^{2+} mobilization. *J.Biol.Chem.* **267:** 4479–4488.

Gurevich VV, Dion SB, Onorato JJ, Ptasienski J, Kim CM, Strene-Marr R, Hosey MM, Benovic JL. (1995) Arrestin interactions with G protein-coupled receptors. *J. Biol. Chem.* **270:** 720–731.

Hawes BE, Luttrell LM, Exum ST, Lefkowitz RJ. (1994) Inhibition of G protein-coupled receptor signaling by expression of cytoplasmic domains of the receptor. *J. Biol. Chem.* **269:** 15776–15785.

Hayashida W, Horiuchi M, Dzau VJ. (1996) Intracellular third loop domain of angiotensin II type-2 receptor. *J. Biol. Chem.* **271:** 21985–21992.

Heckert LL, Daley IJ, Griswold MD. (1992) Structural organisation of the follicle-stimulating hormone receptor gene. *Mol. Endocrinol.* **6:** 70–80.

Hille B. (1994) Modulation of ion-channel function by G-protein-coupled receptors. *Trends Neurosci.* **17:** 531–536.

Horne WC, Shyu J-F, Chakraborty M, Baron R. (1994) Signal transduction by calcitonin: multiple ligands, receptors, and signaling pathways. *Trends Endocrinol. Metab.* 5: 395–401.

Hunter T. (1995) Protein kinases and phosphatases: the Yin and Yang of protein phosphorylation and signaling. *Cell* 80: 225–236.

Hunter T, Karin M. (1992) The regulation of transcription by phosphorylation. *Cell* **70:** 375–387.

Hunyady L, Balla T, Catt KJ. (1996) The ligand binding site of the angiotensin AT1 receptor. *Trends Pharmacol. Sci.* **17:** 135–140.

Inglese J, Freedman NJ, Koch WJ, Lefkowitz RJ. (1993) Structure and mechanism of the G-protein-coupled receptor kinases. *J. Biol. Chem.* **268:** 23735–23738.

Irvine RF. (1992) Inositol phosphates and Ca^{2+} entry: toward a proliferation or a simplification? *FASEB J.* **6:** 3085–3091.

Kalkbrenner F, Degtiar VE, Schenker M, Brendel S, Zobel A, Heschler J, Wittig B, Schultz G. (1995) Subunit composition of G_0 proteins functionally coupling galanin receptors to voltage-gated calcium channels. *EMBO J.* **14:** 4728–4737.

Kaupp UB. (1991) The cyclic nucleotide-gated channels of vertebrate photoreceptors and olfactory epithelium. *Trends Neurosci.* **14:** 150–157.

Kleuss C, Scherübl H, Hescheler J, Schultz G, Wittig B. (1993). Selectivity in signal transduction determined by γ subunits of heterotrimeric G proteins. *Science* **259:** 832–834.

Koch WJ, Milano CA, Lefkowitz RJ. (1996) Transgenic manipulation of myocardial G-protein-coupled receptors and receptor kinases. *Circ. Res.* **78:** 511–516.

Krystek SR, Patel PS, Rose PM *et al.* (1994) Mutation of peptide binding site in transmembrane region of a G protein-coupled receptor accounts for endothelin receptor subtype selectivity. *J. Biol. Chem.* **269:** 12383–12386.

Leeb-Lundberg LMF, Mathis SA, Herzig MCS. (1994) Antagonists of bradykinin that stabilise a G-protein-uncoupled state of the B2 receptor act as inverse agonists in rat myometrial cells. *J. Biol. Chem.* **269:** 25970–25973.

Lefkowitz RJ. (1993) G protein-coupled receptor kinases. *Cell* **74:** 409–412.

Lefkowitz RJ, Cotecchia S, Samama P, Costa T. (1993) Constitutive activity of receptors coupled to guanine nucleotide regulatory proteins. *Trends Pharmacol. Sci.* **14:** 303–307.

Lemmon MA, Ferguson KM, Schlessinger J. (1996) PH domains: diverse sequences with a common fold recruit signaling molecules to the cell surface. *Cell* **85:** 621–624.

LeRoith D, Werner H, Beitner-Johnson D, Roberts CT. (1995) Molecular and cellular aspects of the insulin-like growth factor I receptor. *Endocr. Rev.* **16:** 143–163.

Li B, Jyk T, Oare D, Yen R, Fairbrother WJ, Wells JA, Cunningham BC. (1995) Minimization of a polypeptide hormone. *Science* **270:** 1657–1660.

Liu M, Simon MI. (1996) Regulation by cAMP-dependent protein kinase of a G-protein-mediated phospholipase C. *Nature* **382:** 83–87.

Lohse MJ, Benovic JL, Caron MG, Lefkowitz RJ. (1990) Multiple pathways of rapid β_2-adrenergic receptor desensitization: delineation with specific inhibitors. *J. Biol. Chem.* **265:** 3202–3209.

Mende U, Schmidt CJ, Yi F, Spring DJ, Neer EJ. (1995) The G protein γ subunit: requirements for dimerization with β subunits. *J. Biol. Chem.* **270:** 15892–15898.

Meyer DJ, Campbell GS, Cochran BH, Argetsinger LS, Larner AC, Finbloom DS, Carter-Su C, Schwartz J. (1994) Growth hormone induces a DNA binding factor related to the interferon-stimulated 91kDa transcription factor. *J. Biol. Chem.* **269:** 4701–4704.

Milligan G, Wise A, MacEwan DJ *et al.* (1995) Mechanisms of agonist-induced G-protein elimination. *Biochem. Soc. Trans.* **23:** 166–170.

Müller S, Lohse MJ. (1995) The role of G-protein βγ subunits in signal transduction. *Biochem. Soc. Trans.* **23:** 141–148.

Nahmias C, Strosberg AD. (1995) The angiotensin AT_2 receptor: searching for signal-transduction pathways and physiological function. *Trends Pharmacol. Sci.* **16:** 223–225.

Namba T, Sugimoto Y, Negishi M, Irie A, Ushikubi F, Kakizuka A, Ito S, Ichikawa A, Narumiya S. (1993) Alternative splicing of C-terminal tail of prostaglandin E receptor subtype EP3 determines G-protein specificity. *Nature* **365:** 166–170.

Neer EJ. (1995) Heterotrimeric G proteins: organizers of transmembrane signals. *Cell* **80:** 249–257.

Neubig RR. (1994) Membrane organisation in G-protein mechanisms. *FASEB J.* **8:** 939–946.

Offermanns S, Schultz G. (1994) Complex information processing by the transmembrane signaling system involving G proteins. *Naunyn-Schmiedeberg's Arch. Pharmacol.* **350:** 329–338.

Okamoto T, Nishimoto I. (1992) Detection of G protein-activator regions in M4 subtype muscarinic, cholinergic, and alpha 2-adrenergic receptors based upon characteristics in primary structure. *J. Biol. Chem.* **267:** 8342–8346.

Oppermann M, Freedman NJ, Alexander RW, Lefkowitz RJ. (1996) Phosphorylation of the type 1A angiotensin II receptor by G-protein-coupled receptor kinases and protein kinase C. *J. Biol. Chem.* **271:** 13266–13272.

Parmentier M, Libert F, Vassart G. (1995) La famille des récepteurs couplés aux protéines G et ses orphelins. *Médecine Sciences* **11:** 222–231.

Patti ME, Sun XJ, Bruening JC, Araki E, Lipes MA, White MF, Kahn CR. (1995) 4PS/insulin receptor substrate (IRS)-2 is the alternative substrate of the insulin receptor in IRS-1-deficient mice. *J. Biol. Chem.* **270:** 24670–24673.

Pitcher JA, Payne ES, Csortos C, DePaoli-Roach AA, Lefkowitz RJ. (1995) The G-protein-coupled receptor phosphatase: a protein phosphatase type 2A with a distinct subcellular distribution and substrate specificity. *Proc. Natl Acad. Sci. USA* **92:** 8343–8347.

Post GR, Brown JH. (1996) G protein-coupled receptors and signaling pathways regulating growth responses. *FASEB J.* **10:** 741–749.

Postel-Vinay MC, Finidori J. (1995) Growth hormone receptor: structure and signal transduction. *Eur. J. Endocrinol.* **133:** 654–659.

Premont RT, Inglese J, Lefkowitz RJ. (1995) Protein kinases that phosphorylate activated G proteins-coupled receptors. *FASEB J.* **9:** 175–182.

Probst WC, Snyder LA, Schuster DI, Brosius J, Sealfon SC. (1992) Sequence alignment of the G-protein coupled receptor superfamily. *DNA Cell Biol.* **11:** 1–20.

Rands E, Candelore MR, Cheung AH, Hill WS, Strader CD, Dixon RAF. (1990). Mutational analysis of β-adrenergic receptor glycosylation. *J. Biol. Chem.* **265:** 10759–10764.

Reinhart J, Mertz LM, Catt KJ. (1992) Molecular cloning and expression of cDNA encoding the murine gonadotropin-releasing hormone receptor. *J. Biol. Chem.* **267:** 21281–21284.

Rens-Domiano S, Hamm HE. (1995) Structural and functional relationships of heterotrimeric G-proteins. *FASEB J.* **9:** 1059–1066.

Robbins DJ, Zhen E, Cheng M, Xu S, Ebert D, Cobb MH. (1994) Map kinases ERK1 and ERK2: pleiotropic enzymes in a ubiquitous signaling network. *Adv. Cancer Res.* **63:** 93–115.

Ross EM. (1995) G protein GTPase-activating proteins: regulation of speed, amplitude, and signaling selectivity. *Recent Prog. Horm. Res.* **50:** 207–221.

Samama P, Cotecchia S, Costa T, Lefkowitz RJ. (1993) A mutation-induced activated state of the beta 2-adrenergic receptor. Extending the ternary complex model. *J. Biol. Chem.* **268:** 4625–4636.

Sandberg K. (1994) Structural analysis and regulation of angiotensin II receptors. *Trends Endocrinol. Metab.* 5: 28–35.

Savarese TM, Fraser CM. (1992) In vitro mutagenesis and the search for structure-function relationships among G protein-coupled receptors. *Biochem. J.* **283:** 1–19.

Schindler C, Darnell JE Jr. (1995) Transcriptional responses to polypeptide ligands: the JAK-STAT pathway. *Ann. Rev. Biochem.* **64:** 621–651.

Schlosser SF, Almeida OF, Patchev VK, Yassouridis A, Elands J. (1994) Oxytocin-stimulated release of adrenocorticotropin from the rat pituitary is mediated by arginine vasopressin receptors of the V1b type. *Endocrinology* **135:** 2058–2063.

Schneider H, Feyen JHM, Seuwen K. (1994) A C-terminally truncated human parathyroid hormone receptor is functional and activates multiple G proteins. *FEBS Lett.* **351:** 281–285.

Schütz W, Freissmuth M. (1992) Reverse intrinsic activity of antagonists on G protein-coupled receptors. *Trends Pharmacol. Sci.* **13:** 376–380.

Segaloff DL, Ascoli M. (1993) The lutropin/choriogonadotropin receptor ... 4 years later. *Endocr. Rev.* **14:** 324–347.

Skolnik EY, Lee C-H, Batzer A *et al.* (1993) The SH2/SH3 domain containing protein GRB2 interacts with tyrosine phosphorylated IRS-1 and Shc: implications of insulin control of ras signaling. *EMBO J.* **12:** 1929–1936.

Strader CD, Fong TM, Tota MR, Underwood D. (1994) Structure and function of G-protein-coupled receptors. *Ann. Rev. Biochem.* **63:** 101–132.

Sunahara RK, Dessauer CW, Gilman AG. (1996) Complexity and diversity of mammalian adenylyl cyclases. *Ann. Rev. Pharmacol. Toxicol.* **36:** 461–480.

Surprenant A, Horstman DA, Akbarali H, Limbird LE. (1992) A point mutation of the α_2-adrenoceptor that blocks coupling to potassium but not calcium currents. *Science* **257:** 977–980.

Timmermans PBMWM, Wong PC, Chiu AT, Herblin WF. (1991) Nonpeptide angiotensin II receptor antagonists. *Trends Pharmacol. Sci.* **12:** 55–62.

Ungerer M, Böhm M, Elce JS, Erdmann E, Lohse MJ. (1993) Altered expression of β-adrenergic receptor in the failing human heart. *Circulation* **87:** 454–463.

VanderKuur JA, Allevato G, Billestrup N, Norstedt G, Carter-Su C. (1995) Growth hormone-promoted tyrosyl phosphorylation of SHC proteins and SHC association with Grb2. *J. Biol. Chem.* **270:** 7587–7593.

Vassart G, Dumont JE. (1992) The thyrotropin receptor and the regulation of thyrocyte function and growth. *Endocr. Rev.* **13:** 596–611.

Wakao H, Gouilleux F, Groner B. (1994) Mammary gland factor (MGF) is a novel member of the cytokine regulated transcription factor gene family and confers the prolactin response. *EMBO. J.* **13:** 2182–2191.

Wall MA, Coleman DE, Lee E, Iniguez-Lluhi JA, Posner BA, Gilman AG, Sprang SR. (1995) The structure of the G protein heterotrimer $G_i\alpha_1\beta_1\gamma_2$. *Cell* **83:** 1047–1058.

Wedegaertner PB, Wilson PT, Bourne HR. (1995) Lipid modifications of trimeric proteins. *J. Biol. Chem.* **270:** 503–506.

Wells JA, Cunningham BC, Fuh G, Lowman HB, Bass SH, Mulkerrin MG, Ultsch M, deVos AM. (1993) The molecular basis for growth hormone-receptor interactions. *Recent Prog. Horm. Res.* **48:** 253–275.

Wess J. (1993) Molecular basis of muscarinic acetylcholine receptor function. *Trends Pharmacol. Sci.* **14:** 308–313.

Wong SK, Ross EM. (1994) Chimeric muscarinic cholinergic: beta-adrenergic receptors that are functionally promiscuous among G proteins. *J. Biol. Chem.* **269:** 18968–18976.

Yarden Y, Ullrich A. (1988) Growth factor receptor tyrosine kinases. *Ann. Rev. Biochem.* **57:** 443–478.

Zolnierowicz S, Cron P, Solinas-Toldo S, Friess R, Lin HY, Hemmings BA. (1994) Isolation, characterization, and chromosome localization of the porcine calcitonin receptor gene: identification of two variants of the receptor generated by alternative splicing. *J. Biol. Chem.* **269:** 19530–19538.

4

The steroid hormone superfamily of receptors

V. Krishna K. Chatterjee, Roderick J. Clifton-Bligh, Clare Matthews

4.1 Introduction

The principal mode of action of steroid hormones is to modulate the transcription of target genes in the nucleus, leading to alterations in levels of mRNA synthesis. Receptors for steroid hormones are high affinity ligand-binding proteins (with dissociation constants ~ 10^{-10}M) which exist in low concentrations (2–6 x 10^4 molecules) in cells. The biochemical characterization and purification of these proteins initially led to the cloning of cDNAs encoding the glucocorticoid (GR) and oestrogen (ER) receptors. Since then, a number of highly homologous proteins have been identified which comprise a superfamily of nuclear receptors (Evans, 1988) (*Figure 4.1*). Some of these proteins represent receptors for steroid hormones (androgen (AR), progesterone (PR), mineralocorticoid (MR), vitamin D (VDR)) whilst other members of this family bind structurally unrelated ligands such as thyroid hormone and retinoic acid. However, the majority of these novel proteins have been designated 'orphan' receptors as no cognate ligand has yet been identified. Some receptors (GR, ER, AR, MR) are encoded by a single gene locus whereas others (e.g. thyroid (TR), retinoic acid (RAR) receptors) are encoded by two or multiple loci. Moreover, each gene can undergo alternative splicing to generate highly homologous receptor isoforms differing in structure at the amino or carboxylterminus (Gronemeyer and Laudet, 1995). Nuclear receptor homologues have been isolated from *Drosophila, Caenorhabditis elegans*, zebrafish and many higher eukaryotes but not from yeast, suggesting that this family of proteins evolved at the metazoan stage.

The nuclear receptors can be aligned on the basis of primary amino acid sequence homology to delineate distinct domains (*Figure 4.2*). The N-terminal region (A/B domain) varies in length and composition and is poorly conserved between different receptors. A central cysteine-rich region (C domain) exhibits the highest homology and is followed by the C-terminal region (D/E/F domains) which also contains some conserved sequence motifs (Green and Chambon, 1988). These receptors also share a common mode of action as ligand-inducible transcription factors. Following synthesis, the receptors translocate to the nucleus and bind with high affinity to specific regulatory DNA sequences or response elements, usually

Molecular Endocrinology, edited by G. Rumsby and S.M. Farrow.

Class

I Steroid receptors

Receptor	Ligand
GR	Glucocorticoid
PR	Progesterone
ER	Oestrogen
AR	Androgen

II RXR heterodimers

Receptor	Ligand
TRα,β	Thyroid hormone
RARα,β,γ	all-*trans* RA
PPARα,β,γ	Eicosanoids
VDR	1,25-Dihydroxyvitamin D_3
FXR	Farnesoids
LXR	Oxysterols

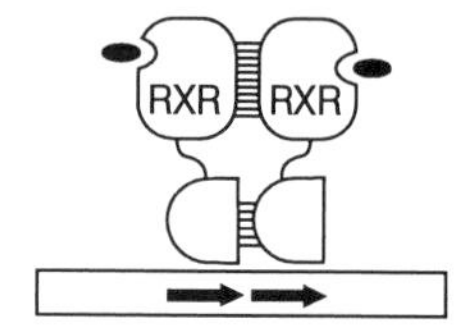

III Dimeric orphan receptors

Receptor	Ligand
RXR	9-*cis* RA
COUP-TF	?
HNF-4	?

IV Monomeric orphan receptors

Receptor	Ligand
SF-1	?
NGF-IB	?
RORα,β	?

Figure 4.1. A functional classification of members of the steroid/nuclear receptor superfamily. Class I includes receptors for steroid hormones which bind as homodimers to palindromically arranged hexanucleotide half-sites as indicated by arrows. Class II contains receptors which form heterodimers with the retinoid X receptor (RXR) on half-sites in a direct repeat configuration. Class III represents orphan receptors which form homodimers and includes RXR which can form homodimers in some contexts when liganded. Class IV are receptors which can bind monomerically to response elements consisting of a slightly extended single hexanucleotide motif. FXR, farnesoid X-activated receptor; LXR, liver-expressed receptor; COUP-TF, chicken ovalbumin upstream promoter transcription factor. For explanation of other abbreviations for receptors, please see relevant text.

located in the promoter region of target genes. Receptor occupancy by ligand then results in modulation of target gene transcription. This review will focus on three aspects of nuclear hormone receptor function:first, the normal pathways of receptor action as well their structural determinants will be considered in detail; second, the role of nuclear receptor defects in human disease will be discussed; finally, the potential physiological functions of orphan receptors will be explored by considering phenotypes associated with targeted disruption of their genes in mice.

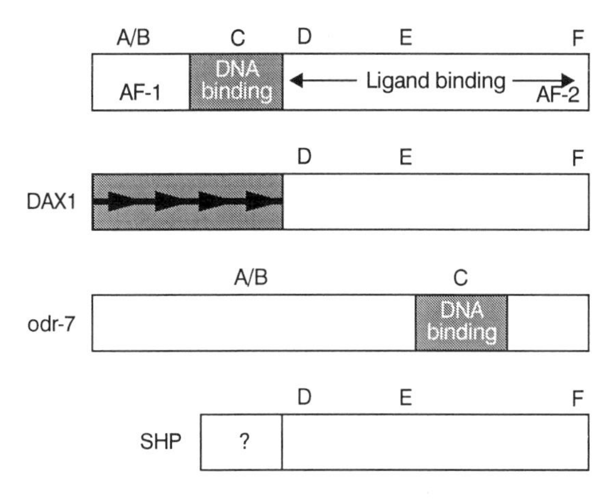

Figure 4.2. Schematic representation of the domain structure of nuclear receptors and some unusual variant receptors. Most steroid and nuclear receptors can be subdivided into six domains (A–F) mediating distinct functions:the A/B domains may contain a constitutive transcriptional activation function (AF-1); the C domain contains zinc-finger motifs which mediate binding to regulatory DNA sequences in target genes; the D/E/F domains encompass ligand binding, dimerization and hormone-dependent transcription activation (AF-2) functions.The C-terminal part of DAX1 (dosage-sensitive sex-reversal–AHC critical region on the X chromosome) shares homology with nuclear receptors, but is coupled to a distinct N-terminal DNA-binding domain which does not contain zinc-finger motifs. The orphan receptor odr-7 from *C.elegans* contains recognizable A/B and C domains but no ligand-binding domain. SHP (small heterodimer partner) is an orphan receptor which contains D/E/F domains but no recognized DNA-binding domain (Seol *et al.,* 1996).

4.2 Structure–function relationships in nuclear receptors

4.2.1 Subcellular location

Following synthesis, some receptor proteins (GR, PR, MR) remain predominantly cytosolic and interact via their C-terminal D/E domains with heat-shock proteins (Hsp90, Hsp70 and others) to form high molecular weight complexes (Pratt, 1993). Occupancy by ligand leads to receptor dissociation from these proteins followed by translocation to the nucleus. In contrast, other receptors (e.g. TR, RAR), do not form complexes with Hsp90 and are constitutively nuclear. For both classes of receptor, transport into the nucleus is dependent on a conserved motif of basic (lysine–arginine) residues in the D domain which acts as a nuclear localization signal (NLS) required for nuclear pore recognition (see *Figure 4.3*), analogous to the NLS in the simian virus 40 T antigen (Guichon-Mantel *et al.,* 1989). Nuclear import is an energy-dependent process and is counterbalanced by passive diffusion of receptors into the cytosol such that receptors shuttle between the two compartments.

4.2.2 Target gene recognition and DNA binding

Inspection of the primary amino acid sequence of the C domain of virtually all nuclear receptors indicates the presence of two cysteine-rich motifs and it was proposed that each of these could coordinate a zinc ion to form a finger-like structure capable of directing sequence-specific DNA binding. Domain-exchange experiments have verified the role of the C domain in DNA binding, and mutational analyses indicate that the receptor–DNA interaction is mediated by a region called the 'P' box at the base of the first finger (*Figure 4.3*), with different amino acids within this region dictating DNA response element recognition. Thus, receptors such as GR, PR, MR and AR, which bind optimally to the hexanucleotide sequence AGAACA, contain the sequence GSCKV within the P box, whereas TR, VDR, RAR and peroxisome-proliferator-activated receptor (PPAR), which recognize the hexanucleotide AGGTCA, contain the residues EGCKG in this region (Umesono and Evans, 1989). When the crystal structures of the DNA-binding domains (DBDs) complexed with their response elements were elucidated for GR (Luisi *et al.*, 1991) and ER (Schwabe *et al.*, 1993), it was confirmed that these P-box residues were encompassed within an α-helix which lay within the major groove of DNA. These structures also showed that the distal part of the second zinc finger forms another α-helix which buttresses the first helix, but does

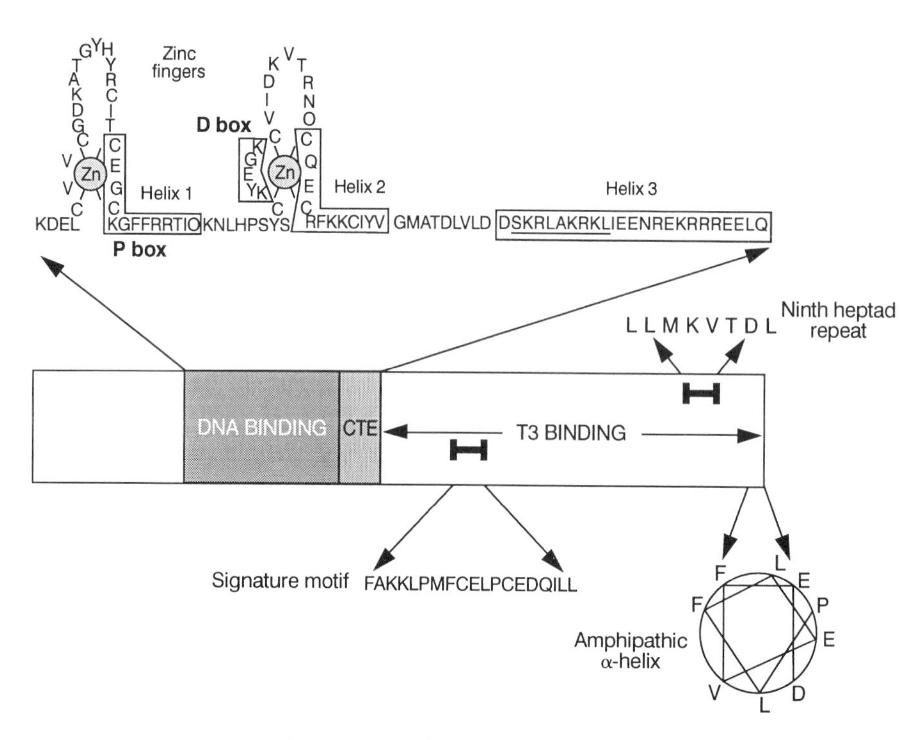

Figure 4.3. A representation of important functional motifs in the thyroid hormone receptor (TR). Two zinc fingers together with a C-terminal extension (CTE) form three α-helices which constitute the DNA-binding domain. The role of the P and D boxes is discussed in the text. A sequence (underlined) within the CTE mediates nuclear localization. Two motifs, the signature and ninth heptad repeat (involved in dimerization), and the amphipathic α-helix (hormone-dependent transcriptional activation) are indicated.

not itself participate in DNA recognition (*Figure 4.3*). Furthermore, it was apparent that the receptors formed homodimers consisting of two DBDs each bound to adjacent hexanucleotide response elements. Protein–protein interactions between residues within the 'D' box at the beginning of the second zinc finger of each DBD (*Figure 4.3*), constituted a dimerization interface. It is reasonable to assume that these structures will be prototypic for other steroid receptors (e.g. PR, MR, AR) which favour binding as homodimers to two hexanucleotide 'half-sites' (AGAACA or AGGTCA), arranged in an inverted repeat configuration (*Figure 4.4*).

However, a subset of nuclear receptors (TR, RAR, VDR, PPAR) were shown to interact with DNA as a complex with an auxiliary nuclear protein (Leid *et al.*, 1992). When isolated, this factor was shown to be the retinoid X receptor – another member of the nuclear receptor family (Kliewer *et al.*, 1992). Thus, RXR is a heterodimeric partner, common to a number of nuclear receptors, which augments their DNA-binding affinity. Another characteristic of this receptor subset is that they bind optimally to response elements consisting of hexanucleotide half-sites arranged tandemly in a 'direct repeat' configuration (*Figure 4.4*). Studies of a large number of natural response elements in target genes have enabled consensus DNA-binding sites to be derived. Such analyses indicate that signalling specificity is conferred by preferential recognition of direct repeats with differing spacings of intervening nucleotides by particular heterodimeric complexes. Thus RXR–VDR binds optimally to a direct repeat (DR) separated by a three nucleotide spacing (DR+3), RXR-TR to a DR+4, and RXR–RAR to either a DR+2 or a DR+5 response element (Umesono *et al.*, 1991). These heterodimeric complexes also exhibit polarity when bound to DNA, such that RXR occupies the 5′ half-site and its partner the 3′ position within a direct repeat (Kurokawa *et al.*, 1994). Crystallographic analyses of the RXR-TR DBDs bound to a DR+4 sequence have confirmed this polarity (Rastinejad *et al.*, 1995). This structure also reveals important differences when

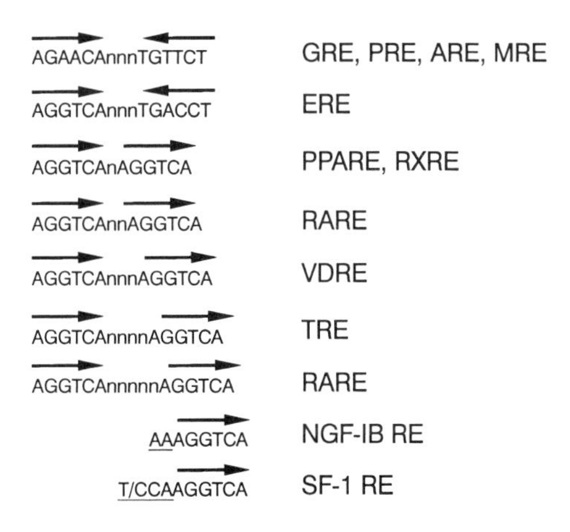

Figure 4.4. The regulatory DNA sequences bound by nuclear receptors have been analysed to derive optimized consensus response elements (RE). An arrow signifies a half-site (AGGTCA or AGAACA) on the coding or non-coding strand, which binds a receptor monomer. Intervening non-conserved 'spacer' nucleotides are denoted 'n'. Some receptors recognize additional nucleotides (underlined) 5′ to the half-site.

compared with ER or GR DBD structures: first, the D box of RXR interacts asymmetrically with residues in the tip of the first zinc finger of TR to constitute a dimer interface; second, a third α-helix extending into the hinge domain of TR (*Figure 4.3*) is also involved in DNA-binding and interacts with the minor groove of spacer nucleotides in the direct repeat. This additional DNA interaction together with the asymmetric dimer interface probably dictates the selectivity of RXR–TR heterodimers for a DR+4 response element – with structural modelling suggesting that alternative nucleotide spacings would be sterically unfavourable.

A third type of receptor–DNA interaction is exhibited by some orphan receptors such as NGF-IB, SF-1, and ROR (*Figure 4.1*), which have been shown to bind monomerically to response elements that are slightly extended half-sites and which include additional nucleotides 5′ to the AGGTCA motif (*Figure 4.4*) (Wilson *et al.*, 1993). Analogous to the third α-helix in TR, it has been suggested that residues distal to the two zinc finger modules, constitute a carboxylterminal extension (CTE) of the DBD which interacts with these additional nucleotides. Lastly, some orphan receptors (COUP-TF, HNF-4, RXR) have been shown to form homodimers on DNA.

Although optimized consensus response elements have been derived for many receptors, some promiscuity has been identified in both: for example, the palindromic response element containing two AGAACA motifs is recognized by a number of steroid receptors and a DR+1 response element can not only mediate transcriptional activation by a PPAR–RXR heterodimer but also by an RXR–RXR homodimer depending on the promoter context (Mangelsdorf *et al.*, 1991). Conversely, receptors such as TR and RAR can also interact with response elements arranged either as an inverted palindrome or an everted repeat with an eight nucleotide spacing (Tini *et al.*, 1994). Future studies are likely to increase the repertoire of known interactions between individual receptors as well as configurations of response elements. This multiplicity represents a mechanism for generating diversity, enabling nuclear receptors to regulate a large number of target genes in a temporal and tissue-specific manner.

4.2.3 Dimerization

The ability of nuclear receptors to form homo- and heterodimers is mediated by at least two protein–protein interactions between adjacent monomers: first, the interface between adjacent DNA-binding domains discussed above; second, regions within the C-terminal ligand-binding domain (LBD) which constitute another dimer interface. The primary amino acid sequence of the TR LBD contains a series of nine heptad repeats of conserved hydrophobic residues (*Figure 4.3*) (Forman and Samuels, 1990), and their homology with the 'leucine zipper' dimerization motif in Fos or Jun led to the hypothesis that they might play a similar role in nuclear receptors. Subsequent mutational analyses have confirmed that the ninth heptad repeat (*Figure 4.3*) is important for dimerization by thyroid and retinoic acid receptors (Au-Fliegner *et al.*, 1993) and a similar region mediates homodimer formation in the ER (Fawell *et al.*, 1990). A number of naturally occurring TR mutants in the first and second heptads identified in thyroid hormone resistance syndrome (see Section 4.3.4) dimerize normally, suggesting that these heptads are not critical for this function (Collingwood *et al.,* 1994). Mutagenesis of residues in another domain, termed the 'thyroid receptor-associated protein (TRAP) domain', in the hinge region of the LBD, indicates that they are also involved in heterodimerization by TR, RAR and VDR (Rosen *et al.,* 1993).

Some insights into the nature of the dimer interface can also be gained from the crystal structures of nuclear receptor LBDs: the RXR LBD crystallized as a dimer with an interface formed mainly by helices 9 and 10 (Bourguet *et al.,* 1995); the TR and RAR LBDs were monomeric, but the hydrophobic residues of the ninth heptad are on the surface of helix 11, with the potential to participate in protein–protein interactions (Renaud *et al.,* 1995; Wagner *et al.,* 1995); finally, the TRAP domain involved in dimerization has been shown to contain residues which are so highly conserved as to constitute a 'signature motif' characteristic of most nuclear receptors. Comparison of the crystal structures indicates that most of these conserved residues mediate interactions between helices 3, 4, 5 and 8 (*Figure 4.5*), which stabilize the hydrophobic core, resulting in the common canonical fold of LBDs (Wurtz *et al.,* 1996). Mutation of these probably impairs dimerization by disrupting the hydrophobic core.

4.2.4 Ligand binding

For most nuclear receptors, the residues involved in ligand binding are widely distributed across the D/E/F domains. These amino acids are thought to form a

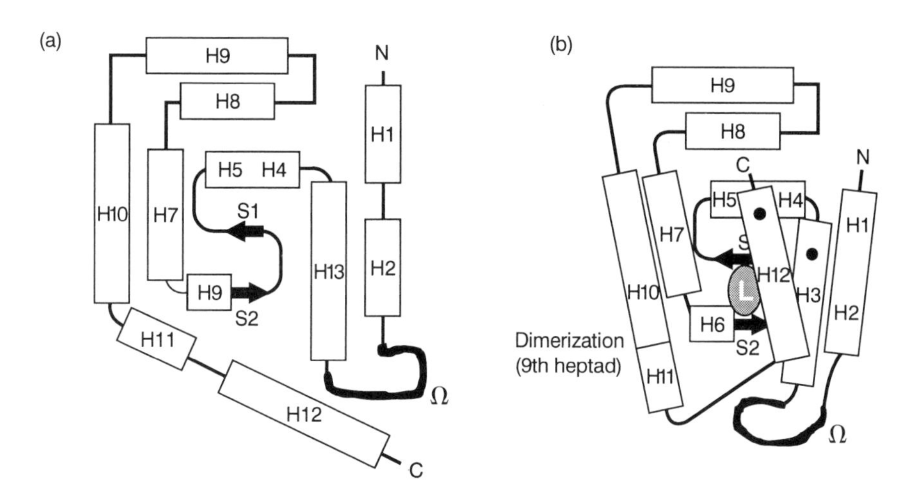

Figure 4.5. Schematic representation of the nuclear ligand-binding domain in the absence (a) and presence (b) of hormone, based on the unliganded RXR and liganded RAR crystal structures (Bourguet *et al.,* 1995; Renaud *et al.,* 1995). On the basis of these structures, it was proposed that ligand binding induces conformational changes, reorienting helices 10 and 11 and releasing helix 12 from the omega loop, such that the latter seals the aperture of the hormone-binding cavity. This also brings helix 12 close to helices 3 and 4, to constitute an activation or co-activator interacting surface. A similar mechanism has been proposed for other nuclear receptors, based on sequence alignments suggesting a common canonical fold (Wurtz *et al.,* 1996). H1–H12, α-helices; S1 and S2, β-strands; L, ligand; Ω, omega loop.

pocket that accomodates ligand, which is usually hydrophobic in nature. Specific cysteine residues have been shown to be important for ligand binding in GR and ER, but determinants in other receptors have not been mapped. For some receptors (GR, PR), ligand binding leads to dissociation from heat-shock proteins, promoting nuclear translocation. The binding of ligand is also associated with marked conformational changes within the LBD, as has been shown by circular dichroism spectroscopy for TR or altered protease sensitivity for PR or RXR.

Many of these biochemical observations have been confirmed by the crystallization of unliganded RXR and liganded RAR and TR LBDs (Bourguet *et al.*, 1995; Renaud *et al.*, 1995; Wagner *et al.*, 1995). The overall topology of these receptor structures is highly homologous, consisting of a triple-layered fold of 12 α-helices. A predominantly hydrophobic pocket within this forms the hormone-binding cavity, and for RAR it has been suggested that ligand is drawn into this by electrostatic forces between a carboxylate group in retinoic acid and positively charged side chains of the amino acids lysine and arginine within the pocket. Comparison of the unliganded RXR and hormone-bound RAR structures indicates marked differences in conformation, leading the authors to propose a 'mousetrap' model, where realignment of helix 12 seals ligand within its binding cavity (*Figure 4.5*). Future crystallization of other receptors in the unliganded state, will determine whether this model is more generally applicable.

Recent studies also suggest that allosteric effects may influence ligand binding – particularly with the subset of receptors which heterodimerize with RXR. There is evidence to suggest that, in a TR–RXR or RAR–RXR complex bound to DNA, the retinoid X receptor is unable to bind its cognate ligand (9-*cis* RA) and remains a 'silent partner' (Forman *et al.*, 1995). However, added complexity is provided by the observation that when RXR forms heterodimers with some orphan receptors (LXR, NGF-IB), it does retain the ability to bind 9-*cis* RA and activate gene transcription (Perlmann and Jansson, 1995; Willy *et al.*, 1995). For example, the oxysterol 22(R)-hydroxycholesterol has recently been shown to activate LXR and this compound acts synergistically with 9-*cis* RA on an LXR–RXR heterodimer (Janowski *et al.*, 1996), indicating that transcriptional signalling via both components of a heterodimer is possible in some contexts.

4.2.5 Repression of basal transcription

A subset of nuclear receptors (e.g. TR, RAR) but not steroid receptors (e.g. GR, PR, ER) exhibit the ability to inhibit or 'silence' basal gene transcription in the absence of ligand. It seems likely that this phenomenon is physiologically relevant as these receptors are constitutively nuclear and have been shown by '*in vivo*' footprinting to bind response elements even when unliganded. Silencing has been studied most intensively with TR, where it has been mapped to the C-terminal D/E/F domains (Baniahmad *et al.*, 1992). One hypothesis for this effect is that the receptor might interact with and inhibit components of the basal transcription complex – and indeed TR does interact with factors such as transcription factor IIB (TF-IIB) (Fondell *et al.*, 1993) or TATA-binding protein (TBP) (Tone *et al.*, 1994).

An alternative hypothesis postulated the existence of an additional cellular factor(s) containing an intrinsic repression function, which could be recruited by unliganded TR, but dissociates from the receptor in the presence of hormone (Casanova *et al.*, 1994). Two cellular factors (nuclear receptor co-repressor (N-CoR) and SMRT/TRAC), have recently been isolated which fulfil criteria as candidate co-repressors. These proteins interact optimally with unliganded TR or RAR heterodimers bound to DNA which results in strong inhibition of basal promoter activity (Chen and Evans, 1995; Horlein *et al.*, 1995). Following ligand binding, the co-repressor dissociates, relieving repression and allowing recruitment of other co-factors which mediate transcriptional activation (*Figure 4.6*).

4.2.6 Activation of transcription

The hallmark of steroid and nuclear receptors is their ability to activate transcription by enhancing the function of the transcription initiation complex. Although studies indicate a direct interaction between some receptors and general transcription factors (e.g. ER and TBP-associated factor II30 (TAFII30) (Jacq *et al.*, 1994); TR and TBP-associated factor 110 (TAF110) (Petty *et al.*, 1996)), there is increasing evidence to suggest that specific 'adaptor' or 'co-activator' proteins may mediate the activation process. Domain-exchange experiments have established that, in general, nuclear receptors contain two types of transcription activation function:their N-terminal A/B regions may contain a weak transactivation function (AF-1) which is constitutive and manifests when unliganded receptor is bound to DNA; in contrast, the C-terminal D/E/F domains encode a powerful ligand-dependent activation function (AF-2).

Constitutive transcription activation function (AF-1). Some members of the nuclear receptor family exhibit the ability to activate target gene transcription in a ligand-independent manner. This activation function (AF-1) has been mapped to the N-terminal A/B domains in most cases, and varies in strength, being very potent in some receptors (e.g. GR) (Hollenberg and Evans, 1988), but weak (e.g. ER) or even absent (e.g. VDR) in others. Another characteristic of AF-1 activity is that, even for a given receptor (e.g. ER), its effects are dependent on target gene and cell type (Tora *et al.*, 1989). Finally, the AF-1 function of some receptors such as ER or PPARγ (Adams *et al.*, 1997) has been shown to be modulated by phosphorylation within the A/B domain. However, little is known about the interaction of the A/B domain with co-activators which might mediate AF-1 activity. A recent study suggested that the A/B domain and D/E/F domains of ER might be capable of interacting *in vivo*, raising the possibility that this might occur via binding of the two regions to a shared common co-activator(s) (Kraus *et al.*, 1995).

Hormone-dependent activation of transcription (AF-2). Previous studies of *v-erbA*, the oncogenic counterpart of TR, had shown that it was unable to regulate transcription in a hormone-dependent manner. Sequence comparisons indicated that this function mapped to nine amino acids at the C-terminus of TR which are deleted in v-erbA (Zenke *et al.*, 1990). A helical wheel plot indicated that this C-terminal motif, made up of hydrophobic and acidic residues, could form an amphipathic α-helix (*Figure 4.3*), analogous to transcription activation domains present in transcription factors such as GAL4 and VP16. This sequence is highly con-

(a) Silencing of basal transcription

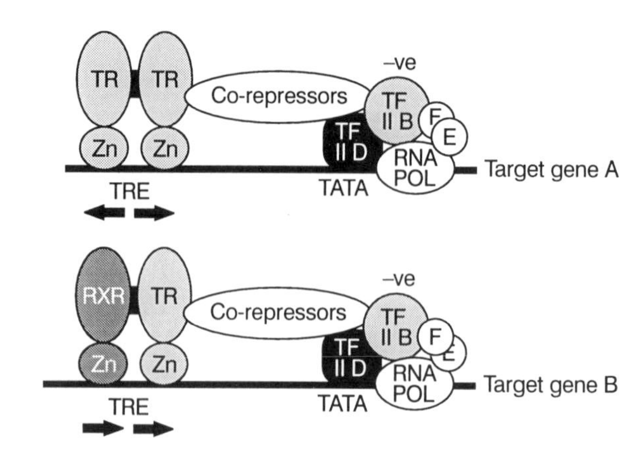

(b) Hormone-induced changes

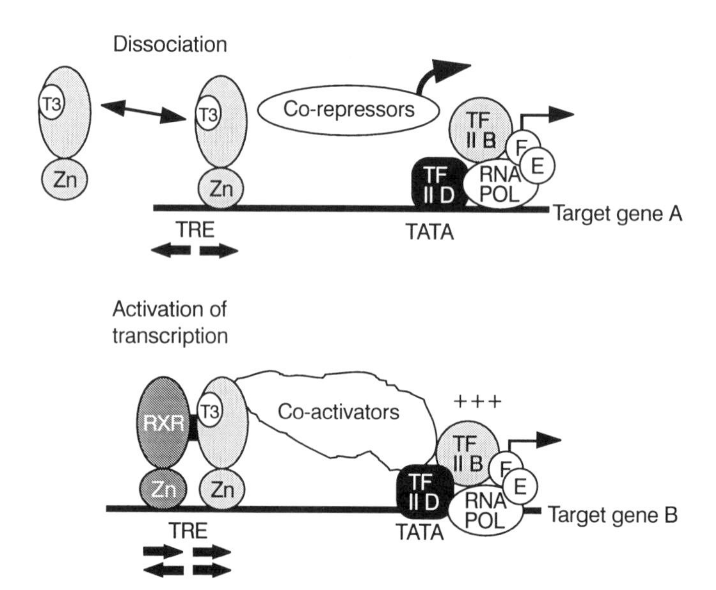

Figure 4.6. Mechanism of action of a Class II nuclear receptor (TR), which is a prototype for other members of this class. (a) In the absence of ligand, the TR binds to response elements as either a homodimer or a heterodimer with RXR. Basal gene transcription is inhibited via intermediary co-repressors. (b) Following T3 binding, TR homodimers dissociate and release of co-repressor results. The TR–RXR heterodimer complex remains DNA bound and recruits co-activators to enhance target gene transcription. The transcription initiation complex is assembled from a number of general transcription factors (TF-IIB, TF-IID, TF-IIE, TF-IIF) and RNA polymerase II (RNA POL).

served amongst nuclear receptors and mutational analyses indicate that hydrophobic and negatively charged (glutamic acid) residues within this motif are critical for AF-2 function in a number of them including ER, TR, RAR and RXR (Danielian *et al.*, 1992; Durand *et al.*, 1994; Tone *et al.*, 1994). The crystal structure of TRα (Wagner *et al.*, 1995) confirms the presence of an amphipathic α-helix (helix 12) at the C-terminus of the receptor, which is also present in the crystal structures of the RARγ and RXRα LBDs. Comparison of the unliganded RXRα versus liganded RARγ crystal structures indicates that the position of helix 12 differs markedly (Renaud *et al.*, 1995), further supporting the notion that ligand binding realigns this α-helix, resulting in co-repressor dissociation as a prelude to transcription activation.

The observation that residues within the amphipathic α-helix which are critical for transcription activation are solvent exposed (Collingwood *et al.*, 1997), supported the hypothesis that this process involved the recruitment of co-activators to the receptor by protein–protein interaction. Far-Western blotting of mammalian cell extracts with radiolabelled receptor LBDs in the presence and absence of ligand has identified several high molecular weight proteins (ER-associated protein (ERAP160, ERAP140), receptor-interacting protein (RIP140)) which bind the ER in a ligand-dependent manner (Cavailles *et al.*, 1995; Halachmi *et al.*, 1994). Cloned RIP140 binds ER in the presence of agonists but not antagonists, fails to interact with a transcriptionally defective ER mutant and enhances transactivation by wild-type ER, fulfilling some criteria for designation as a transcriptional co-activator. Other receptor-interacting proteins (SUG1, transcription intermediary factor (TIF1) and steroid receptor co-activator (SRC-1)), have been identified using the yeast two-hybrid system (Section 2.7.2). TR and RXR bind SUG1 and its human homologue Trip1 (vom Baur *et al.*, 1996). These proteins were initially thought to be part of the RNA polymerase II holoenzyme complex, but may be components of the 26S proteasome and modulate receptor function by altering their turnover. TIF1, which interacts with RXR, RAR and ER, contains a RING finger domain and bromodomain and may mediate chromatin remodelling (vom Baur *et al.*, 1996). SRC-1 was isolated using PR in a yeast two-hybrid assay (Onate *et al.*, 1995), but also interacts with ER, TR and RXR. It is a strong candidate co-activator, as it is recruited to receptor upon binding agonist but not antagonist, enhances receptor AF-2 activity and relieves transcriptional interference or 'squelching' between different receptors. ERAP160 is probably synonymous with SRC-1, and another recently isolated receptor co-factor (TIF2) shares some homology with this protein. Finally, phospho-CREB-binding protein (CBP), a transcription factor which can alter chromatin structure via its histone acetylation function, has also been found to interact with some liganded nuclear receptors (TR, RAR, RXR, ER) in a ternary complex with SRC-1 (Kamei *et al.*, 1996). The precise role of these putative co-activators in mediating transactivation by nuclear receptors remains to be elucidated.

4.2.7 Ligand-dependent transcriptional inhibition

Nuclear receptors can also mediate negative transcriptional regulation by repressing transcription in a ligand-dependent manner. In contrast to response elements mediating transcription activation, such negative response elements are poorly defined and of varying configuration, such that common consensus sequences have not been defined. Furthermore, the mechanism of transcription inhibition by nuclear receptors is also poorly understood. One model suggests that receptor binding to a site which overlaps the DNA sequence recognized by another transcription factor blocks activation by the latter (Akerblom *et al.*, 1988). A second model is exemplified by the transcriptional interference between the GR and the Fos–Jun complex (AP-1), where the activity of the latter appears to be repressed by a direct protein–protein interaction with GR, not requiring sequence-specific DNA-binding by the receptor (Saatcioglu *et al.*, 1994). The recent finding that nuclear receptors can bind proteins such as CBP or signal transducer activator of transcription (Stat5) (Stocklin *et al.*, 1996), which are involved in other signalling pathways, provides a potential third inhibitory mechanism.

4.2.8 Post-translational modification

The transcriptional activity of nuclear receptors can also be modulated by phosphorylation and a number have been shown to exist as phosphoproteins (Orti *et al.*, 1992). The glucocorticoid receptor undergoes phosphorylation at a number of serine residues within the A/B domain which probably contribute to the AF-1 activity of this receptor. The progesterone receptor is phosphorylated on serine and threonine residues at several stages in its pathway of action. Some residues are phosphorylated in the basal state, with others following the binding of ligand and/or interaction with DNA (mediated by a DNA-dependent protein kinase). Phosphorylation of TRα2 – a non-hormone-binding splice variant of TR – regulates its ability to bind DNA and form heterodimers with RXR (Katz *et al.*, 1995).

It has also been suggested that phosphorylation might represent a mechanism whereby some receptors are activated in the absence of ligand. For example, dopamine, acting through its cell surface receptor, phosphorylates and activates PR in the rodent CNS to modulate sexual behaviour (Mani *et al.*, 1994). Phosphorylation of a serine in the A/B domain of the ER by mitogen-activated protein (MAP) kinase augments its AF-1 function, and the receptor also undergoes tyrosine phosphorylation within the LBD, representing a pathway whereby growth factors (e.g. epidermal growth factor) might regulate its action in breast or uterine tissue (Kato *et al.*, 1995).

4.3 Nuclear hormone receptor mutations and disease

Mutations in nuclear receptor genes form the basis of a number of inherited human diseases and defects have been identified using two principal lines of investigation. First, in the 'candidate gene' approach, nuclear receptor defects have been anticipated in cases of hormone resistance, characterized by reduction in target organ responsiveness to circulating hormone. Characterization of the various receptor genes has enabled direct sequencing approaches to dissect the molecular abnormality present in many cases of hormone resistance. The evidence for the pathological role of these defects has then been supported by studies documenting a decreased receptor number or impaired function in each case.

Second, in a 'reverse genetic' approach, linkage studies have localized the molecular basis of a given disease to a chromosomal locus, and positional cloning has revealed a gene encoding a nuclear receptor.

Nuclear receptor defects may be broadly categorized into germline (inherited or sporadic) or somatic. The latter class includes mosaic expression of a receptor defect, and acquired mutations, for instance in tumour cells. Most inherited receptor defects are recessive with disease manifesting only when two defective alleles are present. Dominantly inherited disorders are due to heterozygous defects with haploinsufficiency or inhibition of wild-type receptor action by its mutant counterpart – the phenomenon of the dominant-negative effect.

4.3.1 Androgen receptor defects

Testosterone and its metabolite dihydrotestosterone are critical to male sexual development *in utero* and during puberty (Griffin and Wilson, 1989). The spectrum of androgen resistance ranges from phenotypic females with complete androgen insensitivity (CAIS), to varying degrees of partial insensitivity (PAIS), resulting in ambiguous genitalia or hypospadias and gynaecomastia in males (Quigley *et al.*, 1995). It has also been suggested that some cases of male infertility might represent mild androgen insensitivity. Localization of the AR gene to the Xq11–12 locus and its subsequent isolation led to the pursuit of gene defects in cases of androgen insensitivity. More than 190 mutations have since been described (Quigley *et al.*, 1995) and functional studies have supported the conclusion that the defects identified are the cause of androgen resistance in most cases. The predominance of AR mutations amongst nuclear receptor defects is likely to be due to ascertainment bias; that is, they cause dramatic but not lethal phenotypes. Furthermore, androgen resistance appears to be either recessive or male-specific, since no abnormalities of development or reproductive fitness are known to occur in carrier females (Griffin, 1992).

AR defects have been classified on the basis of androgen-binding assays in genital skin fibroblasts, and have been categorized as receptor positive or receptor negative (absent, decreased or qualitatively abnormal binding) (Griffin and Wilson, 1989). Three broad groups of defects may be defined: (i) mutations leading to reduced receptor number; (ii) mutations in the hormone-binding domain of the AR which disrupt hormone–receptor binding and impair subsequent interaction with transcriptional machinery; and (iii) mutations in the DBD where the receptor is able to bind androgen normally but subsequent interaction with target gene response elements is impaired. In general, 'receptor-positive' cases are associated with DBD mutations whereas receptor-negative resistance involves mutations in the hormone-binding domain. However, mutations in the hormone- or DNA-binding domains that primarily lead to reduced receptor number and, conversely, mutations in the N-terminus or DBD which affect the receptor's affinity for ligand have been described (McPhaul *et al.*, 1993; Zoppia *et al.*, 1993).

The types and position of AR mutations have been comprehensively reviewed (Quigley *et al.*, 1995). Structural defects of the AR gene – complete or partial gene deletions or small insertions – comprise 8% of mutations. Single nucleotide substitutions form the remaining 92%. These are most commonly missense mutations but some premature stop codons and splice junction mutations have been described. Premature stop codons are expected to lead to a receptor truncated of part or all of the C-terminus. Correlating the position or type of mutation with the phenotype it produces has proved difficult owing to phenotypic variability both within and between families harbouring the same mutation (Batch *et al.*, 1993). Nevertheless, complete receptor deletions or mutations which abolish ligand binding generally result in CAIS, whereas less severe reduction in ligand affinity may lead to milder phenotypic abnormalities consistent with PAIS (Quigley *et al.*, 1995). Mutations in the DNA-binding region have been described in both CAIS and PAIS whereas mutations in the N-terminus have only been found in cases of CAIS (Quigley *et al.*, 1995). Interestingly, a premature stop mutation in the N-terminus, described in two siblings with CAIS, was found to lead to downstream initiation and the production of an *N-terminally* truncated receptor (Zoppi *et al.*, 1993). Functional studies *in vitro* demonstrated reduced function for the abnormal receptor, suggesting that, at least in the AR, the N-terminus plays an important role in modulating ligand-dependent transcriptional function (AF-2). A similar phenomenon has been noted in an animal model for androgen resistance, the *Tfm* mouse (Charest *et al.*, 1991).

Germline AR defects have also been reported in Kennedy's syndrome – an X-linked adult-onset form of motor neurone disease characterized by spinal and bulbar muscular atrophy (SBMA) and mild androgen insensitivity evidenced by gynaecomastia, testicular atrophy, oligospermia and raised serum gonadotrophins (Trifiro *et al.*, 1994). In all cases of this syndrome, a trinucleotide (CAG) repeat encoding a polyglutamine tract within the N-terminal A/B region of the AR has been found to be expanded from the normal number (11–33) to between 40 and 62 residues (La Spada *et al.*, 1991). Earlier onset and increased severity of disease correlate with longer repeat expansion (Igarashi *et al.*, 1992), and a tendency toward increased repeat number with transmission from parent to child has been well documented in SBMA families. Functional studies indicate that the polyglutamine expansion can result in altered AR function either due to a decrease in receptor expression (Choong *et al.*, 1996) or its transactivation function (Mhatre *et al.*, 1993), which may account for the mild androgen insensitivity observed. However, neurological features have not been recorded in CAIS or PAIS, suggesting that androgen resistance *per se* is not responsible for neuronal degradation. An alternative model proposes that the excessive number of repeated glutamine residues may promote receptor interaction either with itself or heterologous proteins. This could occur either by the mutant receptor acting as a preferred substrate for protein cross-linking by cellular transglutaminases (Green, 1993) or by the formation of a polar zipper motif by the polyglutamine tract, assisting protein interaction through hydrogen bonds (Perutz *et al.*, 1994). Such inappropriate receptor–protein interactions may generate toxic products resulting in motor neurone loss.

AR defects have also been implicated in oncogenesis. The development of breast cancer has been described in two brothers with PAIS and a germline mutation in the DBD of AR (Wooster *et al.*, 1992). Somatic AR defects have been reported in prostate cancer and, although they are not thought to be a factor in its early pathogenesis, they may contribute to the development of metastatic disease and tumour insensitivity to antiandrogens (Culig *et al.*, 1993; Taplin *et al.*, 1995). Curiously, mutant ARs identified in several cases of metastatic prostate cancer exhibit altered ligand responsiveness such that they can activate transcription in response to progesterone (Culig *et al.*, 1993) and/or oestrogen (Taplin *et al.*, 1995). It has been suggested that such promiscuity for ligand may play a significant role in tumour progression.

4.3.2 Glucocorticoid receptor defects

Generalized inherited glucocorticoid resistance (GIGR) is characterized by hypercortisolaemia with excess adrenocorticotrophic hormone (ACTH) production due to impaired negative feedback by glucocorticoid on the hypothalamus and pituitary. Peripheral stigmata of cortisol excess (Cushing's syndrome) are typically absent and elevations of adrenal androgens and mineralocorticoids, secondary to the excess ACTH drive, are responsible for the clinical features of the syndrome: isosexual precocious pseudopuberty in boys, virilization and menstrual irregularities in women (adrenal androgen excess), and hypertension with hypokalaemia in either sex (excess mineralocorticoid effect). The elevated levels of cortisol and its metabolites may also overwhelm renal 11β-hydroxysteroid dehydrogenase, aggravating the excess activation of the mineralocorticoid receptor. The degree of glucocorticoid insensitivity in this disorder is partial, with the compensatory increase in circulating cortisol levels preventing features of adrenal insufficiency. Complete GIGR has not been documented in man as the phenotype is presumed to be lethal. This notion is supported by the observation that GR-deficient knockout mice die shortly after birth with pulmonary atelectasis (despite the presence of surfactant), and adrenal cortical enlargement and defective hepatic gluconeogenesis (Cole *et al.*, 1995). Biochemical abnormalities in the knockout mice mimicked GIGR with elevated serum corticosterone and aldosterone levels and raised ACTH, reflecting a loss of the negative feedback effects of glucocorticoid in the pituitary–adrenal axis.

Thirty-five cases of GIGR have been described (Arai and Chrousos, 1994) and receptor mutations have been documented in three individuals (reviewed by Brönnegård and Carlstedt-Duke, 1995; Malchoff and Malchoff, 1995). In two cases, homozygosity for different mutations in the ligand-binding domain of GR has been documented. Both mutations were associated with reduced ligand-binding affinities and impaired reporter gene transactivation. A splice-site mutation in GR has also been reported in a large kindred with GIGR, and affected individuals are heterozygous for the receptor defect such that tissue receptor mRNA and protein expression (presumably from the residual normal allele) is 50% that of normal controls (Karl *et al.*, 1993). In this context, it is interesting to note that heterozygous GR-deficient mice also have elevated ACTH and corticosterone levels (Cole *et al.*, 1995), suggesting that loss of one functional GR allele results in glucocorticoid resistance, which contrasts with the lack of phenotype associated with loss of a single allele of the TR gene (see below). In a number of other cases of GIGR, functional receptor abnormalities, including decreased cortisol binding, decreased receptor number, decreased DNA binding and increased thermolability, have been documented, but their molecular basis has not been elucidated.

4.3.3 Oestrogen receptor defects

The long-held view that ER defects would be lethal has been disproved by the description of viable mice homozygous for an insertional disruption of the ER gene (Lubahn *et al.*, 1993). Female ER-deficient mice had hypoplastic uteruses and hyperaemic ovaries and these tissues failed to respond to oestrogens. Surprisingly, male mice were also abnormal with impaired spermatogenesis. Bone mineral density was reduced by 75% in both sexes. A human counterpart of this unexpected male phenotype has also been described recently. A male individual with osteoporosis, tall stature, unfused epiphyses, and biochemical evidence of oestrogen resistance (elevated serum luteinizing hormone and follicle-stimulating hormone in the presence of high serum oestradiol levels) was found to be homozygous for a premature stop mutation in the ER gene (Smith *et al.*, 1994), generating a truncated product predicted to lack both the DNA- and hormone-binding domains. The clinical features of this case and the male knockout mice highlight the importance of oestrogens in epiphyseal maturation, bone mineralization and the regulation of pituitary gonadotrophin secretion in males.

Somatic point mutations or alternate splice variants of the ER are well described in breast cancer although their pathogenic role is disputed. Some of these mutations may result in constitutive activation of the receptor and may allow oestrogen-independent tumour growth (Dickson and Lippman, 1995). Somatic ER defects have also been linked to colorectal cancer – a possible mechanism for this relationship is that inactivation of the ER gene occurs by methylation in ageing colonic mucosa (Issa *et al.*, 1994).

Finally, residual oestradiol binding activity (~5% of wild-type levels) has been observed in tissues from ER-deficient mice. Readthrough transcription of the receptor gene distal to the site of disruption was initially thought to account for this, but an intriguing alternative possibility is the recent description of a second ER gene (ERβ) in the rat. This receptor binds oestradiol and is mainly expressed in prostate and ovary, but its physiological role is not understood (Kuiper *et al.*, 1996).

4.3.4 Thyroid hormone receptor defects

The cloning of TRs and identification of two thyroid hormone genes (TRα and TRβ) in man has led to the elucidation of the pathogenetic defect responsible for the syndrome of resistance to thyroid hormone (RTH). This disorder, usually inherited as an autosomal dominant, is characterized by elevated circulating levels of thyroid hormone together with a failure to suppress pituitary thyroid-stimulating hormone (TSH) secretion and variable peripheral refractoriness to hormone action (Refetoff *et al.*, 1993). Affected individuals are heterozygous for diverse mutations in the TRβ gene which localize to the C-terminal hormone-binding domain (Chatterjee and Beck-Peccoz, 1994). The clinical manifestations of RTH are highly variable ranging from asymptomatic individuals deemed to have generalized resistance (GRTH), to thyrotoxic features such as failure to thrive, low body mass index, tachycardia or dysrhythmia suggesting predominant pituitary resistance (PRTH). Both GRTH and PRTH are associated with TRβ mutations, indicating that the two disorders represent phenotypic variants of a single genetic entity (Adams *et al.*, 1994). In keeping with their location, mutant receptors in RTH exhibit impaired hormone binding and transcriptional activity. In addition, the mutant receptors are capable of inhibiting the action of their wild-type counterparts when co-expressed (Chatterjee *et al.*, 1991). The 'dominant-negative' inhibitory effects of mutant receptors *in vivo* are supported by the finding of a unique family with recessively inherited RTH (Refetoff *et al.*, 1967). Individuals who were heterozygous for a deletion of one allele of the TRβ gene and therefore had no mutant receptor expressed were clinically and biochemically normal, and this has also been confirmed in heterozygote TRβ knockout mice (Forrest *et al.*, 1996). Interestingly, another unusual feature shared by individuals homozygous for a TRβ gene deletion and homozygous TRβ knockout mice is profound sensorineural deafness, suggesting that TR plays an important role in auditory development.

A splice variant of the β gene (TRβ2), is most highly expressed in the pituitary and hypothalamus. Accordingly, it has been suggested that the dominant-negative action of mutant TRs in these tissues impairs negative feedback within the pituitary–thyroid axis, leading to the characteristic abnormalities in thyroid function (high serum free T4, T3; normal TSH) which lead to the diagnosis of RTH (Collingwood *et al.*, 1994). Studies *in vitro* indicate that artificial mutations which impair DNA binding or heterodimerization by mutant TRs abolish their dominant-negative activity (Nagaya and Jameson, 1993). This correlates with the clustering of all natural RTH mutations described hitherto to three regions in the LBD of TRβ, outside areas known to be important for DNA-binding or dimerization. No naturally occurring defects in the human TRα receptor gene have been described although, based on its tissue distribution (brain, myocardium, skeletal muscle), these might not manifest the classical biochemical features of RTH.

4.3.5 Vitamin D receptor defects

Hereditary vitamin D_3 resistant rickets type II (HVDRR II) or calcitriol-resistant rickets (CRR) (Hochberg and Weisman, 1995) is inherited in autosomal recessive fashion and characterized by early onset of severe rickets, impaired dentition and alopecia, together with hypocalcaemia and elevated circulating 1,25-dihydroxyvitamin D_3 levels. About 50 cases of CRR have been reported (Hewison and O'Riordan, 1994) and a number of different mutations in the VDR gene have been described to date. The majority are point mutations in the DBD of the receptor (Hochberg and Weisman, 1995), occurring either at the tips or the base of each zinc finger. In keeping with this location, functional studies with nuclear extracts from patient skin fibroblasts show normal ligand binding but reduced receptor binding to a DNA–cellulose column (Hughes *et al.*, 1988).

Mutations in the C-terminal domain of VDR have also been described which reduce the receptor's ability to bind ligand or activate transcription (Kristjansson *et al.*, 1993). Most recently, two novel LBD mutations have been described. The first (I314S) exhibited reduced hormone-dependent transactivation *in vitro* which could be restored with excess ligand, and this correlated with virtual resolution of clinical disease with pharmacological doses of 1,25-dihydroxyvitamin D_3. The second mutation (R391C) involved a residue adjacent to the 'ninth heptad' which mediates dimerization, and functional analyses indicate that interaction of this mutant receptor with RXR is indeed impaired (Whitfield *et al.*, 1996). By analogy with RTH (see above), the majority of natural VDR mutants with impaired DNA binding or dimerization are predicted to lack dominant-negative activity and no natural dominant-negative VDR mutations have been described hitherto. In keeping with this, there is no recognized clinical phenotype associated with heterozygote VDR mutant individuals.

Considerable controversy has arisen following an initial report that allelic variants in the 3′-untranslated region (3′-UTR) of the VDR gene are associated with osteoporosis (Morrison *et al.*, 1994). Support for the association with bone mineral density has been found in some subsequent studies but not in others. A recent meta-analysis of data from 16 studies cautiously favoured a small effect of VDR genotype on bone mineral density (Cooper and Umbach, 1996). Two hypotheses have been put forward to explain the association: the first is that certain alleles are in linkage with the changes in the 3′-UTR of the gene which leads to reduced stability of the VDR mRNA; the second is the possibility that the VDR alleles are in linkage disequilibrium with another gene that influences bone mineral density. However, it is also clear that the VDR gene is not the sole genetic determinant of bone mineral density and further studies are underway to identify other genes involved (Econs and Speer, 1996). The debate has been widened by the finding of an association between VDR genotype and primary hyperparathyroidism (Carling *et al.*, 1995); this requires confirmation in further studies.

4.3.6 DAX1 and adrenal hypoplasia congenita

Adrenal hypoplasia congenita (AHC) is a disorder of adrenal gland development resulting in early-onset adrenal insufficiency characterized by a reduction in all adrenocortical hormone levels and failure to respond to ACTH. After localization to the Xp21 locus, the gene underlying AHC has recently been cloned (Zanaria *et al.*, 1994) and termed *DAX1* (dosage-sensitive sex-reversal – AHC critical region on the X chromosome). Deletion, missense and premature stop mutations in *DAX1* have been described in 20 cases of AHC (Muscatelli *et al.*, 1994). Although DAX1 is a member of the nuclear receptor superfamily, its similarity to other members is restricted to the LBD which shares 50% homology. Notably, the protein lacks N-terminal A/B or zinc-finger DBDs which are replaced by a novel region containing four tandem repeats of a glycine/alanine rich motif (*Figure 4.2*). Although the molecular function of DAX1 is unknown, it has been shown to bind retinoic acid response elements, resulting in antagonism of RAR action in transfection studies (Zanaria *et al.*, 1994). Hypogonadotropic hypogonadism may also be associated with AHC (see Section 10.2.2), and it seems probable that the DAX1 mutations are also responsible for this feature (Zanaria *et al.*, 1994). While the same chromosomal locus is also involved in sex determination, in that patients with a normal sex-determining region Y chromosome (SRY) gene and autosomal duplications of this region develop as phenotypic females (hence the term 'dosage-sensitive sex-reversal' (DSS)), DAX1 mutations are not thought to be responsible for DSS, which may involve a defect in another gene which maps to the same interval.

4.3.7 HNF-4α defects

Hepatocyte nuclear factor 4 (HNF-4) was first characterized as a transcription factor in liver extracts and interacts with regulatory DNA elements in the transthyretin and apolipoprotein CIII genes. When isolated, it was found to be an orphan member of the steroid/nuclear receptor superfamily (Sladek *et al.*, 1990). The mammalian receptor is most highly expressed in liver, intestine and kidney and its *Drosophila* homologue plays a role in development of visceral endoderm. This receptor may function as a homodimer (*Figure 4.1*) and appears unable to interact with RXR. A homozygous knockout of the murine gene resulted in embryonic lethality with cell death and failure of gastrulation (Chen *et al.*, 1994).

Recently, a rare monogenic form of maturity-onset diabetes of the young (MODY-1), characterized by early onset of glucose intolerance (at age <25 years), was linked to 20q12–13 encompassing the HNF-4α gene locus. Sequence analysis in a single large MODY-1 pedigree indicated that affected individuals were heterozygous for a nonsense mutation (Q268X) in the HNF-4α gene, consistent with the dominant mode of inheritance of this disorder (Yamagata *et al.*, 1996). The mutation is predicted to generate a truncated receptor protein, containing a DBD but lacking C-terminal dimerization and transcription activation regions. The relationship of this receptor defect to the pathogenesis of diabetes remains unclear; HNF-4α is expressed in the pancreatic islet and also regulates the expression of another transcription factor (HNF-1α) which weakly modulates insulin gene expression and which is defective in another form of MODY (MODY-3). It has been suggested that haploinsufficency for HNF-4α has a deleterious effect on pancreatic β-cell function resulting in impaired insulin secretion. It remains to be established whether the phenotype of heterozygous HNF-4α knockout mice, which are known to be viable, accords with this hypothesis.

4.3.8 Nuclear receptor defects without a recognized human counterpart

Targeted disruption of the PR gene in the mouse suggests that it may have pleiotropic effects in the reproductive axis (Lydon *et al.*, 1995). Both female and male homozygous mice developed to adulthood with no gross defects and male fertility was not compromised. Female mice exhibited several abnormalities:a failure to ovulate due to incomplete follicular rupture in oogenesis, uterine hyperplasia and inflammation, impaired mammary gland development, and altered sexual behaviour. Heterozygous mice were fully fertile. Although corresponding mutations in the human PR gene might be expected to impair fertility which would reduce their prevalence, we can anticipate the eventual description of PR defects in cases with the appropriate phenotype.

The physiological importance of MR function is underscored by the recent observation that MR-deficient mice die 10 days after birth with weight loss and elevated levels of renin, angiotensin II and aldosterone (Berger *et al.*, 1996). Some of these features are shared by the human syndrome of pseudohypoaldosteronism type I (PHA1) (see Section 9.5.5) (Komesaroff *et al.*, 1994), characterized by salt-wasting, hyperreninaemia and elevated serum aldosterone. However, despite exhaustive studies, no mutations in MR have been found in such cases. Recent linkage studies indicate genetic heterogeneity, mapping PHA1 to either chromosome 16p or 12p, and also support the notion that downstream pathways may be involved as the human MR gene is located on chromosome 4q (Strautnieks *et al.*, 1996).

4.4 Nuclear receptor knockouts

For many of the growing number of orphan nuclear receptors, no human disease associations have been made as yet and the identification of cognate ligands has proved elusive, such that the physiological role of these factors is poorly understood. An approach which has been successful in some cases is the targeted disruption of the receptor gene by homologous recombination in mice. The phenotypes of some receptor knockouts are described below.

4.4.1 Steroidogenic factor 1

SF-1 (see also Sections 8.3.2 and 10.2.1) was first isolated as a transcription factor interacting with the promoters of adrenal steroidogenic enzyme genes, although it has since been shown to regulate gonadal (anti-Müllerian hormone) and pituitary (glycoprotein hormone α-subunit) gene expression as well (Parker and Schimmer, 1996). The receptor is highly expressed in the adrenal glands, gonads and placenta and its *Drosophila* homologue FTZ-F1 regulates expression of the fushi-tarazu gene.

Mice homozygous for a null mutation in the SF-1 gene fail to develop adrenal glands and gonads and their early postnatal death from adrenal insufficency can be prevented by corticosteroid administration (Luo *et al.*, 1994). Subsequent studies also showed an absence of pituitary gonadotrophin (LH, FSH) expression, which could be induced by administration of exogenous gonadotrophin-releasing hormone (GnRH) (Ikeda *et al.*, 1995). The latter observation indicates the presence of viable pituitary gonadotrophs in SF-1 mutant mice, suggesting that the gonadotrophin deficiency is more likely to be due to impaired release of endogenous GnRH from hypothalamic neurones, possibly because of absent input from the ventromedial hypothalamic nucleus which is lacking in these mice (Ikeda *et al.*, 1995). Thus, SF-1 appears to be critical for development and function at multiple levels of the hypothalamic–pituitary–gonadal axis. Whether some cases of adrenal hypoplasia and hypogonadism which are not due to DAX1 gene mutations (see above) are associated with mutations in the human SF-1 gene remains to be elucidated.

4.4.2 Nerve growth factor inducible receptor

Exposure of PC12 cells to nerve growth factor (NGF) rapidly induces expression of an orphan receptor called NGF-IB and a closely related nuclear receptor Nur-related factor 1 (NURR1) is also growth factor inducible. Like SF-1, NGF-IB has also been shown to regulate steroidogenic enzyme gene expression and its transcript is highly expressed in the hypothalamic–adrenal axis following stress (Crawford *et al.*, 1995). Receptor expression is also upregulated in apoptotic T cells and this pathway can be inhibited by expression of a dominant-negative mutant NGF-IB (Liu *et al.*, 1994). In this context, it was surprising that NGF-IB null mutant mice exhibited no discernible defect, particularly with respect to adrenocortical and T-cell function (Crawford *et al.*, 1995). However, NGF-IB-deficient mice express NURR1 more highly, suggesting some compensatory functional redundancy between these two orphan receptors.

4.4.3 Retinoid orphan receptor α

Although this orphan receptor (RORα) shares some homology with the retinoic acid receptors, until recently its function was poorly understood. It exhibits constitutive transcriptional activity when bound monomerically to a response element (Giguere *et al.*, 1994). The homozygous *staggerer* (sg) mouse exhibits severe cerebellar ataxia due to a defect in Purkinje cell maturation, morphology, gene expression and number. The sg locus was mapped to mouse chromosome 9 encompassing the RORα gene. Sequence analysis indicated that the mutant mice were homozygous for a 122 bp intragene deletion causing a shift in reading frame at residue 272 leading to a premature stop codon 27 residues later. The predicted RORα protein is truncated and lacks most of the C-terminal LBD. *In situ* hybridization confirmed that RORα is highly expressed in developing normal Purkinje cells and midbrain with much lower levels of expression in mutant mice. The histology in sg mice (reduced dendritic arborization of Purkinje cells and decreased granule cell number) is similar to that seen in hypothyroidism, except that it is not alleviated by thyroid hormone treatment. This has led to the suggestion that RORα and TR act cooperatively to facilitate cerebellar Purkinje cell maturation (Hamilton *et al.*,1996).

4.4.4 Retinoic acid and retinoid X receptors

As morphogens, the retinoids play an important role in embryonic development and postnatally they are indispensable for growth, reproduction and vision. These physiological effects are mediated by a diverse array of receptors:there are three retinoic acid receptor genes (RARα, β, γ) with multiple isoforms at each locus which are activated by all *trans* and 9-*cis* RA; three retinoid X receptors (RXRα, β, γ) respond only to 9-*cis* RA. RARs form heterodimers with RXR, which also serves as a heterodimer partner for many other nuclear receptors. RXR also homodimerizes in some response elements. One implication of such diversity is that the effect of receptor gene knockouts is likely to be complex; this has been substantiated experimentally. A detailed account of the various phenotypes has been published elsewhere (Kastner *et al.*, 1995) and is beyond the scope of this review. Nevertheless, some overall conclusions can be drawn. A number of knockouts (RARα1, RARβ, RARγ2) were not associated with a discernible histological phenotype; RARα and RARγ knockouts exhibited decreased viability, poor growth and male sterility while double RAR gene null mutants showed many congenital malformations (lung hypoplasia, cardiac and genitourinary anomalies, eye defects). RXRα null mutations were embryonic lethal due to cardiac ventricular hypoplasia (Sucov *et al.*, 1994) whereas RXRβ mutants were viable with impaired spermatogenesis; a combination of particular RXR/RAR double knockouts, resulted in an exaggeration of some developmental defects (e.g. persistent truncus arteriosus with RXRα plus RARα; anterior segment eye defects with RXRα plus RARγ or β2). Many of these developmental anomalies were highly homologous to those seen in fetal vitamin A deficiency, further supporting the notion that RARs and RXRs mediate the physiological effects of retinol derivatives. The absence of a phenotype with some single retinoid receptor knockouts is also indicative of some functional redundancy between receptor isoforms. Finally, the enhanced severity of malformations in compound receptor knockouts may provide direct evidence for the involvement of particular RAR/RXR heterodimer combinations in morphogenesis in particular tissues.

Human acute promyelocytic leukaemia (APML) is associated with a characteristic balanced chromosomal translocation t(15;17) (q22;q21) and all *trans* RA is a recognized adjunctive therapy for APML and can also induce differentiation of HL-60 myeloid leukaemic cells *in vitro*. These seemingly unrelated observations were unified by the finding that the 15:17 translocation generates a chimaeric transcript, fusing the DBD and LBD of the RARα gene to a novel gene (PML), whose function is unknown (Dyck *et al.,* 1994). Unlike wild-type PML, which localizes to discrete nuclear structures, it has been shown that the hybrid PML–RAR fusion protein complexes with its normal counterpart in an aberrant microparticulate distribution which interferes with normal myeloid differentiation. These morphological changes are reversed by all *trans* RA treatment, relieving the block in differentiation.

4.4.5 Peroxisome-proliferator-activated receptor

The α isoform of this orphan receptor was first isolated from mouse liver and shown to be activated by the hypolipidaemic agent clofibrate which causes hepatic peroxisome proliferation, hence PPAR. Since then, PPARα has been shown to regulate a number of target genes (e.g. acyl-CoA oxidase, cytochrome P450 4A) that are involved in fatty acid metabolism, suggesting that this receptor plays a role in hepatic peroxisomal β-oxidation and microsomal ω-oxidation of fatty acids and detoxification of xenobiotics. PPARα can also be activated by other agents such as the hypolipidaemic compound Wy14,643 and eicosatetraynoic acid (Keller and Wahli, 1993). However, targeted disruption of the PPARα gene resulted in viable homozygous null mice with no gross phenotypic defects apart from an absence of biochemical and peroxisome-proliferative responses to fibrates (Lee *et al.,* 1995). More recently, it has been shown that the arachidonic acid metabolite 8S-HETE can activate PPARα and that leukotriene B_4 binds the receptor (Devchand *et al.,* 1996). The observation that leukotriene B_4 induces a greater inflammatory response in PPARα knockout mice has led to the intriguing suggestion that this receptor may mediate hepatic inactivation of pro-inflammatory mediators to regulate the inflammatory response (Devchand *et al.,* 1996).

The γ isoform of PPAR was first identified as part of a transcription factor complex which interacts with a major enhancer in the adipocyte P2 gene. The receptor is induced early in murine pre-adipocyte differentiation and expressed at high levels in mature adipocytes. Furthermore, retroviral overexpression of PPARγ in murine fibroblasts promotes their conversion to adipocytes, inducing adipocyte-specific gene expression and lipid accumulation. Together, these observations signify the importance of PPARγ as a key transcription factor in adipocyte-specific gene expression and differentiation (Spiegelman and Flier, 1996). Progress towards identifying the ligand for PPARγ was made by the unexpected observation that thiazolidenediones bind this receptor. These compounds were initially synthesized as potentially hypolipidaemic derivatives of clofibrate, but developed subsequently due to their unexpected antidiabetic properties, related to their ability to enhance insulin action *in vivo*. Troglitazone and BRL49653 are thiazolidenediones which have been shown to ameliorate insulin resistance in clinical trials and are near to therapeutic use in non-insulin-dependent diabetes. These agents are also powerful inducers of murine pre-adipocyte differentiation and their rank order potency for this mirrors their insulin-sensitizing action, suggesting that both effects are PPARγ mediated. It has been proposed recently that 15 deoxy Δ12,14 prostaglandin J_2 may represent a natural endogenous PPARγ ligand (Kliewer *et al.*, 1995). These findings indicate that PPARγ will be an important target for drug development in diabetes and suggest that this gene may also contribute to the pathogenesis of obesity and insulin-resistant states.

4.5 Conclusions

In the decade since the isolation of the first cDNAs for steroid receptors, an exponential increase in our understanding of their action has been accompanied by a similar expansion in members of this family which now number over 150 distinct proteins. The challenges for the future are twofold.

The first is to gain a better understanding of the action of known receptors at the cellular level; how does the apparent promiscuity of receptor interactions *in vitro* translate into specificity of signalling *in vivo*? Conversely, how do a finite number of receptors and ligands mediate a diverse pattern of biological responses; what are the specific roles of the seemingly myriad array of receptor co-activators and co-repressors and what is the structural basis for their organization into higher order complexes with receptors? How in turn do these complexes interact with the basic transcriptional machinery and chromatin?

The second challenge is to assign a physiological role for the diverse number of orphan receptors. Gene knockouts in mice can be valuable but delineation of the phenotype can sometimes be limited by early embryonic lethality or functional redundancy of receptor isoforms. Accordingly, the chromosomal mapping of novel receptor genes via the Human Genome Project and linkage of these loci to human diseases of unknown aetiology will provide a valuable complementary approach to elucidating receptor function. These genetic approaches will be fuelled by pharmacological quests for novel orphan receptor ligands, using combinatorial chemistry and high-throughput screening assays. In this context, it is interesting to note that at least one leading pharmaceutical company has adopted the nuclear receptor family as part of a systematic approach to drug discovery. Perhaps this is added testament – were it necessary – to the physiological importance of this family of proteins and their therapeutic potential.

Acknowledgements

Owing to space constraints, we apologize that numerous primary references could not be cited directly, but are subsumed within other reviews cited in the text. R.J.C.-B. is a Commonwealth Scholar and C.M. is a Wellcome Advanced Training Fellow. V.K.K.C is supported by the Wellcome Trust and Medical Research Council (UK).

References

Adams M, Matthews C, Collingwood TN, Tone Y, Beck-Peccoz P, Chatterjee VKK. (1994) Genetic analysis of 29 kindreds with generalized and pituitary resistance to thyroid hormone. *J. Clin. Invest.* **94**: 506–515.

Adams M, Reginato M, Shao D, Lazar MA, Chatterjee VKK. (1997) Transcriptional activation by peroxisome-proliferator activated receptor γ is inhibited by phosphorylation at a consensus mitogen-activated protein kinase site. *J. Biol. Chem.* **272**: 5128–5132.

Akerblom IE, Slater EP, Beato M, Baxter JD, Mellon PM. (1988) Negative regulation by glucocorticoids through interference with a cAMP responsive enhancer. *Science* **241**: 350–353.

Arai K, Chrousos GP. (1994) Glucocorticoid resistance. In: *Hormones, Enzymes and Receptors* (eds MC Sheppard, PM Stewart). Baillière Tindall, London, pp. 317–331.

Au-Fliegner M, Helmer E, Casanova J, Raaka BM, Samuels HH. (1993) The conserved ninth C-terminal heptad in thyroid hormone and retinoic acid receptors mediates diverse responses by affecting heterodimer but not homodimer formation. *Mol. Cell. Biol.* **13**: 5725–5737.

Baniahmad A, Kohne AC, Renkawitz R. (1992) A transferable silencing domain is present in the thyroid hormone receptor, in the v-erbA oncogene product and in the retinoic acid receptor. *EMBO J.* **11**: 1015–1023.

Batch JA, Davies HR, Evans BAJ, Hughes IA, Patterson MN. (1993) Phenotypic variation and detection of carrier status in the PAIS syndrome. *Arch. Dis. Child.* **68**: 453–457.

Berger S, Cole TJ, Scmid W, Schutz G. (1996) Analysis of glucocorticoid and mineralocorticoid signalling by gene targeting. *Endocr. Res.* **22**: 641–652.

Bourguet W, Ruff M, Chambon P, Gronemeyer H, Moras D. (1995) Crystal structure of the ligand binding domain of the human nuclear receptor RXRα. *Nature* **375**: 377–382.

Brönnegård M, Carlstedt-Duke J. (1995) The genetic basis of glucocorticoid resistance. *Trends Endocrinol. Metab.* **6**: 160–164.

Carling T, Kindmark A, Hellman P, Lundgren S, Rastad J, Åkerstrom G, Melhus H. (1995) Vitamin D receptor genotypes in primary hyperparathyroidism. *Nature Med.* **1**: 1309–1311.

Casanova J, Helmer E, Selmi-Ruby S, Qi J-S, Au-Fliegner M, Desai-Yajnik V, Koudinova N, Yarm F, Raaka BM, Samuels HH. (1994) Functional evidence for ligand-dependent dissociation of thyroid hormone and retinoic acid receptors from an inhibitory cellular factor. *Mol. Cell. Biol.* **14**: 5756–5765.

Cavailles V, Dauvois S, L'Horset F, Lopez G, Hoare S, Kushner PJ, Parker MG. (1995) Nuclear factor RIP140 modulates transcriptional activation by the estrogen receptor. *EMBO J.* **14**: 3741–3751.

Charest NJ, Zhou Z-X, Lubahn DB, Olsen KL, Wilson EM, French FS. (1991) A frameshift mutation destabilises AR messenger RNA in the *Tfm* mouse. *Mol. Endocrinol.* **5**: 573–581.

Chatterjee VKK, Beck-Peccoz. (1994) Thyroid hormone resistance. In: *Hormones, Enzymes and Receptors* (eds MC Sheppard, PM Stewart). Baillière Tindall, London, pp. 267–283.

Chatterjee VKK, Nagaya T, Madison LD, Datta S, Rentoumis A, Jameson JL. (1991) Thyroid hormone resistance syndrome:inhibition of normal receptor function by mutant thyroid hormone receptors. *J. Clin. Invest.* **87**: 1977–1984.

Chen JD, Evans RM. (1995) A transcriptional corepressor that interacts with nuclear hormone receptors. *Nature* **377**: 454–457.

Chen WS, Manova K, Weinstein DC, Duncan SA, Plump AS, Prezioso VR, Bacharova RF, Darnell JE. (1994) Disruption of the HNF-4 gene, expressed in visceral endoderm, leads to cell death in embryonic ectoderm and impaired gastrulation of mouse embryos. *Genes Devel.* **8**: 2466–2477.

Choong CS, Kemppainen JA, Zhou Z-X, Wilson EM. (1996) Reduced androgen receptor gene expression with first exon CAG repeat expansion. *Mol. Endocrinol.* **10:** 1527–1535.

Cole TJ, Blendy JA, Monaghan AP, Krieglstein K, Schmid W, Aguzzi A, Fantuzzi G, Hummler E, Unsicker K, Schütz G. (1995) Targeted disruption of the glucocorticoid receptor gene blocks adrenergic chromaffin cell development and severely retards lung maturation. *Genes Devel.* **9**: 1608–1621.

Collingwood TN, Adams M, Tone Y, Chatterjee VKK. (1994) Spectrum of transcriptional dimerization and dominant-negative properties of twenty different mutant thyroid hormone β receptors in thyroid hormone resistance syndrome. *Mol. Endocrinol.* **8**: 1262–1277

Collingwood TN, Rajanayagam O, Adams M *et al.* (1997) A natural transactivation mutation in the thyroid hormone receptor b:impaired interaction with putative transcriptional mediators. *Proc. Natl Acad. Sci. USA* **94:** 248–253.

Cooper GS, Umbach DM. (1996) Are vitamin D receptor polymorphisms associated with bone mineral density? A meta-analysis. *J. Bone. Min. Res.* **11:** 1841–1849.

Crawford PA, Sadovsky Y, Woodson K, Lee SL, Millbrandt J. (1995) Adrenocortical function and regulation of the 21-hydroxylase gene in NGF-IB deficient mice. *Mol. Cell. Biol.* **15:** 4331–4336.

Culig Z, Hobisch A, Cronauer MV, Cato ACB, Hittmair A, Radmayr C, Eberle J, Bartsch G, Klocker H. (1993) Mutant androgen receptor detected in an advanced-stage prostatic carcinoma is activated by adrenal androgens and progesterone. *Mol. Endocrinol.* **7**: 1541–1550.

Danielian PS, White R, Lees JA, Parker MG. (1992) Identification of a conserved region required for hormone dependent transcriptional activation by steroid hormone receptors. *EMBO J.* **11:** 1025–1033.

Devchand PR, Keller H, Peters JM, Vazquez M, Gonzalez FJ, Wahli W. (1996) The PPARα-leukotriene B_4 pathway to inflammation control. *Nature* **384:** 39–43.

Dickson RB, Lippman ME. (1995) Growth factors in breast cancer. *Endocr. Rev.* **16:** 559–589.

Durand B, Saunders M, Gaudon C, Roy B, Losson R, Chambon P. (1994) Activation function 2. (AF-2) of retinoic acid receptor and 9-cis retinoic acid receptor:presence of a conserved autonomous constitutive activating domain and influence of the nature of the response element on AF-2 activity. *EMBO J.* **13:** 5370–5382.

Dyck JA, Maul GG, Miller WH, Chen JD, Kakizuka A, Evans RM. (1994) A novel macromolecular structure is a target of the promyelocyte-retinoic acid receptor oncoprotein *Cell* **76:** 333–343.

Econs MJ, Speer MC. (1996) Genetic studies of complex diseases:let the reader beware. *J. Bone Min. Res.* **11:** 1835–1840.

Evans RM. (1988) The steroid and thyroid hormone receptor superfamily. *Science* **240:** 889–895.

Fawell SE, Lees JA, White R, Parker MG. (1990) Characterization and colocalization of steroid binding and dimerization activities in the mouse estrogen receptor. *Cell* **60:** 953–962.

Fondell JD, Roy AL, Roeder RG. (1993) Unliganded thyroid hormone receptor inhibits formation of a functional preinitiation complex: implications for active repression. *Genes Devel.* **7:** 1400–1410.

Forman BM, Samuels HH. (1990) Interactions among a subfamily of nuclear hormone receptors; the regulatory zipper model. *Mol. Endocrinol.* **4:** 1293–1301.

Forman BM, Umesono K, Chen J, Evans RM. (1995) Unique response pathways are established by allosteric interactions among nuclear hormone receptors. *Cell* **81:** 541–550.

Forrest D, Hanebuth E, Smeyne RJ, Everds N, Stewart CL, Wehner JM, Curran T. (1996) Recessive resistance to thyroid hormone in mice lacking thyroid hormone receptor β:evidence for tissue-specific modulation of receptor function. *EMBO J.* **15:** 3006–3015.

Giguere V, Tini M, Flock G, Ong E, Evans RM, Otulakowski G. (1994) Isoform-specific aminoterminal domains dictate DNA-binding properties of RORα, a novel family of orphan hormone receptors. *Genes Devel.* **8:** 538–553.

Green H. (1993) Human genetic diseases due to codon reiteration:relationship to an evolutionary mechanism. *Cell* **74:** 955–956.

Green S, Chambon P. (1988) Nuclear receptors enhance our understanding of transcriptional regulation. *Trends Genet.* **4:** 309–314.

Griffin JE. (1992) Androgen resistance – the clinical and molecular spectrum. *New Engl. J. Med.* **326:** 611–618.

Griffin JE, Wilson JD. (1989) The androgen receptor syndromes:5α-reductase deficiency, testicular feminization, and related syndromes. In: *The Metabolic Basis of Inherited Disease* (eds CR Scriver, AL Beaudet, WS Sly, D Valle). 6th Edn, McGraw-Hill, New York, pp. 1919–1944.

Gronemeyer H, Laudet V. (1995) Sequences of nuclear receptors. In: *Protein Profile* (ed. P Sheterline). Academic Press, London, pp. 1184–1197.

Guichon Mantel A, Loosfelt H, Lescop P, Sar S, Atger M, Perrot-Aplanat M, Milgrom E. (1989) Mechanisms of nuclear localization of the progesterone receptor:evidence for receptor interaction between monomers. *Cell* **57:** 1147–1154.

Halachmi S, Marden E, Martin G, MacKay H, Abbondanza C, Brown M. (1994) Estrogen receptor-associated proteins:possible mediators of hormone-induced transcription. *Science* **264:** 1455–1458.

Hamilton BA, Frankel WN, Kerrebrock AW *et al.* (1996) Disruption of the nuclear hormone receptor RORα in staggerer mice. *Nature* **379:** 736–739.

Hewison M, O'Riordan JLH. (1994) Vitamin D resistance. In: *Hormones, Enzymes and Receptors* (eds MC Sheppard, PM Stewart). Baillière Tindall, London, pp. 305–315.

Hochberg Z, Weisman Y. (1995) Calcitriol-resistant rickets due to vitamin D receptor defects. *Trends Endocrinol. Metab.* **6:** 216–220.

Hollenberg S, Evans RM. (1988) Multiple and cooperative transactivation domains of the human glucocorticoid receptor *Cell* **55:** 899–906.

Horlein AJ, Naar AM, Heinzel T *et al.* (1995) Ligand-dependent repression by the thyroid hormone receptor mediated by a nuclear receptor corepressor. *Nature* **377:** 397–404.

Hughes MR, Malloy PM, Kieback DG, Kesterson RA, Pike JW, Feldman D, O'Malley BW. (1988) Point mutations in the human vitamin D receptor gene associated with hypocalcaemic rickets. *Science* **242:** 1702–1705.

Igarashi S, Tanno Y, Onodera O *et al.* (1992) Strong correlation between the number of CAG repeats in androgen receptor genes and the clinical onset of features of spinal and bulbar muscular atrophy. *Neurology* **42:** 2300–2302.

Ikeda Y, Luo X, Abbud R, Nilson JH, Parker KL. (1995) The nuclear receptor steroidogenic factor 1 is essential for the formation of the ventromedial hypothalamic nucleus. *Mol. Endocrinol.* **9:** 478–486.

Issa J-PJ, Ottaviano YL, Celano P, Hamilton SR, Davidson NE, Baylin SB. (1994) Methylation of the oestrogen receptor CpG island links ageing and neoplasia in human colon. *Nature Genet.* **7:** 536–540.

Jacq X, Brou C, Lutz Y, Davidson I, Chambon P, Tora L. (1994) Human TAFII30 is present in a distinct TFIID complex and is required for transcriptional activation by the estrogen receptor. *Cell* **79:** 107–117.

Janowski BA, Willy PJ, Devi TR, Falck JR, Mangelsdorf DJ. (1996) An oxysterol signalling pathway mediated by the nuclear receptor LXRα. *Nature* **383:** 728–731.

Kamei Y, Xu L, Heinzel T *et al.* (1996) A CBP integrator complex mediates transcriptional activation and AP-1 inhibition by nuclear receptors. *Cell* **85:** 403–414.

Kastner P, Mark M, Chambon P. (1995) Nonsteroid nuclear receptors:what are genetic studies telling us about their role in real life. *Cell* **83:** 859–869.

Karl M, Lamberts SW, Detera-Wadleigh SD, Encio IJ, Stratakis CA, Hurley DM, Accili D, Chrousos GP. (1993) Familial glucocorticoid resistance caused by a splice site deletion in the human glucocorticoid receptor gene. *J. Clin. Endocrinol. Metab.* **76:** 683–689.

Kato S, Endoh H, Masuhiro Y *et al.* (1995) Activation of the estrogen receptor through phosphorylation by mitogen-activated protein kinase. *Science* **270:** 1491–1494.

Katz D, Reginato MJ, Lazar MA. (1995) Functional regulation of thyroid hormone receptor variant TRα2 by phosphorylation *Mol. Cell. Biol.* **15:** 2341–2348.

Keller H, Wahli W. (1993) Peroxisome-proliferator activated receptors:a link between endocrinology and nutrition. *Trends Endocrinol. Metab.* **4:** 291–296.

Kliewer SA, Umesono K, Manglesdorf DJ, Evans RM. (1992) Retinoid X receptor interacts with nuclear receptors in retinoic acid, thyroid hormone and vitamin D_3 signalling. *Nature* **355:** 446–449.

Kliewer SA, Lenhard JM, Willson TM, Patel I, Morris DC, Lehmann JM. (1995) A prostaglandin J_2 metabolite binds peroxisome proliferator-activated receptor gamma and promotes adipocyte differentiation. *Cell* **83:** 813–819.

Komesaroff PA, Funder JW, Fuller PJ. (1994) Mineralocorticoid resistance. In: *Hormones, Enzymes and Receptors* (eds MC Sheppard, PM Stewart). Baillière Tindall, London, pp. 333–355.

Kraus WL, McInerney EM, Katzenellenbogen BS. (1995) Ligand-dependent transcriptionally productive association of the amino and carboxyl-terminal regions of a steroid hormone nuclear receptor. *Proc. Natl Acad. Sci. USA* **92:** 12314–12318.

Kristjansson K, Rut AR, Hewison M, O'Riordan JLH, Hughes MR. (1993) Two mutations in the hormone binding domain of the vitamin D receptor cause tissue resistance to 1, 25 dihydroxyvitamin D_3. *J. Clin. Invest.* **92:** 12–16.

Kuiper GGJM, Enmark E, Pelto-Huikko M, Nilsson S, Gustafsson J-A. (1996) Cloning of a novel estrogen receptor expressed in rat prostate and ovary. *Proc. Natl Acad. Sci. USA* **93:** 5925–5930.

Kurokawa R, DiRenzo J, Boehm M, Sugarman J, Gloss B, Rosenfeld MG, Heyman R, Glass CK. (1994) Regulation of retinoid signalling by receptor polarity and allosteric control of ligand binding. *Nature* **371:** 528–531.

La Spada AR, Wilson EM, Lubahn DB, Harding AE, Fischbeck KH. (1991) Androgen receptor gene mutations in X-linked spinal and bulbar muscular atrophy. *Nature* **352:** 77–79.

Lee SS, Pineau T, Drago J, Lee EJ, Owens JW, Kroetz DL, Fernandez-Salguero PM, Westphal H, Gonzalez FJ. (1995) Targeted disruption of the α isoform of the peroxisome proliferator-activated receptor gene in mice results in abolishment of the pleiotropic effects of peroxisome proliferators. *Mol. Cell. Biol.* **15:** 3012–3022.

Leid M, Kastner P, Lyons R *et al.* (1992) Purification, cloning and RXR identity of the HeLa cell factor with which RAR or TR heterodimerises to bind target sequences efficiently. *Cell* **68:** 377–395.

Lubahn DB, Moyer JS, Golding TS, Couse JF, Korach KS, Smithies O. (1993) Alteration of reproductive function but not prenatal sexual development after insertional disruption of the mouse estrogen receptor gene. *Proc. Natl Acad. Sci. USA* **90:** 11162–11166.

Liu Z-G, Smith SW, McLaughlin KA, Schwartz LM, Osborne BA. (1994) Apoptotic signals delivered through the T cell receptor of a T cell hybrid require the immediate early gene nur 77. *Nature* **367:** 281–284.

Luisi BF, Xu WX, Otwinowski Z, Freedman LP, Yamamoto KR, Sigler PB. (1991) Crystallographic analysis of the interaction of the glucocorticoid receptor with DNA. *Nature* **352:** 497–505.

Luo X, Ikeda Y, Parker KL. (1994) A cell-specific nuclear receptor is essential for adrenal and gonadal development and sexual differentiation. *Cell* **77:** 481–490.

Lydon JP, DeMayo FJ, Funk CR, Mani SK, Hughes AR, Montgomery CA, Shyamala G, Conneely OM, O'Malley BW. (1995) Mice lacking progesterone receptor exhibit pleiotropic reproductive abnormalities. *Genes Devel.* **9:** 2266–2278.

Malchoff CD, Malchoff DM. (1995) Glucocorticoid resistance in humans. *Trends Endocrinol. Metab.* **6:** 89–95.

Mangelsdorf DJ, Umesono K, Kliewer SA, Borgmeyer U, Ong ES, Evans RM. (1991) A direct repeat in the cellular retinol-binding protein type II gene confers differential regulation by RAR and RXR. *Cell* **66:** 555–561.

Mani S, Allen J, Clark J, O'Malley B. (1994) Convergent pathways for steroid hormone and neurotransmitter induced rat sexual behaviour. *Science* **265:** 1246–1249.

McPhaul MJ, Marcelli M, Zoppi S, Griffin JE, Wilson JD. (1993) Genetic basis of endocrine disease 4:the spectrum of mutations in the androgen receptor gene that causes androgen resistance. *J. Clin. Endocrinol. Metab.* **76:** 17–23.

Mhatre AN, Trifiro MA, Kaufman M, Kazemi-Esfarajani P, Figlewicz D, Rouleau G, Pinsky L. (1993) Reduced transcriptional regulatory competence of the androgen receptor in X-linked spinal and bulbar muscular atrophy. *Nature Genet.* **5:** 184–187.

Morrison NA, Qi JC, Tokita A, Kelly PJ, Crofts L, Nguyen TV, Sambrook PN, Eisman JA. (1994) Prediction of bone density from vitamin D receptor alleles. *Nature* **367:** 284–287.

Muscatelli F, Strom TM, Walker AP *et al.* (1994) Mutations in the *DAX-1* gene give rise to both X-linked adrenal hypoplasia congenita and hypogonadotropic hypogonadism. *Nature* **372:** 672–676.

Nagaya T, Jameson JL. (1993) Thyroid hormone receptor dimerization is required for dominant-negative inhibition by mutations that cause thyroid hormone resistance. *J. Biol. Chem.* **268:** 15766–15771.

Onate SA, Tsai SY, Tsai M-J, O'Malley BW. (1995) Sequence and characterization of a coactivator for the steroid hormone receptor superfamily. *Science* **270:** 1354–1357.

Orti E, Bodwell JE, Munck A. (1992) Phosphorylation of steroid hormone receptors *Endocr. Rev.* **13:** 105–128.

Parker KL, Schimmer BP. (1996) The roles of the nuclear receptor steroidogenic factor 1 in endocrine differentiation and development. *Trends Endocrinol. Metab.* **7:** 203–207.

Perlmann T, Jansson L. (1995) A novel pathway for vitamin A signaling mediated by RXR heterodimerization with NGF-IB and NURR1 *Genes Devel.* **9:** 769–782.

Perutz MF, Johnson T, Suzuki M, Finch JT. (1994) Glutamine repeats as polar zippers:their possible role in inherited neurodegenerative diseases. *Proc. Natl Acad. Sci. USA* **91:** 5355–5358.

Petty KJ, Krimkevich YI, Thomas D. (1996) A TATA binding protein-associated factor functions as a coactivator for thyroid hormone receptors *Mol. Endocrinol.* **10:** 1632–1645.

Pratt WB. (1993) The role of heat shock proteins in regulating the function, folding and trafficking of the glucocorticoid receptor. *J. Biol. Chem.* **268:** 21455–21458.

Quigley CA, De Bellis A, Marschke KB, El-Awady MK, Wilson EM, French FS. (1995) Androgen receptor defects:historical, clinical, and molecular perspectives. *Endocr. Rev.* **16:** 271–321.

Rastinejad F, Perlmann T, Evans RM, Sigler PB. (1995) Structural determinants of nuclear receptor assembly on DNA direct repeats. *Nature* **375:** 203–211.

Refetoff S, DeWind LT, DeGroot LJ. (1967) Familial syndrome combining deaf-mutism, stippled epiphyses, goiter and abnormally high PBI:possible target organ refractoriness to thyroid hormone. *J. Clin. Endocrinol. Metab.* **27:** 279–294.

Refetoff S, Weiss RE, Usala SJ. (1993) The syndromes of resistance to thyroid hormone *Endocr. Rev.* **14:** 348–399.

Renaud JP, Rochel N, Ruff M, Vivat V, Chambon P, Gronemeyer H, Moras D. (1995) Crystal structure of the RARγ binding domain bound to all-trans retinoic acid. *Nature* **378:** 681–689.

Rosen ED, Beninghof EG, Koenig RJ. (1993) Dimerization interfaces of thyroid hormone, retinoic acid, vitamin D and retinoid X receptors. *J. Biol. Chem.* **268:** 11534–11541.

Saatcioglu F, Claret F-X, Karin M. (1994) Negative transcriptional regulation by nuclear receptors *Semin. Cancer Biol.* 5: 347–359.

Schwabe JWR, Chapman L, Finch JT, Rhodes D. (1993) The crystal structure of the estrogen receptor DNA-binding domain bound to DNA:how receptors discriminate between their response elements. *Cell* **75:** 567–578.

Seol W, Choi H-S, Moore DD. (1996) An orphan nuclear hormone receptor that lacks a DNA-binding domain and heterodimerises with other receptors. *Science* **272:** 1336–1339.

Sladek FM, Zhong W, Lai E, Darnell JE. (1990) Liver-enriched transcription factor HNF-4 is a novel member of the steroid hormone receptor superfamily. *Genes Devel.* **4:** 2353–2365.

Smith EP, Boyd J, Frank GR, Takahashi H, Cohen RM, Specker B, Williams TC, Lubahn DB, Korach KS. (1994) Estrogen resistance caused by a mutation in the estrogen-receptor gene in a man. *New Engl. J. Med.* **331:** 1056–1061.

Spiegelman BM, Flier JS. (1996) Adipogenesis and obesity; rounding out the big picture. *Cell* **87:** 377–389.

Stocklin E, Wissler M, Gouilleux F, Groner B. (1996) Functional interactions between Stat5 and the glucocorticoid receptor *Nature* **383:** 726–728.

Strautnieks SS, Thompson RJ, Hanukoglu A, Dillon MJ, Hanukoglu I, Kuhnle U, Seckl J, Gardiner RM, Chung E. (1996) Localisation of pseudohypoaldosteronism genes to chromosome 16p12.2–13.11 and 12p13.1-pter by homozygosity mapping. *Hum. Mol. Genet.* 5: 293–299.

Sucov HM, Dyson E, Gumeringer CL, Price J, Chien KR, Evans RM. (1994) RXRα mutant mice establish a genetic basis for vitamin A signalling in heart morphogenesis. *Genes Devel.* **8:** 1007–1018.

Taplin ME, Bubley GJ, Sunster TD, Frantz ME, Spooner AE, Ogata GK, Kew H, Balk SP. (1995) Mutation of the androgen-receptor gene in metastatic androgen-independent prostate cancer. *New Engl. J. Med.* **332:** 1293–1295.

Tini M, Tsui L-C, Giguere V. (1994) Heterodimeric interaction of the retinoic acid and thyroid hormone receptors in transcriptional regulation on the γF-crystallin everted retinoic acid response element. *Mol. Endocrinol.* **8:** 1494–1506.

Tone Y, Collingwood TN, Adams M, Chatterjee VKK. (1994) Functional analysis of a transactivation domain in the thyroid hormone β receptor *J. Biol. Chem.* **269:** 31157–31161.

Tora L, White J, Brou C, Tasset D, Webster N, Scheer E, Chambon P. (1989) The human estrogen receptor has two independent nonacidic transcriptional activation functions. *Cell* **59:** 477–487.

Trifiro MA, Kazemi-Esfajani P, Pinsky L. (1994) X-linked muscular atrophy and the androgen receptor. *Trends Endocrinol. Metab.* **5:** 416–421.

Umesono K, Evans RM. (1989) Determinants of target gene specificity for steroid/thyroid hormone receptors. *Cell* **57:** 1139–1146.

Umesono K, Murakami KK, Thompson CC, Evans RM. (1991) Direct repeats as selective response elements for the thyroid hormone, retinoic acid and vitamin D3 receptors. *Cell* **65:** 1255–1266.

vom Baur E, Zechel C, Heery D, Heine M, Garnier JM, Vivat V, Le Douarin B, Gronemeyer H, Chambon P, Losson R. (1996) Differential ligand-dependent interactions between the AF-2 activating domain of nuclear receptors and the putative transcriptional intermediary factors mSUG1 and TIF1. *EMBO J.* **15:** 110–124.

Wagner RL, Apriletti JW, McGrath ME, West BL, Baxter JD, Fletterick RJ. (1995) A structural role for hormone in the thyroid hormone receptor. *Nature* **378:** 690–697.

Whitfield GK, Selznick SH, Haussler CA, Hsieh J-C, Galligan MA, Jurutka PW, Thompson PD, Lee SM, Zerwekh JE, Haussler MR. (1996) Vitamin D receptors from patients with resistance to 1,25-dihydroxyvitamin D3:point mutations confer reduced transactivation in response to ligand and impaired interaction with the retinoid X receptor heterodimeric partner. *Mol. Endocrinol.* **10:** 1617–1631.

Willy PJ, Umesono K, Ong ES, Evans RM, Heyman RA, Mangelsdorf DJ. (1995) LXR, a nuclear receptor that defines a distinct retinoid response pathway. *Genes Devel.* **9:** 1033–1045.

Wilson TE, Fahrner TJ, Millbrandt J. (1993) The orphan receptors NGFI-B and steroidogenic factor 1 establish monomer binding as a third paradigm of nuclear-receptor DNA interactions. *Mol. Cell. Biol.* **13:** 5794–5804.

Wooster R, Mangion J, Eeles R ***et al.*** (1992) A germline mutation in the androgen receptor gene in two brothers with breast cancer and Reifenstein syndrome. *Nature Genet.* **2:** 132–134.

Wurtz J-M, Bourguet W, Renaud J-P, Vivat V, Chambon P, Moras D, Gronemeyer H. (1996) A canonical structure for the ligand binding domain of nuclear receptors *Nature Struct. Biol.* **3:** 87–94.

Yamagata K, Furuta H, Oda N, Kaisaki PJ, Menzel S, Cox NJ, Fajans SS, Signorini S, Stoffel M, Bell GI. (1996) Mutations in the hepatocyte nuclear receptor factor-4α gene in maturity-onset diabetes of the young. (MODY1). *Nature* **384:** 458–460.

Zanaria E, Muscatelli F, Bardoni ***et al.*** (1994) An unusual member of the nuclear hormone receptor superfamily responsible for X-linked adrenal hypoplasia congenita. *Nature* **372:** 635–641.

Zenke M, Munoz A, Sap J, Vennstrom B, Beug H. (1990) v-erbA oncogene activation entails loss of the hormone-dependent regulator activity of c-erbA. *Cell* **61:** 1035–1049.

Zoppi S, Wilson CM, Harbison MD, Griffin JE, Wilson JD, McPhaul MJ, Marcelli M. (1993) Complete testicular feminization caused by an amino-terminal truncation of the androgen receptor with downstream initiation. *J. Clin. Invest.* **91:** 1105–1112.

5

Molecular insights into mechanisms regulating glucose homeostasis

Peter R. Shepherd

5.1 Introduction

Many tissues of the body such as the brain and blood cells have an absolute requirement for glucose as an energy source such that if blood glucose levels fall below 3 mM the functioning of these tissues is severely impaired. Conversely, prolonged elevation of blood glucose levels also has major deleterious effects on many tissues. Therefore a system of counterbalancing hormones has evolved to ensure that blood glucose levels are maintained at around 5 mM. The major hormonal regulators of glucose homeostasis are glucagon and insulin.

In the absorptive state following a meal, glucose is rapidly absorbed into the blood creating a need to lower blood glucose levels and to store some of the glucose for later use. As blood glucose rises above about 5 mM, a glucose-sensing mechanism in the β cells of the endocrine pancreas causes secretion of insulin into the portal vein. The secreted insulin acts on liver, muscle and fat to rapidly decrease hepatic glucose output and to promote glucose uptake and storage in all three tissues. Following binding to target tissue, insulin is rapidly degraded and blood glucose and insulin levels both fall. The lower blood glucose level results in lower insulin secretion and hence 3–4 h after a meal the concentration of insulin in the blood returns to fasting levels. As the glucose bolus from the meal is absorbed the body must switch to other sources to maintain adequate blood glucose levels. The falling blood glucose concentration stimulates glucagon secretion from the α cells of the pancreas. In the liver the combination of low insulin and increased glucagon act to increase gluconeogenesis and breakdown of liver glycogen stores. This results in a net output of glucose from the liver which is used to supply glucose required for neural tissues and blood cells in the post-absorptive state.

Defects in the mechanism controlling glucose homeostasis are relatively common in humans, the most widespread of these being diabetes. Diabetes has been broadly classified into two categories: (i) type 1 or insulin-dependent diabetes

Molecular Endocrinology, edited by G. Rumsby and S.M. Farrow.

(IDDM) arises from autoimmune destruction of the β cells of the pancreas (Atkinson and Maclaren, 1994). The resulting lack of insulin secretion has profound effects on glucose metabolism and will ultimately result in death without insulin replacement therapy. (ii) Type 2 or non-insulin-dependent diabetes (NIDDM) is a slow onset disease arising most often in older individuals (DeFronzo, 1992; Granner and O'Brien, 1992; Kahn *et al.*, 1996; Moller and Flier, 1991). This form of diabetes is almost always preceded by insulin resistance and is often associated with obesity. Insulin resistance is initially characterized by hyperinsulinaemia and decreased effectiveness of insulin in stimulating glucose metabolism in muscle and fat. Subsequently, hyperglycaemia is observed and, as the disease progresses to frank diabetes, dysregulation of hepatic glucose output is observed along with impaired glucose stimulation of insulin release. Evidence strongly suggests that NIDDM arises from a combination of multiple genetic and environmental factors (Kahn *et al.*, 1996; Moller and Flier, 1991). In a small minority of cases, distinct genetic factors have been identified which cause insulin-resistant conditions linked with NIDDM. These experiments of nature have often provided important insights into the mechanisms by which glucose metabolism is regulated.

This chapter describes recent advances in our understanding of how whole body glucose homeostasis is maintained including the molecular mechanisms by which glucose regulates insulin secretion and the mechanisms by which insulin and glucagon regulate glucose metabolism in liver, muscle and adipose tissue. How derangements in these processes contribute to diabetes is also discussed.

5.2 The role of the endocrine pancreas

The endocrine pancreas is made up of four major cell types (α, β, γ, δ) and the regulated secretion of peptide hormones from these cells provides the main mechanism by which the body achieves glucose homeostasis. The major role of the α cells is to secrete glucagon in response to low blood glucose levels while that of the β cells is to secrete insulin in response to rising glucose. The δ cells produce somatostatin which inhibits secretion of many peptide hormones including insulin and glucagon. As somatostatin secretion is increased in response to glucagon it is likely to play a physiologically relevant role in regulating hormone secretion in the pancreas (Unger and Orci, 1981).

5.2.1 Glucagon expression and secretion in the α cells

Glucagon is a 29-amino-acid peptide hormone with a high degree of homology to vasoactive intestinal protein, gastric inhibitory protein and secretin (Lefebvre, 1995; Philippe, 1991). The mammalian glucagon gene encodes for a 180 amino-acid preproglucagon molecule which is expressed in pancreatic α cells as well as in other specific locations such as the L cells of the gut. Complete processing of preproglucagon yields the four related peptide hormones glucagon, glucagon-related pancreatic peptide (GRPP) and two isoforms of glucagon-like peptide (GLP-1 and GLP-2). However, differential processing mechanisms exist for preproglucagon in different tissues. Therefore, in the α cells of the pancreas only glucagon and small amounts of GLP-1 are produced while the L-cells of the gut do not produce glucagon. Glucagon gene expression is decreased by hyperinsulinaemia and increased by hypoinsulinaemia (Philippe, 1991). This has physiological importance in human type-1 and type-2 diabetics, where the pancreatic α cell is still fully functional. In these subjects increased rates of glucagon production are observed which can be suppressed by administration of counteracting doses of insulin (Unger and Orci, 1981). However, the intracellular mechanisms by which glucose and insulin levels regulate glucagon secretion are still poorly understood (Lefebvre, 1995).

5.2.2 Insulin synthesis and storage

Insulin is a member of a gene family which includes insulin-like growth factor (IGF)-I, IGF-II and relaxin (Steiner *et al.*, 1985). However, unlike the other members of this gene family, expression of insulin is restricted exclusively to the β cells of the pancreas. Furthermore, while circulating reservoirs of IGF-I exist in complex with specific binding proteins, no such mechanism exists for insulin. The insulin gene is transcribed and translated as preproinsulin. This contains a 24-amino-acid signal sequence which is cleaved during transit through the rough endoplasmic reticulum (ER) to yield the 81-amino-acid proinsulin molecule. This is a single polypeptide chain with three internal disulphide bonds. From the trans-Golgi network proinsulin is sorted into vesicles destined for secretion (Hutton, 1994). These vesicles go through a process of maturation during which time a series of specific endopeptidases sequentially cleave the proinsulin. The processing of proinsulin to insulin is thought to be largely effected by the subtilisin-related proteases prohormone convertase-1 and -2 (PC1 and PC2) which are also expressed in neural tissues where they are associated with the processing of several neuroendocrine prohormones including proopiomelanocortin (POMC) (Hutton, 1994; Rouille *et al.*, 1995). These Ca^{2+} dependent soluble proteases are able to specifically cleave proinsulin in the low pH environment of the maturing insulin storage granule. Pathogenic alterations in endopeptidase action causing defective proinsulin processing have recently been identified in a single human subject and in the fat/fat obese insulin-resistant mouse model. The human subject, who was obese but had normal fasting glycaemia, was found to have extremely high levels of circulating proinsulin (O'Rahilly *et al.*, 1995) and virtually no fully processed insulin. It is assumed that the high level of proinsulin compensated for the lack of insulin. The defect in this case was attributed to a lack of functional PC1 and consistent with defective POMC processing in the subject. In the fat/fat mouse a defect in carboxypeptidase E is associated with hyperproinsulinaemia (Naggert *et al.*, 1995). Mutations in the insulin gene have also been identified which result in defective processing of proinsulin and the secretion of partially processed insulin (Steiner *et al.*, 1990).

Fully processed insulin consists of two chains, the 30-amino-acid A-chain and the 21-amino-acid B-chain, which are linked by two disulphide bonds. The central portion of the proinsulin molecule which is released is known as the C-chain (or more commonly as C-peptide) and is largely retained in the mature storage granule and secreted along with insulin. C-peptide does not seem to have any major physiological role in maintaining glucose homeostasis as type-1 diabetics supplemented only with insulin are able to control glucose metabolism reasonably well. However, unlike insulin, C-peptide persists in the bloodstream and is often used as a method for estimating rates of insulin secretion.

Glucose controls the release of insulin at multiple levels (Steiner *et al.*, 1985). The acute effect is to regulate the release of preformed insulin storage granules from the pancreatic β cells. However, glucose also increases the biosynthesis of insulin by effects on transcription and translation of the gene (Philippe, 1991; Steiner *et al.*, 1985). In addition, the rates of biosynthesis of the processing enzymes PC2, and in particular PC1 are raised (Hutton, 1994). These effects occur over a period of several hours and the main function appears to be to ensure that β cell insulin stores are replenished ready for the next glucose challenge.

5.2.3 Glucose-sensing mechanisms in the β cell

Insulin secretion is exquisitely linked to blood glucose levels in the physiological range and the mechanism by which the pancreatic β cells sense changes in glucose concentrations has been a major focus of research efforts. Therefore, the rate-limiting steps regulating glucose influx and metabolism in the β cell are likely to function as glucose sensors. One candidate for this role is GLUT2 which is the major glucose transporter isoform found in the plasma membrane of the β cell (Gould and Holman, 1993). GLUT2 is a passive glucose transporter with a relatively high V_{max}. The K_m for glucose is high therefore the rate of glucose transport into the cell will vary as the concentration of glucose fluctuates over the physiological range. GLUT2 is required for efficient glucose-induced insulin secretion as demonstrated by studies in transgenic mice in which antisense RNA reduced GLUT2 levels in β cells by 80% (Valera *et al.*, 1994b). These mice develop impaired insulin secretion and elevated blood glucose levels. A limited number of mutations in the *GLUT2* gene have been identified in humans (Matsubara *et al.*, 1995; Mueckler *et al.*, 1994). While most of these do not appear to be pathogenic, a single NIDDM patient was found to have a V197I mutation in one allele of the *GLUT2* gene. This mutation abolished glucose transport and the reduced expression of functional GLUT2 is likely to contribute to the pathogenesis of diabetes in this subject. However, in normal individuals the total capacity for glucose transport into a β cell is far greater than the cellular glycolytic capability so, in reality, it is unlikely that GLUT2 plays a major glucose-sensing role (Efrat *et al.*, 1994).

A better candidate for the glucose sensor is glucokinase (Efrat *et al.*, 1994; Matschinsky *et al.*, 1993). Glucokinase is the major enzyme responsible for phosphorylation of free glucose in the β cell as hexokinase is suppressed by glucose-6-phosphate levels under *in vivo* conditions (Matschinsky, 1996; Matschinsky *et al.*, 1993). Glucokinase has a K_m for glucose of ≈ 10 mM which means that rising concentrations of blood glucose in the physiological range will result in a proportional increase in the rate of glycolysis in the β cell (Matschinsky, 1996; Matschinsky *et al.*, 1993). Phosphorylation of glucose appears to be a rate-limiting step in the glycolytic pathway in the β cell and, as discussed below, it is proposed that the increased rate of glycolysis ultimately regulates insulin secretion. The importance of glucokinase in this regulation is highlighted by transgenic approaches using β-cell-specific antisense RNA (Efrat *et al.*, 1994b) and homologous recombination to knockout the glucokinase gene (Grupe *et al.*, 1995). In these experiments the homozygote knockout causes death *in utero* while, in the heterozygotes and animals treated with antisense RNA, a 50–70% reduction in β-cell glucokinase levels was associated with reduced capacity to secrete insulin in response to glucose. However, the most convincing evidence for the role of glucokinase comes from the finding that mutations which reduce glucokinase function in humans are associated with impaired glucose-stimulated insulin secretion (Byrne *et al.*, 1994) (*Figure 5.1*) and the development of a subclass of NIDDM known as maturity onset diabetes of the young (specifically MODY2) (Bell *et al.*, 1996).

5.2.4 Signals responsible for insulin secretion

The metabolism of glucose in pancreatic β cells proceeds largely by glycolysis and very little glucose is stored as glycogen or converted to lactate. The metabolism of glucose is an absolute requirement for the stimulation of insulin release. Various agents such as amino acids (particularly arginine and leucine), fatty acids and gut hormones [such as GLP-1 (Orskov, 1992)] can increase the secretion of insulin but only in the presence of facilitating concentrations of glucose (above 3 mM). While these potentiation effects may have important physiological roles in fine-tuning the release of insulin, the primary signal remains glucose. This suggests that a metabolite(s) of glucose acts to stimulate insulin secretion while the other secretagogues are only able to stimulate a subset of the pathways required for insulin secretion.

Recently, there has been significant progress in understanding the mechanisms by which glucose stimulates insulin release (*Figure 5.2*). It has long been established that the glucose-stimulated release of insulin requires the rapid closing of an inward rectifying K^+ channel resulting in depolarization of the β-cell plasma membrane (Dukes and Philipson, 1996). The potassium channel in question is inhibited by ATP and it was originally postulated that the rise in cellular ATP subsequent to increased glycolysis caused the closing of the K^+ channel. However, subsequent data suggested that the K^+ channel would remain shut throughout the physiological range of ATP concentrations. It is now thought more likely that it

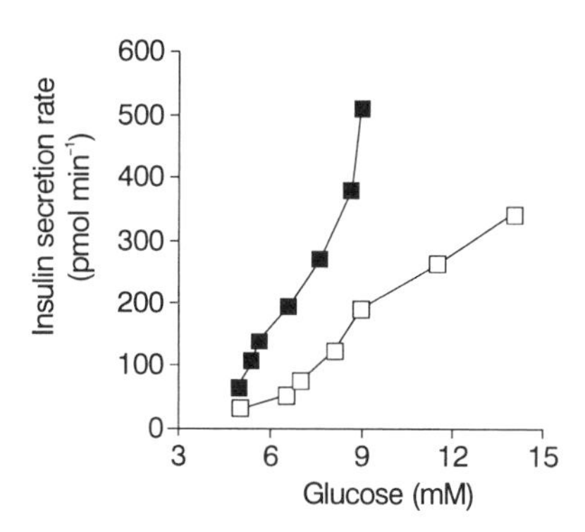

Figure 5.1. The ability of glucose to stimulate insulin secretion is impaired in subjects with glucokinase mutations. Insulin secretion rates at indicated glucose concentrations in control subjects (closed squares) or subjects with a pathogenic mutation in the glucokinase gene (open squares). Adapted from Byrne *et al.* (1994) by copyright permission of The American Society for Clinical Investigation.

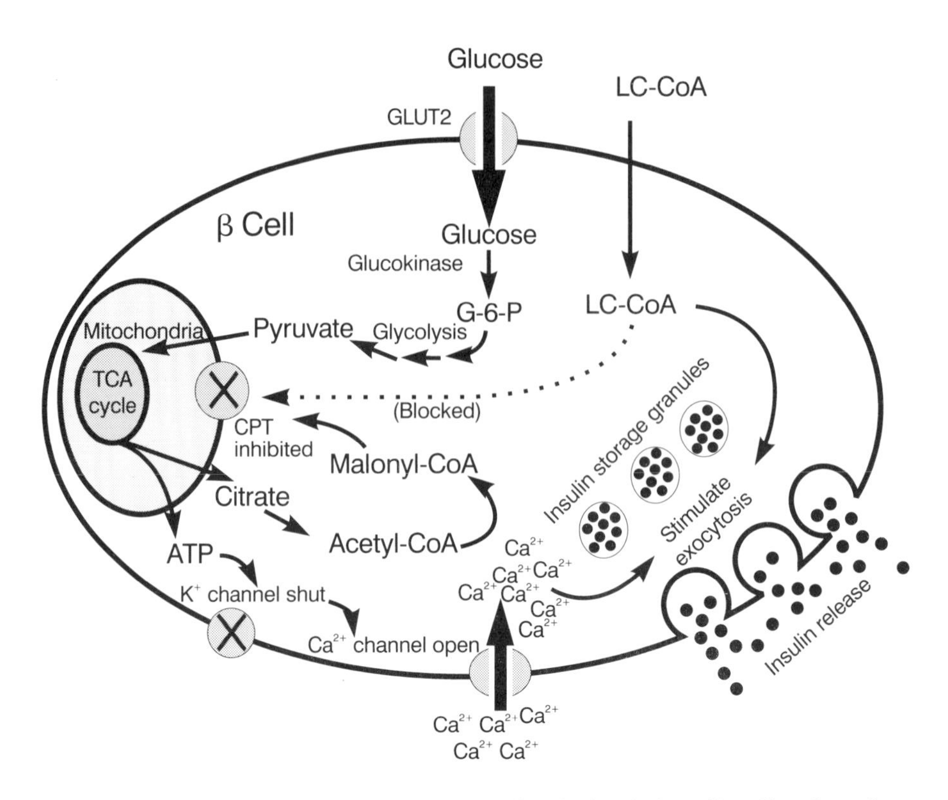

Figure 5.2. Mechanisms involved in glucose-mediated stimulation of insulin release from pancreatic β cells. CPT, carnitine palmitoyl CoA transferase; G-6-P, glucose-6-phosphate; LC-CoA, long-chain fatty-acyl-CoA esters; TCA, tricarboxylic acid.

is in fact the change in ATP/ADP ratio during glucose metabolism that regulates the K^+ channel as it is known that ADP antagonizes the effects of ATP in this channel. The importance of the ATP/ADP ratio in maintaining appropriate insulin secretion is demonstrated in subjects who have a subtype of NIDDM known as maternally inherited diabetes and deafness (Maasen and Kadowaki, 1996). These subjects are characterized by having a high ratio of mitochondria with gene mutations which reduce oxidative phosphorylation capacity resulting in a lower ATP/ADP ratio. The pancreatic β cells of these subjects have a reduced capacity to secrete insulin. Impairment of mitochondrial number or function may also contribute to defects of insulin secretion associated with ageing.

The association between inhibition of transmembrane K^+ transport and insulin secretion *in vivo* is clearly demonstrated in a recent transgenic mouse model in which a non-ATP-sensitive K^+ channel was overexpressed in the β cell (Philipson *et al.*, 1994). The result is constitutive K^+ transport and decreased insulin secretion in the face of a glucose challenge. Furthermore, the sulphonylurea class of antidiabetic drugs increase insulin secretion by causing closure of the β-cell K^+ channel (Dukes and Philipson, 1996). The target for the sulphonylureas is the recently cloned sulphonylurea receptor which is thought to act as a regulatory protein for ion channels including the ATP-sensitive K^+ channel. Mutations have recently been described in this receptor which are far more prevalent in NIDDM patients than in controls (Inoue *et al.*, 1996). Mutations in the sulphonylurea receptor have also been associated with the human syndrome of familial persistent hyperinsulinaemia and hypoglycaemia in infancy (PHHI) (Thomas *et al.*, 1995). The β cells from these subjects lack functional ATP-sensitive K^+-channel activity and display unrestrained insulin secretion. These results suggest that the sulphonylurea receptor plays an important role in regulating K^+-channel activity and hence has an important role in regulating insulin secretion.

The rapid depolarization of the β cell induced by K^+-channel closure activates an L-type voltage-sensitive calcium channel which causes a rapid influx of Ca^{2+} ions into the cell (Dukes and Philipson, 1996). Intracellular calcium stores can also be mobilized and the result is an oscillation in intracellular Ca^{2+} concentrations. The rise in intracellular Ca^{2+} levels is necessary for glucose-induced insulin release although how this rise in intracellular Ca^{2+} participates in secretion of insulin storage granules is not known.

However, while the Ca^{2+} influx is necessary for insulin secretion, it alone is not sufficient. A large amount of evidence is now accumulating to suggest that glucose-induced production of malonyl CoA plays a key role in regulating insulin secretion (Prentki and Corkey, 1996). In this model, the malonyl CoA inhibits carnitine palmitoyl CoA transferase (CPT), the protein which is responsible for the transport of long-chain fatty-acyl-CoA esters (LC-CoA) into the mitochondria for subsequent oxidation. This results in an increase in cytosolic LC-CoA concentrations and LC-CoA is known to have stimulatory effects on exocytosis. It has therefore been postulated that the concurrent increases in intracellular free Ca^{2+} and LC-CoA combine to regulate insulin secretion (Prentki and Corkey, 1996).

5.2.5 Amylin

Amylin (or islet amyloid polypeptide) is one of the peptides that remains in mature insulin storage granules and is thus co-secreted with insulin (Westermark *et al.*, 1992). Amylin is structurally related to the calcitonin gene-related peptide and it has been speculated that it may have some hormonal function in glucose metabolism. However, studies in humans (Bennet *et al.*, 1994) and transgenic mice which overexpress amylin in the pancreas (Hoppener *et al.*, 1993; Yagui *et al.*, 1995) have failed to find any major glucoregulatory role for amylin. The physiological role of amylin remains unclear but the human peptide (but not rodent amylin) is highly prone to the formation of amylin fibrils which are a major component of the amyloid deposits that are often found in islets of NIDDM patients (Westermark *et al.*, 1992). These deposits could therefore play a role in the diminution of β-cell numbers in NIDDM (Westermark *et al.*, 1992).

5.3 The role of liver, muscle and adipose tissue in the regulation of glucose homeostasis

The effects of insulin on glucose homeostasis are largely mediated through a reduction in hepatic glucose output with simultaneous stimulation of glucose uptake into muscle and adipose tissue.

5.3.1 Glucose metabolism in liver

In the post-absorptive state, the bulk of the glucose utilized by the body is supplied by the liver (*Figure 5.3*). Unlike many other tissues, the liver possesses the machinery for both the import and export of glucose. Glucose influx and efflux across the liver plasma membrane is regulated by the high capacity GLUT2 glucose transporter (Gould and Holman, 1993). Like the pancreas, the liver contains glucokinase which rapidly phosphorylates glucose to glucose-6-phosphate. Therefore, the mutations in glucokinase and GLUT2 described above which affect insulin secretion from the pancreas are also likely to have effects on liver glucose metabolism. However, unlike the pancreas, the liver also contains glucose-6-phosphatase (Burchell *et al.*, 1994). This enzyme is found in the ER and catalyses the release of phosphate from glucose-6-phosphate. The glucose is then transported into the cytosol via the GLUT7 glucose transporter isoform in the membranes of the ER (Waddell *et al.*, 1992) thus increasing the cytosolic concentration of free glucose. As this rises, glucose is transported across the plasma membrane by GLUT2.

High levels of glucose-6-phosphate are maintained in the post-absorptive state by a combination of glycogenolysis and gluconeogenesis. This in turn leads to increased intracellular levels of free glucose and, as extracellular glucose levels are relatively low, the net flow of glucose is out of the liver.The importance of hepatic glycogenolysis and glucose-6-phosphatase in regulating whole body glucose homeostasis is illustrated by the pathophysiology of type-1 glycogen storage disease. This is an autosomal recessive condition caused by mutations in the coding sequence of the glucose-6-phosphatase gene which severely impairs or abolishes

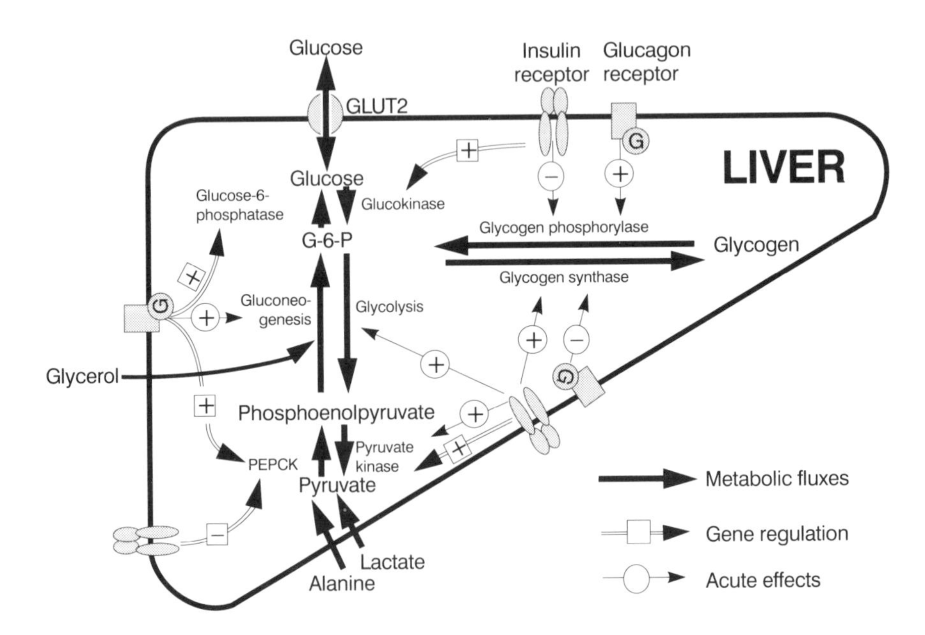

Figure 5.3. Mechanisms involved in regulating glucose metabolism in the liver. G-6-P, glucose-6-phosphate; PEPCK, phosphoenolpyruvate carboxykinase.

enzyme activity (Lei *et al.*, 1995). This results in an inability to mobilize liver glycogen stores to free glucose leading to post-absorptive hypoglycaemia and lactic acidosis (Nordlie and Sukalski, 1986).

Gluconeogenesis and glycogenolysis in the liver are stimulated by glucagon when plasma glucose and insulin concentrations are low (Lefebvre, 1995) (*Figure 5.3*). The liver is the major site of glucagon action and the hepatocyte membranes are rich in glucagon receptors which are members of the G-protein-linked seven-transmembrane-spanning class of receptors (see Section 3.2.3) (Christophe, 1995). Glucagon stimulation increases cAMP levels resulting in activation of protein kinase A which activates the glycogen-bound form of protein phosphatase-1 (PP-1G). This enzyme then causes inactivation of glycogen synthase but activates glycogen phosphorylase thus promoting glycogenolysis (Lawrence, 1992). Glucagon also upregulates expression of phosphoenolpyruvate carboxykinase (PEPCK) which is a key step in gluconeogenesis from precursors such as lactate and amino acids. Circulating levels of lactate and glycerol are elevated in insulin-resistant and diabetic subjects, and there is some debate as to whether alterations in the level of these gluconeogenic precursors can increase hepatic glucose output.

Insulin has a powerful effect on liver metabolism as this tissue is the first recipient of the insulin secreted into the portal vein and thus sees a higher concentration than peripheral tissues. In addition, the liver is more sensitive to insulin than muscle and fat. The main action of insulin is to reduce glucose output by causing a shift from glucose production (glycogenolysis and gluconeogenesis) to glucose utilization and storage (glycolysis and glycogen synthesis) (Barrett and Liu, 1993; Granner and Pilkis, 1990) (*Figure 5.3*). However, unlike the situation in muscle and fat, insulin does not directly affect the transport of glucose across the liver plasma membrane. Insulin effects on the liver generally oppose those of glucagon. For example, insulin inhibits PP-1G and therefore glycogen phosphorylase while simultaneously activating glycogen synthase. The acute inhibitory effect of the hormone on hepatic gluconeogenesis is largely mediated by increasing fructose-2,6-bisphosphate which activates 6-phosphofructokinase and stimulates the glycolytic utilization of glucose. In addition, insulin stimulates pyruvate kinase.

Over longer time frames (1–2 h) insulin stimulates glucokinase and inhibits glucose-6-phosphatase expression. Glucose and insulin also combine to stimulate pyruvate kinase gene expression (Granner and Pilkis, 1990) and studies in transgenic mice have recently defined the elements in the pyruvate kinase gene promoter which mediate the effects of insulin and glucose (Noguchi *et al.*, 1993). Importantly, insulin also counteracts the effects of glucagon and downregulates expression of PEPCK. However, insulin is unable to suppress PEPCK levels in the liver of type-2 diabetics and the resulting increase in PEPCK activity is thought to contribute to raised hepatic glucose production in these subjects (Barrett and Liu, 1993; Granner and Pilkis, 1990). The importance of this counter-regulatory effect of insulin on PEPCK levels to whole body glucose homeostasis is demonstrated most clearly in two transgenic rodent models which constitutively overexpress PEPCK in liver. In both of these models, relatively modest elevations in hepatic PEPCK cause significant hyperglycaemia and hyperinsulinaemia (Rosella *et al.*, 1995; Valera *et al.*, 1994a; *Figure 5.4*). Elements in the *PEPCK* gene promoter region responsible for downregulation of gene

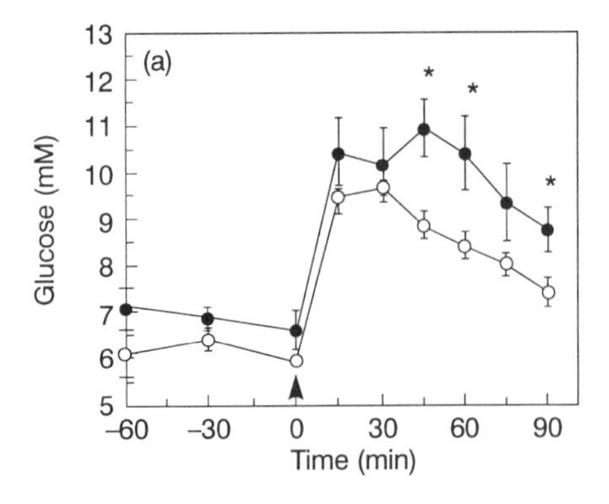

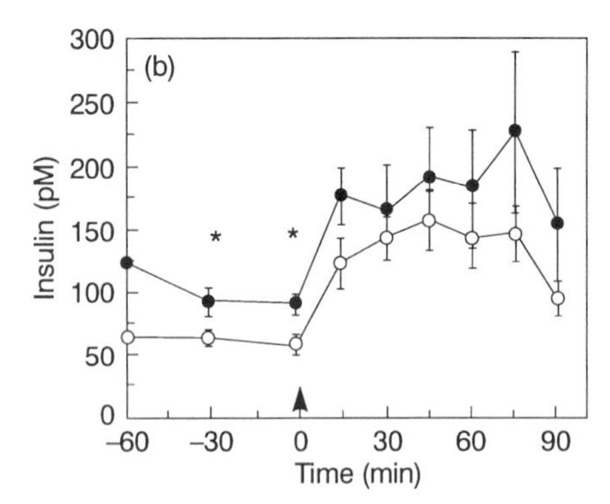

Figure 5.4. PEPCK overexpression in liver causes whole body insulin resistance. Plasma blood glucose (a) and insulin (b) levels were determined in transgenic rats overexpressing PEPCK in liver (closed circle) or control animal (open circle) before and after an intraperitoneal glucose load (injected at time 0). Results are means ± SEM of five separate rats; *, $P<0.05$. Redrawn with permission from Rosella *et al.* (1995); © The Endocrine Society.

expression by insulin have been identified (Granner and O'Brien, 1992) but little is known of the signal transduction cascade leading from the insulin receptor to changes in *PEPCK* gene transcription. However, one recent study indicates that insulin stimulation of the enzyme phosphatidylinositol-3-kinase (PI3-kinase) is a required step in this process (Sutherland *et al.*, 1995).

5.3.2 Glucose metabolism in muscle and adipose tissue

In the post-absorptive state, uptake of glucose into muscle and adipose tissue is low and the major metabolic fuel for muscle is fatty acids and, to a lesser extent, glucose from glycogen stores. However, the rate of glucose uptake into muscle and adipose tissue is increased dramatically by the post-prandial increase in circulating insulin. A number of studies indicate that muscle is the major site of insulin-stimulated glucose disposal *in vivo* and that fat plays a lesser role (DeFronzo, 1988). This is partly due to the fact that muscle makes up a much larger proportion of total body mass than fat in a normal individual. In the presence of insulin, muscle uses glucose as a metabolic fuel via glycolysis (oxidative pathways) but the largest proportion of glucose taken up into muscle is stored as glycogen (non-oxidative pathways). In adipocytes glycogen synthesis is minimal but the glucose may be used to form the glycerol backbone during triacylglycerol synthesis, a process which is also stimulated by insulin. Another major fate of glucose in adipose tissue is to be metabolized to lactate (Digirolamo *et al.*, 1992).

Key role of muscle and adipocyte glucose transport in regulation of glucose homeostasis. The ability of muscle and adipose tissue to dramatically alter their rate of glucose uptake in response to insulin is due to the unique properties of GLUT4 which is the predominant glucose transporter isoform expressed in these tissues (Gould and Holman, 1993). GLUT4 is a transmembrane-facilitated glucose transporter with a high level of sequence homology and predicted membrane topology to GLUT2. However, unlike GLUT2, it has a low K_m which means that it is functioning at its maximal rate in the physiological range of glucose concentrations. A further difference between GLUT2 and GLUT4 is their subcellular localization. While GLUT2 resides constitutively in the plasma membrane, GLUT4 recycles between the plasma membrane and an intracellular storage pool. In unstimulated muscle and adipocytes GLUT4 is largely located in the intracellular pool. Insulin stimulation causes the rapid recruitment of GLUT4 to the plasma membrane which results in a rapid increase in the V_{max} for glucose transport into the tissue (Gould and Holman, 1993). Acute exercise (or experimentally induced contraction or hypoxia) also stimulates the translocation of GLUT4 and increases the rate of glucose transport in muscle (Goodyear *et al.*, 1991; Lund *et al.*, 1995). The effects of insulin and exercise on glucose transport are additive indicating that distinct intracellular pools of GLUT4 are recruited in response to each of these stimuli. The intracellular signalling mechanisms used by these two stimuli to induce GLUT4 translocation from these two pools appear to differ (Lund *et al.*, 1995). Furthermore, while insulin stimulation of glucose transport into muscle is severely impaired in NIDDM (Zierath, 1995) it appears that the mechanism stimulating glucose transport in response to acute exercise remains intact (Azevedo *et al.*, 1995). Long-term exercise training also increases GLUT4 levels and this has been implicated in the increased insulin sensitivity observed with exercise in humans (Houmard *et al.*, 1991), and explains the observation that exercise has a glucose-lowering effect in insulin-resistant states.

Several lines of evidence point to glucose transport across the plasma membrane as the rate-limiting step for glucose metabolism in muscle and adipocytes *in vivo* (Rothman *et al.*, 1992; Yki-Jarvinen *et al.*, 1987), even in the metabolically perturbed states associated with IDDM and NIDDM (Butler *et al.*, 1990; Yki-Jarvinen *et al.*, 1990). Overexpression of GLUT4 in muscle increases glucose uptake and glycogen synthesis (Deems *et al.*, 1994; Treadway *et al.*, 1994; Tsao *et al.*, 1996). Basal and insulin-stimulated glucose transport and oxidative glucose metabolism are also dramatically increased in adipocytes from these animals overexpressing GLUT4 (Deems *et al.*, 1994; Tozzo *et al.*, 1995).

The mouse models overexpressing GLUT4 in muscle are more insulin responsive with lower circulating insulin and glucose levels both in the fasting and fed states (Ikemoto *et al.*, 1995; Leturque *et al.*, 1996; Liu *et al.*, 1993; Tsao *et al.*, 1996). Post-prandial insulin and glucose levels are also significantly lower in mice expressing GLUT4 selectively in adipose tissue, despite the fact that muscle is thought to be the major site of insulin-stimulated glucose disposal (Shepherd *et al.*, 1993) (*Figure 5.5*). Thus, the *in vivo* disposal of glucose into muscle and adipocytes is increased significantly in animals overexpressing GLUT4 and is presumably the major factor responsible for lowering blood glucose levels as hepatic glucose output is unchanged despite lower insulin levels and significantly higher glucagon levels (Ren *et al.*, 1995; Treadway *et al.*, 1994). The lower insulin levels are presumably secondary to the lower blood glucose levels. The potential importance of GLUT4 levels in regulating glucose homeostasis under insulin-resistant conditions is demonstrated in the db/db obese mouse model which has leptin resistance (see Section 10.3.1). In these mice, overexpression of GLUT4 in

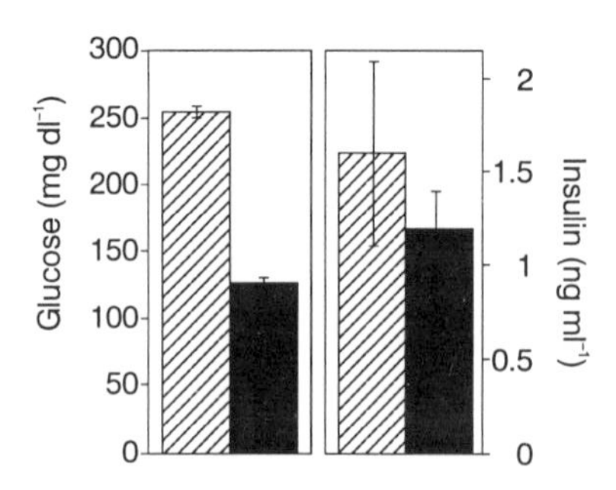

Figure 5.5. Improved insulin sensitivity in transgenic mice overexpressing GLUT4 specifically in adipose tissue. Plasma glucose and insulin levels 30 min after an intraperitoneal glucose load in transgenic mice overexpressing GLUT4 selectively in adipocytes (solid bars) or non-transgenic control litter mates (hatched bars). Results are means ± SEM. Adapted from Shepherd *et al.* (1993a) with permission from the American Society for Biochemistry and Molecular Biology.

muscle and fat reduced fasting glycaemia to normal and greatly reduced hyperglycaemia after a glucose load (Gibbs *et al.*, 1995). Furthermore, GLUT4 overexpression was able to significantly reduce hyperglycaemia in mice that had been made insulinopaenic by streptozotocin-induced pancreatic ablation (Leturque *et al.*, 1996). Conversely lowering GLUT4 levels in muscle reduces the overall ability of the body to dispose of glucose. This has been clearly demonstrated in mice where one allele of the *GLUT4* gene has been disrupted. In these animals GLUT4 levels are reduced and severe insulin resistance develops (Rosetti *et al.*, 1996). However, decreases in GLUT4 levels do not appear to play a significant role in the impairment of insulin sensitivity in human diabetics as GLUT4 levels are reduced in adipocytes but not in muscle which is the major site of insulin-stimulated glucose disposal (Shepherd and Kahn, 1993b).

Therefore, the level of GLUT4 expression has important implications for overall control of glucose homeostasis. The factors responsible for tissue-specific expression of GLUT4 and its differential regulation in muscle and adipose tissue have been studied. Transgenic mouse models using various elements of the *GLUT4* gene promoter linked to reporter genes have been used to dissect promoter elements that are responsible for this tissue-specific expression and for regulation of expression of GLUT4 in response to the *in vivo* milieu (Olson and Pessin, 1996).

Other major regulatory steps in glucose metabolism in muscle and fat. The first committed step of glucose metabolism in muscle and fat is phosphorylation and hexokinase-2 is the major enzyme catalysing this step (Wilson, 1995). The activity of hexokinase-2 is not acutely regulated by insulin and most evidence suggests that under normal circumstances it is not rate limiting for glucose metabolism in these tissues (Rothman *et al.*, 1992). Insulin-stimulated glucose transport into muscle *in vitro* was increased in transgenic mice overexpressing hexokinase-2 specifically in muscle (Chang *et al.*, 1996). However, unlike transgenic models overexpressing GLUT4 in muscle there was no effect on *in vivo* glucose clearance rates or insulin levels indicating that glucose transport is a more important determ... than glucose phosphorylation of whole body glucose disposal. It has been suggested that hexokinase could become a rate-limiting step in glucose metabolism in diabetes where lower glucose-6-phosphate levels are observed (Rothman *et al.*, 1995).

Insulin acutely stimulates the activities of pyruvate dehydrogenase (PDH) and acetyl CoA carboxylase (ACC) (Lawrence, 1992). The increased PDH activity diverts more glucose-derived carbon units into the citric acid cycle and hence encourages glucose oxidation. The increase in ACC activity increases levels of malonyl CoA thus inhibiting CPT and blocking fatty acid oxidation in the mitochondria. The reduction in mitochondrial fatty acid oxidation in turn encourages the oxidative metabolism of glucose.

Glycogen storage is the most important pathway for non-oxidative metabolism of glucose in muscle. *In vivo* insulin stimulation of glycogen synthesis is in part mediated by the allosteric activation of glycogen synthase by increased levels of glucose-6-phosphate that arise from increased glucose transport into muscle (Shulman *et al.*, 1995). As in the liver, insulin also stimulates the rate of glycogen synthesis by activating protein kinase cascades which ultimately activate the glycogen-bound form of PP-1G resulting in the activation of glycogen synthase and the inactivation of glycogen phosphorylase (Lawrence, 1992).

Other mechanisms by which adipose tissue may affect glucose homeostasis. The close links between obesity and insulin resistance suggest that events occurring in adipose tissue may influence whole body glucose homeostasis. In fact, recent evidence suggests that white adipose tissue can function as an endocrine organ and play an active role in regulating energy balance (Flier, 1995). For example, leptin (the *ob* gene product; see Section 10.3.1) (Billington and Levine, 1996; Zhang *et al.*, 1994) is expressed solely in adipocytes and secreted into the circulation. Leptin is thought to act largely on the hypothalamus to restrict appetite, and mice containing a homozygous nonsense mutation in this gene (ob/ob mice) are hyperphagic and develop severe obesity with insulin resistance. However, human obesity is typified by an increase in leptin and therefore if leptin is playing a role in the development of human obesity it is likely to be due to resistance of leptin receptors in the hypothalamus. White adipose tissue also secretes tumour necrosis factor (TNF)-α and expression of this protein is increased in obese and insulin-resistant animals (Flier, 1995). TNF-α causes insulin-resistance at the cellular level and is therefore postulated to be a major factor in inducing insulin resistance associated with obesity. Indeed, in obese rodent models systemic neutralization of TNF-α is associated with an improvement in whole body insulin sensitivity (Hotamisligil *et al.*, 1993). However, similar studies in obese diabetic human subjects have so far failed to demonstrate similar benefits (Ofei *et al.*, 1996).

Brown adipose tissue may also play a special role in glucoregulatory mechanisms. This tissue has a high capacity for glucose metabolism and is distinguished from white adipose tissue by the presence of large numbers of mitochondria which uniquely express uncoupling protein (UCP). UCP is a proton transporter which destroys the proton gradient required to drive ATP production in the mitochondria, glucose metabolism being instead directed toward heat production. Brown adipose tissue depots are clearly identifiable in rodents where they are thought to be involved in non-shivering thermogenesis. An interesting transgenic mouse model has been developed in which the UCP promoter region was fused to a diphtheria toxin resulting in the ablation of brown adipose tissue while white adipose tissue remained normal (Hamann *et al.*, 1995). These animals become hyperphagic, severely obese and insulin resistant suggesting a key role for brown adipose tissue in glucose homeostasis in rodents. However, whether or not brown fat plays a role in energy metabolism in the adult human remains controversial.

5.4 Role of insulin-signalling pathways in maintaining glucose homeostasis

5.4.1 Insulin receptor

The effects of insulin on target cells are mediated by binding of the hormone to transmembrane insulin receptors which trigger a series of intracellular signalling events, the nature of which are only now beginning to be understood (see Section 3.2.1). The insulin receptor is expressed at high levels in muscle, adipose tissue and liver (Lee and Pilch, 1994) where it is transcribed from a single gene. The pro-receptor is cleaved into a 130 kDa α-chain and a 95 kDa β-chain. Two variants of the α-chain exist resulting from alternative splicing of exon 11 of the insulin receptor gene although the functional significance of this is poorly understood. The fully functional insulin receptor comprises a tetramer of two α subunits, which are largely extracellular and which contain the insulin-binding sites, and two largely intracellular β subunits, containing the tyrosine kinase activity critical for signal transduction. Binding of insulin to the receptor has two major effects: (i) the tyrosine kinase activity contained in the β subunit is activated and (ii) the insulin–insulin receptor complex is internalized. Internalization results in degradation of the insulin and this represents the major mechanism for downregulation of circulating insulin. The vast majority of the insulin receptors are not degraded and are able to translocate back to the plasma membrane.

The insulin receptor shares a high degree of sequence and structural homology with the IGF-I receptor (DeMeyts *et al.*, 1994). They also share significant functional abilities and, when expressed at similar levels in cell culture, background activation of either insulin or IGF-I receptors induces similar effects on glucose metabolism (Weiland *et al.*, 1991). Further complexity is introduced by the fact that insulin receptor αβ heterodimers can form hybrid receptors with IGF-I receptor αβ heterodimers (Siddle *et al.*, 1994). However, IGF-I receptor levels are low in insulin-responsive tissues and both the IGF-I receptor and the hybrid receptor have a low affinity for insulin. Therefore these receptor systems are not thought to play major roles in whole body glucoregulation.

The insulin receptor has a crucial role in glucose homeostasis and decreasing levels of insulin receptor function are inversely related to the degree of insulin resistance. Transgenic mice have been produced in which both alleles of the insulin receptor gene have been disrupted. These mice develop severe ketoacidosis and die soon after birth (Accili *et al.*, 1996; Joshi *et al.*, 1996). A similar phenotype is observed in the human syndrome of leprechaunism where various combinations of mutations at the insulin receptor locus result in the lack of functional insulin receptors. These individuals have severe *in utero* growth retardation, severe insulin resistance and invariably die soon after birth (Krook and O'Rahilly, 1996). Additional evidence for the essential role of the insulin receptor comes from the finding that certain severely insulin-resistant subjects have circulating antibodies to the receptor. These auto-antibodies block insulin binding and thus reduce insulin-induced glucose metabolism (Kahn *et al.*, 1976). A variety of mutations have also been identified in the insulin receptor gene in humans which impair insulin signalling to varying degrees and are associated with insulin resistance (Krook and O'Rahilly, 1996). The effect that even modest alterations in insulin receptor function can have on whole body glucose homeostasis is highlighted by a recent transgenic mouse model in which a dominant negative insulin receptor mutant was overexpressed specifically in muscle (Moller *et al.*, 1996). These mice have reduced insulin receptor tyrosine kinase activity in muscle and become insulin resistant. Interestingly, these mice also have increased body fat levels which suggests a link between insulin resistance and the development of obesity.

5.4.2 Intracellular signalling pathways

The tyrosine kinase activity of the insulin receptor is essential for regulation of glucose metabolism. While most receptor tyrosine kinases form stable signalling complexes at the receptor which regulate downstream signalling, the substrates of the insulin receptor only transiently form such complexes. A number of intracellular proteins in liver, muscle and adipose tissue are rapidly phosphorylated on tyrosine residues following stimulation of the insulin receptor. These include Shc and the insulin receptor substrates (IRS)-1 and -2 in all three tissues, a 120 kDa ecto-ATPase in liver, and the 15 kDa aP2 fatty acid-binding protein and a 60 kDa protein of unknown function in adipose tissue (Myers and White, 1996).

The best understood mechanism by which ligand-induced stimulation of receptor tyrosine kinases leads to recruitment of downstream signalling pathways is the *src* homology 2 (SH2) domain paradigm (Pawson and Gish, 1992). SH2 domains are motifs of approximately 100 amino acids found in a range of signalling molecules including the adapter proteins Grb2 and Nck, the SH2-containing protein tyrosine phosphatase (SH-PTP2) and the p85 subunit of PI3-kinase. The SH2 domains enable the formation of specific signalling complexes as they allow binding with other signalling molecules in a tyrosine phosphorylation-dependent manner. These interactions arise because SH2 domains bind with highest affinity to specific tyrosine-containing amino acid motifs and binding affinity is highest when the tyrosine has been phosphorylated.

The basic outline of two major tyrosine kinase-dependent insulin signalling cascades has been identified (*Figure 5.6*). In one scheme, insulin-induced tyrosine phosphorylation of Shc allows the interaction of Shc with the SH2 domain of Grb2 (Bonfini *et al.*, 1996). This triggers a signalling cascade which activates ras and ultimately the mitogen-activated protein (MAP) kinase cascade. However, it has been established that activation of the MAP kinase cascade is neither necessary nor sufficient for insulin action on glucose transport and glycogen synthesis in muscle and adipose tissue (Denton and Tavare, 1995). Evidence suggests that insulin-stimulated tyrosine phosphorylation of the highly homologous adapter

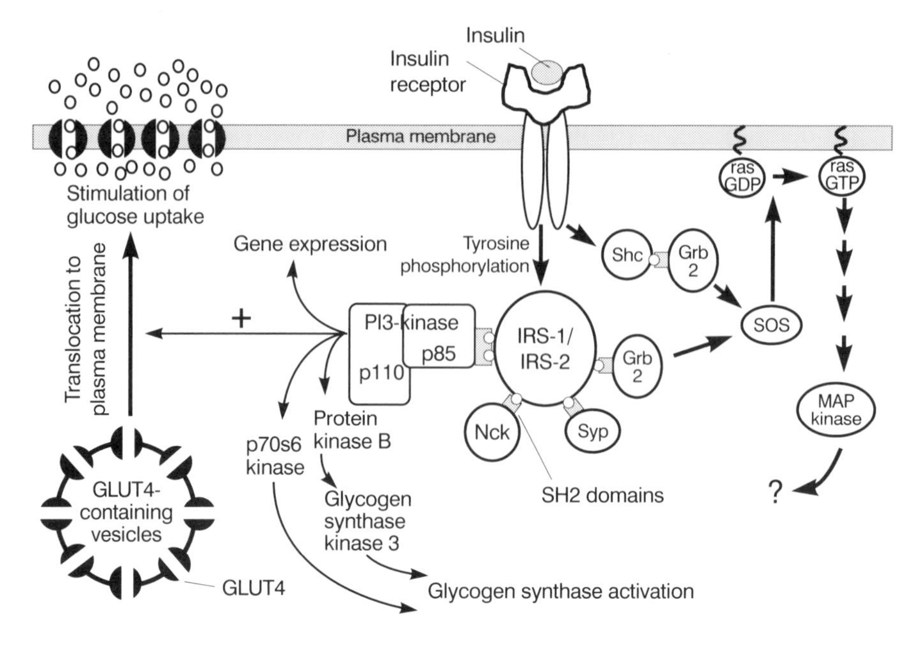

Figure 5.6. Intracellular signalling pathways stimulated by insulin. Shc, *src* homology collagen-like protein; SOS, son of sevenless protein; Syp, SH-PTP2.

proteins IRS-1 and IRS-2 has an important function in glucose metabolism. IRS-1 and IRS-2 are both soluble proteins of approximately 180 kDa and share significant sequence and structural homology. Each contains around 20 potential tyrosine-containing motifs capable of interacting with SH2 domains following insulin-induced tyrosine phosphorylation (Myers and White, 1996). IRS-1 and IRS-2 are able to recruit a number of SH2 domain-containing proteins including Grb2, SH-PTP2 and the p85 subunit of PI3-kinase (Myers and White, 1996). The importance of IRS proteins in insulin signalling has been demonstrated in transgenic mice where both alleles of IRS-1 have been disrupted. These mice become hyperglycaemic and hyperinsulinaemic and the ability of insulin to stimulate glucose transport into muscle and adipocytes is greatly reduced (Araki *et al.*, 1994; Tamemoto *et al.*, 1994). In these mice, continued expression of IRS-2 only partially compensates for the loss of IRS-1 (Patti *et al.*, 1995) which indicates that IRS-1 and IRS-2 have similar but not identical roles in insulin signalling.

Insulin-induced association of IRS proteins with the p85 subunit of PI3-kinase appears to play a central role in insulin effects on glucose metabolism. The p85 is tightly associated with the 110 kDa catalytic subunit of PI3-kinase and the formation of an IRS–p85–p110 complex results in the activation of cellular PI3-kinase activity (Shepherd *et al.*, 1996a). A large amount of evidence is accumulating which demonstrates that PI3-kinase activity is necessary for insulin to stimulate glucose transport (Shepherd *et al.*, 1996a, b; *Figure 5.7*), to activate glycogen synthase (Shepherd *et al.*, 1995) and to stimulate gene expression [e.g. hexokinase-2 (Osawa *et al.*, 1996)] in muscle and adipose tissue. A number of isoforms of the p85 and p110 subunits of PI3-kinase have been identified although

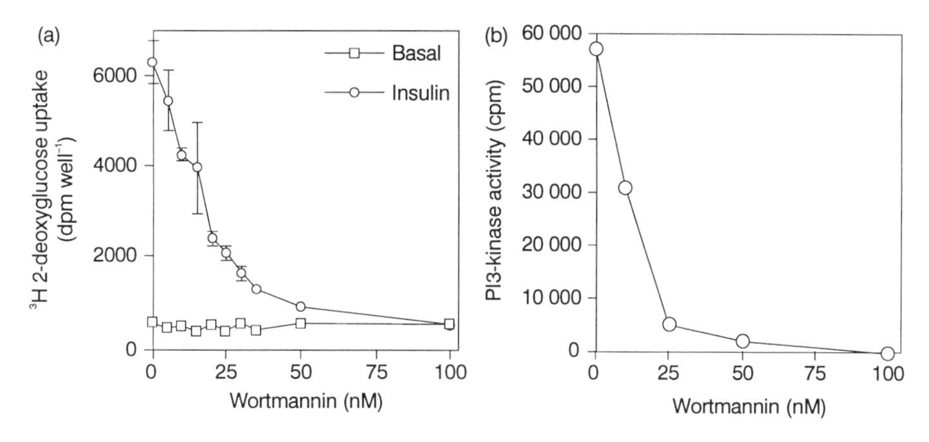

Figure 5.7. The specific PI3-kinase inhibitor wortmannin blocks PI3-kinase activity and insulin stimulation of glucose uptake with similar dose responses in 3T3-L1 adipocytes (P.R. Shepherd, unpublished data).

the individual role of each of these is currently not clear (Shepherd *et al.*, 1996a). However, it is of interest that a polymorphism has been identified in the p85α isoform which is associated with insulin resistance in humans who are homozygous for this sequence variant (Hansen *et al.*, 1996).

5.5 Integration of the individual components regulating glucose homeostasis

The above sections detail the components that are involved in the regulation of glucose metabolism in liver, muscle and adipose tissue. However, the factors governing the ability of an individual to regulate glucose homeostasis are complex. The counteracting effects of insulin and glucagon delicately manipulate glucose metabolism by co-ordinately regulating the activities of a number of enzymes and expression of a range of genes. This suggests that defects at a number of steps regulating insulin secretion or glucose metabolism could alter the set point of this equilibrium. In fact, evidence suggests that in humans there is a high degree of heterogeneity in the efficiency with which individuals control glucose metabolism, even between individuals who are considered to have normal glucose tolerance. This is likely to be due to the large degree of interindividual variation in the activity and expression of molecules important in regulating glucose homeostasis.

As described in previous sections, a number of mutations in key molecules involved in regulating glucose homeostasis have been identified which alone are sufficient to cause severe derangement of glucose metabolism. However, these only explain a small proportion of insulin resistance and a number of polymorphisms have also been identified in such genes which have much more subtle effects on glucose metabolism. These include mutations in GLUT4 (Buse *et al.*, 1992; Choi *et al.*, 1991; O'Rahilly *et al.*, 1992), the glycogen-binding subunit of protein phosphatase-1 (PP-1G) (Hansen *et al.*, 1995), IRS-1(Almind *et al.*, 1996a, b; Clausen *et al.*, 1995) and hexokinase-2 (Echwald *et al.*, 1995; Taylor *et al.*, 1996; Vidalpuig *et al.*, 1995). These polymorphisms alone do not appear to cause derangements in the regulation of glucose homeostasis to levels considered to be insulin resistant but they are likely to influence where an individual lies in the normal spectrum of insulin sensitivity.

Environmental factors (under which broad banner may be included obesity, diet and exercise) and the effects of ageing also contribute to the overall status of glucoregulatory mechanisms. One means by which this occurs is via alterations in the level of expression of molecules important in glucose metabolism. Key examples are the reduction in the levels of IRS-1 and the p85 subunit of PI3-kinase observed in muscle from obese subjects (Goodyear *et al.*, 1995), GLUT4 reductions in muscle associated with ageing (Houmard *et al.*, 1995) and reduction of hexokinase-2 and glycogen synthase in muscle of NIDDM patients (Vestergaard *et al.*, 1995). Heterogeneity in individual insulin sensitivity may also arise from the impact of factors which have inhibitory effects on insulin action. As described above, increased circulating lipid and fatty acid levels in obese subjects have a negative effect on insulin action (Felber *et al.*, 1993) and may directly impair insulin secretion (Prentki and Corkey, 1996). TNF-α may also affect insulin sensitivity and expression of TNF-α is increased in adipose tissue in obese subjects (Hotamisligil *et al.*, 1993). α_2-HS glycoprotein is a 63 kDa plasma protein known to inhibit insulin receptor function (Auberger *et al.*, 1989) as does the membrane-bound glycoprotein PC1 (Maddux *et al.*, 1995), and there is some evidence that elevated levels of these are associated with insulin resistance. Rad is a low molecular-weight GTP-binding protein which is overexpressed in the muscle of NIDDM and which may have inhibitory effects on insulin action (Reynet and Kahn, 1993). These changes in gene expression and the presence of inhibitors of insulin action may combine with the inborn errors in genes regulating glucose metabolism described above further to aggravate insulin resistance. For example, evidence suggests that mutations in PP-1G (Hansen *et al.*, 1995) and IRS-1 (Almind *et al.*, 1996a, b; Clausen *et al.*, 1995) only contribute to insulin resistance when present in obese subjects. Additionally, the elevated glucose levels found in insulin-resistant states may have a compounding effect on the degree of insulin resistance. Hyperglycaemia causes impaired insulin action in muscle although these effects are rapidly reversible when glucose concentrations are lowered (Zierath *et al.*, 1994). One mechanism by which hyperglycaemia impairs insulin action is by increasing flux of glucose through the hexosamine biosynthetic pathway (Robinson *et al.*, 1995). In muscle this has been shown to have inhibitory effects on insulin stimulation of GLUT4 translocation (Baron *et al.*, 1995) and glycogen synthase activation (Crook and McClain, 1996).

In summary, it has not been possible to attribute a single cause to the majority of cases of insulin resistance. However, gene mutations and the influence of environmental factors which reduce the flux through key steps in pathways regulating glucose metabolism are likely to combine over a period of time to cause the slow development of insulin resistance and influence the progression towards NIDDM.

5.6 Conclusion

The major enzymatic pathways regulating glucose metabolism have been understood for some time. However, in recent years most of the important proteins in the insulin regulation of glucose metabolism and many of those involved in insulin secretion have been identified and their genes cloned. These discoveries have allowed the application of techniques for quantification of levels of gene expression and rapid gene scanning techniques to look for gene polymorphisms. Techniques for transgenic gene expression and gene knockout have proved particularly useful for dissecting the role of individual proteins in metabolic pathways. Using these approaches it has been possible to gain powerful insights into the role of particular proteins in the complex metabolic interactions involved in the regulation of whole body glucose homeostasis. These techniques have also allowed a better understanding of how derangements in glucose metabolism can lead to insulin resistance and NIDDM.

References

Accili D, Drago J, Lee EJ *et al.* (1996) Early neonatal death in mice homozygous for a null mutation in the insulin receptor gene. *Nature Genet.* **12:** 106–109.

Almind K, Bjorbaek C, Vestergaard H, Hansen T, Echwald S, Pedersen O. (1996a) Amino acid polymorphisms of IRS-1 in non insulin dependent diabetes mellitus. *Lancet* **342:** 828–832.

Almind K, Inoue G, Pedersen O, Kahn CR. (1996b) A common amino acid polymorphism in IRS-1 causes impaired insulin signalling – evidence from transfection studies. *J. Clin. Invest.* **97:** 2569–2575.

Araki E, Lipes MA, Patti M, Bruning JC, Haag B, Johnson RS, Kahn CR. (1994) Alternative pathway of insulin signalling in mice with targeted disruption of the IRS-1 gene. *Nature* **372:** 186–190.

Atkinson MA, Maclaren NK. (1994) Mechanisms of disease: the pathogenesis of insulin dependent diabetes mellitus. *New Engl. J. Med.* **331:** 1428–1436.

Auberger P, Falquerho L, Contreres JO, Pages G, Lecam G, Rossi B, Lecam A. (1989) Characterization of a natural inhibitor of the insulin receptor tyrosine kinase – cDNA cloning, purification and antimitogenic activity. *Cell* **58:** 631–640.

Azevedo JL, Carey JO, Pories WJ, Morris PG, Dohm GL. (1995) Hypoxia stimulates glucose transport in insulin resistant human skeletal muscle. *Diabetes* **44:** 695–698.

Baron AD, Zhu JS, Weldon H, Maianu L, Garvey WT (1995). Glucosamine induces insulin resistance in vivo by affecting GLUT4 translocation in skeletal muscle – implications for glucose toxicity. *J. Clin. Invest.* **96:** 2792–2801

Barrett EJ, Liu Z. (1993) Hepatic glucose metabolism and insulin resistance in NIDDM and obesity. *Baillieres Clin. Endocrinol. Metab.* **7:** 875–901.

Bell GI, Pilkis SJ, Weber IT, Polonsky KS. (1996) Glucokinase mutations, insulin secretion, and diabetes mellitus. *Ann. Rev. Physiol.* **58:** 171–186.

Bennet WM, Smith DM, Bloom SR. (1994) Islet amyloid polypeptide: does it play a pathophysiological role in the development of diabetes. *Diabetic Med.* **11:** 825–829.

Billington CJ, Levine AS. (1996). Appetite regulation: shedding light on obesity. *Curr. Biol.* **6:** 920–923.

Bonfini L, Migliaccio E, Pelicci G, Lanfrancone L, Pelicci P. (1996) Not all Shc's roads lead to Ras. *Trends Biochem. Sci.* **21:** 257–261.

Burchell A, Allan BB, Hume R. (1994) Glucose 6-phosphatase proteins of the endoplasmic reticulum. *Mol. Membr. Biol.* **11:** 217–227.

Buse JB, Yasuda K, Lay TP, Seo TS, Karam JH, Seino S, Bell GI. (1992) Human GLUT4 muscle/fat glucose transporter gene: sequence, promoter characterization and genetic variations. *Diabetes* **41:** 1436–1446.

Butler PC, Kryshak EJ, Marsh M, Rizza RA. (1990) Effect of insulin on oxidation of intracellularly and extracellularly derived glucose in patients with NIDDM – evidence for primary defect in glucose transport and/or phosphorylation but not oxidation. *Diabetes* **39:** 1373–1380.

Byrne MM, Sturis J, Clement K *et al.* (1994). Insulin secretory abnormalities in subjects with hyperglycemia due to glucokinase mutations . *J. Clin. Invest.* **93:** 1120–1130.

Chang P, Jensen J, Printz RL, Granner DK, Ivy JL, Moller DE. (1996) Overexpression of hexokinase 2 in transgenic mice: evidence that increased phosphorylation augments muscle glucose uptake. *J. Biol. Chem.* **271:** 14834–14839.

Choi WH, O'Rahilly S, Buse JB, Rees A, Morgan R, Flier JS, Moller DE. (1991) Molecular scanning of insulin responsive glucose transporter (GLUT4) gene in NIDDM subjects. *Diabetes* **40:** 1712–1718.

Christophe J. (1995) Glucagon receptors: from genetic structure and expression to effector coupling and biological responses. *Biochim. Biophys. Acta* **1241:** 45–57.

Clausen JO, Hansen T, Bjorbaek C *et al.* (1995) Insulin resistance – interactions between obesity and a common variant of IRS-1. *Lancet* **346:** 397–402.

Crook ED, McClain DA. (1996) Regulation of glycogen synthase and protein phosphatase-1 by hexosamines. *Diabetes* **45:** 322–327.

Deems RO, Evans JL, Deacon RW, Honer CM, Chu DT, Burki K, Fillers WS, Cohen DK, Young DA. (1994) Expression of human GLUT4 in mice results in increased insulin action. *Diabetologia* **37:** 1097–1104.

DeFronzo RA. (1988) The triumverate: beta cell, muscle, liver. *Diabetes* **37:** 667–687.

DeFronzo RA. (1992) Pathogenesis of Type 2 diabetes mellitus: a balanced overview. *Diabetologia* **35:** 389–397.

DeMeyts P, Wallach B, Christoffersen CT, Urso B, Gronskov K, Latus L, Yakushiji F, Ilondo MM. (1994) The insulin like growth factor-1 receptor. *Hormone Res.* **42:** 152–169.

Denton RM, Tavare JM. (1995) Does MAP Kinase have a role in insulin action: the case for and against. *Eur. J. Biochem.* **227:** 597–611.

Digirolamo M, Newby FD, Lovejoy J. (1992) Lactate production in adipose tissue: a regulated function with extra adipose implications. *FASEB J.* **6:** 2405–2412.

Dukes ID, Philipson LH. (1996) K^+ channels: generating excitement in pancreatic beta cells. *Diabetes* **45:** 845–853.

Echwald SM, Bjorbaek C, Hansen T, Clausen JO, Vaestergaard H, Zierath JR, Printz RL, Granner DK. (1995) Identification of a 4 amino acid substitution in hexokinase-2 and studies of relationships to NIDDM, glucose effectiveness and insulin sensitivity. *Diabetes* **44:** 347–353.

Efrat S, Tal M, Lodish HF. (1994a) The pancreatic beta cell glucose sensor. *Trends Biochem. Sci.* **19:** 535–538.

Efrat S, Leiser M, Wu YJ *et al.* (1994b) Ribozyme mediated attenuation of pancreatic beta cell glucokinase expression in transgenic mice results in impaired glucose induced insulin secretion. *Proc. Natl Acad. Sci. USA* **91:** 2051–2055.

Felber JP, Acheson KJ, Tappy L. (1993) *From Obesity to Diabetes.* John Wiley & Sons, Chichester.

Flier JS. (1995) The adipocyte: storage depot or node on the energy information superhighway. *Cell* **80:** 15–18.

Gibbs EM, Stock JL, McCoid SC, Stukenbrok HA, Pessin JE, Stevenson RW, Milici AJ, McNeish JD. (1995) Glycemic improvement in diabetic db/db mice by overexpression of the human insulin regulatable glucose transporter GLUT4. *J. Clin. Invest.* **95:** 1512–1518.

Goodyear LJ, Hirshman MF, Horton ES. (1991) Exercise induced translocation of skeletal muscle glucose transporters. *Am. J. Physiol.* **261:** E795-E799.

Goodyear LJ, Giorgino F, Sherman LA, Carey J, Smith RJ, Dohm GL. (1995) Insulin receptor phosphorylation, IRS-1 phosphorylation, and PI 3-kinase activity are decreased in intact skeletal muscle strips from obese subjects *J. Clin. Invest.* **95:** 2195–2204.

Gould GW, Holman GD. (1993) The glucose transporter family: structure, function and tissue specific expression. *Biochem. J.* **295:** 329–341.

Granner DK, O'Brien RM. (1992) Molecular physiology and genetics of NIDDM. *Diabetes Care* **15:** 369–395.

Granner DK, Pilkis S. (1990) The genes regulating hepatic glucose metabolism. *J. Biol. Chem.* **265:** 10173–10176.

Grupe A, Hultgren B, Ryan A, Ma YH, Bauer M, Stewart TA. (1995) Transgenic knockouts reveal a critical requirement for pancreatic beta cell glucokinase in maintaining glucose homeostasis. *Cell* **83:** 69–78.

Hamann A, Benecke H, LeMarchand-Brustel Y, Susulic VS, Lowell BB, Flier JS. (1995) Characterisation of insulin resistance and NIDDM in transgenic mice with reduced brown fat. *Diabetes* **44:** 1266–1273.

Hansen L, Hansen T, Vestergaard H *et al.* (1995) A widespread amino acid polymorphism at codon 905 of the glycogen associated regulatory subunit of protein phosphatase-1 is associated with insulin resistance and hypersecretion of insulin. *Hum. Mol. Genet.* **4:** 1313–1320.

Hansen T, Andersen CB, Echwald SM, Clausen JO, Urhammer SA, Hansen L, Pedersen O. (1996) A missense mutation at codon 326 met-ile in the gene encoding the PI 3-kinase p85α subunit is associated with decreased glucose tolerance and decreased glucose effectiveness. *Diabetologia* **39** Suppl 1: A5.

Hoppener JWM, Verbeek JS, Dekoning EJP *et al.* (1993) Chronic overproduction of islet amyloid polypeptide amylin in transgenic mice – lysosomal location of human islet amyloid polypeptide and lack of marked hyperglycemia or hyperinsulinemia. *Diabetologia* **36:** 1258–1265.

Hotamisligil GS, Shargill NS, Spiegelman BM. (1993) Adipose expression of tumor necrosis factor a: Direct role in obesity linked insulin resistance. *Science* **259:** 87–89.

Houmard JA, Etgan PC, Neufer PD, Friedman JE, Wheeler WS, Israel RG, Dohm GL. (1991) Elevated skeletal muscle glucose transporter levels in exercise trained middle aged men. *Am. J. Physiol.* **261:** E437-E443.

Houmard JA, Weidner MD, Dolan PL *et al.* (1995) Skeletal muscle GLUT4 protein concentration and aging in humans. *Diabetes* **44:** 555–560.

Hutton JC. (1994) Insulin secretory granule biogenesis and the proinsulin endopeptidases. *Diabetologia* **37** (Suppl 2): S48-S56.

Ikemoto S, Thompson KS, Itakura H, Lane MD, Ezaki O. (1995) Expression of an insulin responsive GLUT4 minigene in transgenic mice – effect of exercise and role in glucose homeostasis. *Proc. Natl Acad. Sci. USA* **92:** 865–869.

Inoue H, Ferrer J, Welling CM *et al.* (1996) Sequence variants in the sulfonylurea receptor gene are associated with NIDDM in caucasians. *Diabetes* **45:** 825–831.

Joshi RL, Lamothe B, Cordonnier N, Mesbah K, Monthioux E, Jami J, Bucchini D. (1996) Targeted disruption of the insulin receptor gene in the mouse results in neonatal lethality. *EMBO J.* **15:** 1542–1547.

Kahn CR, Flier JS, Bar RS, Archer JA, Gorden P, Martin MM, Roth J. (1976) The syndromes of insulin resistance and acanthosis nigricans: insulin receptor disorders in man. *New Engl. J. Med.* **294:** 739–745.

Kahn CR, Vicent D, Doria A. (1996) Genetics of non-insulin dependent diabetes mellitus. *Ann. Rev. Med.* **47:** 509–531.

Krook A, O'Rahilly S. (1996) Mutant insulin receptors in syndromes of insulin resistance. *Baillieres Clin. Endocrinol. Metab.* **10:** 97–122.

Lawrence JC (1992). Signal transduction and protein phosphorylation in the regulation of cellular metabolism. *Ann. Rev. Physiol.* **54:** 177–193.

Lee J, Pilch PF. (1994) The insulin receptor: structure, function and signalling. *Am. J. Physiol.* **266:** C319-C334.

Lefebvre PJ. (1995) Glucagon and its family revisited. *Diabetes Care* **18:** 715–730.

Lei KJ, Pan CJ, Liu JL, Shelley LL, Chou JY. (1995) Structure–function analysis of human glucose-6-phosphatase, the enzyme deficient in glycogen storage disease type 1A. *J. Biol. Chem.* **270:** 11882–11886.

Leturque A, Loizeau M, Vaulont S, Salminen M, Girard J. (1996) Improvement in insulin action in diabetic transgenic mice selectively overexpressing GLUT4 in skeletal muscle. *Diabetes* **45:** 23–27.

Liu ML, Gibbs EM, McCoid SC, Milici AJ, Stukenbrok HA, McPherson RK, Treadway JL, Pessin JE. (1993) Transgenic mice overexpressing the human GLUT4 muscle fat facilitative glucose transporter protein exhibit efficient glycemic control. *Proc. Natl Acad. Sci. USA* **90:** 11346–11350.

Lund S, Holman GD, Schmitz O, Pedersen O. (1995) Contraction stimulates translocation of glucose transporter Glut4 in skeletal muscle through a mechanism distinct from that of insulin. *Proc. Natl Acad. Sci. USA* **92:** 5817–5821.

Maasen JA, Kadowaki T. (1996) Maternally inherited diabetes and deafness: a new diabetes subtype. *Diabetologia* **39:** 375–382.

Maddux BA, Sbraccia P, Kumakura S *et al.* (1995) Membrane glycoprotein PC-1 and insulin resistance in NIDDM. *Nature* **373:** 448–451.

Matschinsky FM. (1996) A lesson in metabolic regulation inspired by the glucokinase glucose sensor paradigm. *Diabetes* **45:** 223–241.

Matschinsky F, Liang Y, Kesavan P *et al.* (1993) Glucokinase as pancreatic beta cell glucose sensor and diabetes gene. *J. Clin. Invest.* **92:** 2092–2098.

Matsubara A, Tanizawa Y, Matsutani A, Kaneko T, Kaku K. (1995) Sequence variations of the pancreatic islet glucose transporter (GLUT2) gene in Japanese subjects with NIDDM. *J. Clin. Endocrinol. Metab.* **80:** 3131–3135.

Moller DE, Flier JS. (1991) Insulin resistance – mechanisms, syndromes and implications. *New Engl. J. Med.* **325:** 938–948.

Moller DE, Chang PY, Yaspelkis BB, Flier JS, Wallberg-Henriksson H, Ivy JL. (1996) Transgenic mice with muscle specific insulin resistance develop increased adiposity, impaired glucose tolerance and dyslipidemia. *Endocrinology* **137:** 2397–2405.

Mueckler M, Kruse M, Strube M, Riggs AC, Chiu KC, Permutt MA. (1994) A mutation in the GLUT2 glucose transporter gene of a diabetic patient abolishes transport activity. *J. Biol. Chem.* **269:** 17765–17767.

Myers MG, White MF. (1996) Insulin signal transduction and the IRS proteins. *Ann. Rev. Pharmacol. Toxicol.* **36:** 615–658.

Naggert JK, Fricker LD, Varlamov O, Nishina PM, Rouille Y, Steiner DF, Carroll RJ, Paigen BJ, Leiter EH. (1995) Hyperproinsulinaemia in obese fat/fat mice associated with a carboxypeptidase-E mutation which reduces enzyme activity. *Nature Genet.* **10:** 135–142.

Noguchi T, Okabe M, Wang Z, Yamada K, Imai E, Tanaka T. (1993) An enhancer unit of L-pyruvate kinase gene is responsible for the transcriptional stimulation by dietary fructose as well as glucose in transgenic mice. *FEBS Lett.* **318:** 269–272.

Nordlie RC, Sukalski KA. (1986) Multiple forms of type 1 glycogen storage disease: underlying mechanisms. *Trends Biochem. Sci.* **11:** 85–88.

Ofei F, Hurel S, Newkirk J, Sopwith M, Taylor R. (1996) Effects of an engineered human anti-TNFalpha antibody (CDP571) on insulin sensitivity and glycemic control in patients with NIDDM. *Diabetes* **45:** 881–885.

Olson AL, Pessin JE. (1996) Transcriptional regulation of GLUT4 gene expression. *Semin. Cell Devel. Biol.* **7:** 287–293.

O'Rahilly S, Rook A, Morgan R, Rees A, Flier JS, Moller DE. (1992) Insulin receptor and insulin responsive glucose transporter (GLUT4) mutations and polymorphisms in a Welsh type 2 (non insulin dependent) diabetic population. *Diabetologia* **35:** 486–489.

O'Rahilly S, Gray H, Humphreys PJ, Krook A, Polonsky KS, White A, Gibson S, Taylor K, Carr C. (1995) Impaired processing of prohormones associated with abnormalities of glucose homeostasis and adrenal function. *New Engl. J. Med.* **333:** 1386–1390.

Orskov C. (1992) Glucagon like peptide-1, a new hormone of the entero-insular axis. *Diabetologia* **35:** 701–711.

Osawa H, Sutherland C, Robey RB, Printz RL, Granner DK. (1996) Analysis of signalling pathways involved in the regulation of hexokinase-II gene transcription by insulin. *J. Biol. Chem.* **271:** 16690–16694.

Patti M, Sun X, Breuning JC, Araki E, Lipes MA, White MF, Kahn CR. (1995) 4PS/insulin receptor substrate 2 (IRS-2) is the alternative substrate of the insulin receptor in IRS-1 deficient mice. *J. Biol. Chem.* **270:** 24670–24673.

Pawson T, Gish GD. (1992) SH2 and SH3 domains: from structure to function. *Cell* **71:** 359–362.

Philippe J. (1991) Structure and pancreatic expression of the insulin and glucagon genes. *Endocr. Rev.* **12:** 252–271.

Philipson LH, Rosenberg MP, Kuznetsov A, Lancaster ME, Worley JF, Roe MW, Dukes ID. (1994) Delayed rectifier K^+ channel overexpression in transgenic islets and beta cells associated with impaired glucose responsiveness. *J. Biol. Chem.* **269:** 27787–27790.

Prentki M, Corkey BE. (1996) Are the beta cell signalling molecules malonyl-CoA and cytosolic long-chain-acyl-CoA implicated in multiple tissue defects of obesity and NIDDM. *Diabetes* **45:** 273–283.

Ren JM, Marshall BA, Mueckler MM, McCaleb M, Amatruda JM, Shulman GI. (1995) Overexpression of GLUT4 protein in muscle increases basal and insulin stimulated whole body glucose disposal in conscious mice. *J. Clin. Invest.* **95:** 429–432.

Reynet C, Kahn CR. (1993) Rad: a member of the ras family overexpressed in muscle of type-2 diabetics. *Science* **252:** 1441–1444.

Robinson KA, Weinstein ML, Lindenmayer GE, Buse MG. (1995) Effects of diabetes and hyperglycemia on the hexosamine synthesis pathway in rat muscle and liver. *Diabetes* **44:** 1438–1446.

Rosella G, Zajac JD, Baker L, Kaczmarczyk SJ, Andrikopolous S, Adams TE, Proietto J. (1995) Impaired glucose tolerance and increased weight gain in transgenic rats overexpressing a non insulin responsive phosphoenolpyruvate carboxykinase gene. *Mol. Endocrinol.* **9:** 1396–1404.

Rosetti L, Stenbit AE, Katz EB, Brazilai N, Chen W, Hu M, Charron MJ. (1996) Disruption of one allele of the murine GLUT4 gene causes a marked resistance to the action of insulin. *J. Invest. Med.* **44:** A265.

Rothman DL, Shulman RG, Shulman GI. (1992). 31-P Nuclear magnetic resonance measurements of muscle glucose-6-phosphate. *J. Clin. Invest.* **89:** 1069–1075.

Rothman DL, Magnusson I, Cline G, Gerard D, Kahn CR, Shulman RG, Shulman GI. (1995) Decreased muscle glucose transport/phosphorylation is an early defect in the pathogenesis of NIDDM. *Proc. Natl Acad. Sci. USA* **92:** 983–987.

Rouille Y, Duguay SJ, Lund K, Furuta M, Gong QM, Lipkind G, Oliva AA, Chan SJ, Steiner DF. (1995) Proteolytic processing mechanisms in the biosynthesis of neuroendocrine peptides – the subtilisin like proprotein convertases. *Frontiers Neuroendocrinol.* **16:** 322–361.

Shepherd PR, Kahn BB. (1993) Cellular defects in glucose transport: lessons from animal models and implications for human insulin resistance In: *Insulin Resistance* (ed. D Moller). John Wiley, Chichester, pp. 253–300.

Shepherd PR, Gnudi L, Tozzo E, Yang H, Leach F, Kahn BB. (1993) Adipose cell hyperplasia and enhanced glucose disposal in transgenic mice overexpressing GLUT4 selectively in adipose tissue. *J. Biol. Chem.* **268:** 22243–22246.

Shepherd PR, Nave BT, Siddle K. (1995) Insulin stimulation of glycogen synthesis and glycogen synthase activity is blocked by wortmannin and rapamycin in 3T3-L1 adipocytes: evidence for the involvement of phosphoinositide 3-kinase and p-70 ribosomal protein-S6 kinase. *Biochem. J.* **305:** 25–28

Shepherd PR, Nave BT, O'Rahilly S. (1996a) The role of phosphoinositide 3-kinase in insulin signalling. *J. Mol. Endocrinol.* **17:** 175–184.

Shepherd PR, Reaves BJ, Davidson HW. (1996b) Essential Role for PI 3-kinase in membrane trafficking. *Trends Cell Biol.* **6:** 92–97

Shulman RG, Boch G, Rothman DL. (1995). In vivo regulation of muscle glycogen synthase and the control of glycogen synthesis. *Proc. Natl Acad. Sci. USA* **92:** 8535–8542.

Siddle K, Soos MA, Field CE, Nave BT. (1994) Hybrid and atypical insulin/IGF-1 receptors. *Hormone Res.* **41** (Suppl 2): 56–65.

Steiner DF, Chan SJ, Welsh JM, Kwok SCM. (1985) Structure and evolution of the insulin gene. *Ann. Rev. Genet.* **19:** 463–484.

Steiner DF, Tager HS, Chan SJ, Nanjo K, Sanke T, Rubenstein AH. (1990) Lessons learned from molecular biology of insulin gene mutations. *Diabetes Care* **13:** 600–609.

Sutherland C, O'Brien RM, Granner DK. (1995) PI 3-kinase but not p70 s6 kinase is required for the regulation of PEPCK gene expression by insulin. *J. Biol. Chem.* **270:** 15501–15506.

Tamemoto H, Kadowaki T, Tobe K *et al.* (1994). Insulin resistance and growth retardation in mice lacking insulin receptor substrate-1. *Nature* **372:** 182–186.

Taylor RW, Printz RL, Armstrong M, Granner DK, Alberti KGMM. (1996) Variant sequences of the hexokinase-2 gene in familial NIDDM. *Diabetologia* **39:** 322–328.

Thomas PM, Cole GJ, Wohllk N, Haddad B, Mathew PM, Rabi W, Aguilar-Bryan L, Gagel RF, Bryan J. (1995) Mutations in the sulfonylurea receptor gene in familial hyperinsulinemic hypoglycemia in infancy. *Science* **268:** 426–429.

Tozzo E, Shepherd PR, Gnudi L, Kahn BB. (1995) Transgenic Glut 4 overexpression in fat enhances glucose metabolism: preferential effect on fatty acid synthesis. *Am. J. Physiol.* **268:** E956-E964.

Treadway JL, Hargrove DM, Nardone NA *et al.* (1994). Enhanced peripheral glucose utilization in transgenic mice expressing the human GLUT4 gene. *J. Biol. Chem.* **269:** 29956–29961.

Tsao TS, Burcelin R, Katz EB, Huang L, Charron MJ. (1996) Enhanced insulin action due to targeted GLUT4 overexpression exclusively in muscle. *Diabetes* **45:** 28–36.

Unger RH, Orci L. (1981) Glucagon and the A-cell. *New Engl. J. Med.* **304:** 1518–1524 and 1575–1580.

Valera A, Pujol A, Pelegrin M, Bosch F. (1994a) Transgenic mice overexpressing phosphoenolpyruvate carboxykinase develop NIDDM. *Proc. Natl Acad. Sci. USA* **91:** 9151–9154.

Valera A, Solanes G, Fernandez-Alvarez J, Pujol A, Ferrer J, Asins G, Gomis R, Bosch F. (1994b) Expression of GLUT2 antisense RNA in beta cells of transgenic mice leads to diabetes. *J. Biol. Chem.* **269:** 28543–28546.

Vestergaard H, Bjorbaek C, Hansen T, Larsen FS, Granner DK, Pedersen O. (1995) Impaired activity and gene expression of hexokinase-II in muscle from NIDDM patients. *J. Clin. Invest.* **96:** 2639–2645.

Vidalpuig A, Printz RL, Stratton IE, Granner DK, Moller DE. (1995) Analysis of the hexokinase-2 gene in subjects with insulin resistance and NIDDM and detection of GLN(142)HIS substitution. *Diabetes* **44**: 340–346.

Waddell ID, Zomerschoe AG, Voice MW, Burchell A. (1992) Cloning and expression of a hepatic microsomal glucose transporter protein. *Biochem. J.* **286**: 173–177.

Weiland M, Bahr F, Hohne M, Schurmann A, Ziehm D, Joost HG. (1991) The signalling potential of the receptors for insulin and IGF-1 in 3T3-L1 adipocytes: comparison of glucose transport activity, induction of oncogene c-fos, glucose transporter mRNA and DNA synthesis. *J. Cell. Physiol.* **149**: 428–435.

Westermark P, Johnson KH, O'Brien TD, Betsholtz C. (1992) Islet amyloid polypeptide – a novel controversy in diabetes research. *Diabetologia* **35**: 297–303.

Wilson JE. (1995) Hexokinases. *Rev. Physiol. Biochem. Pharmacol.* **126**: 65–198.

Yagui K, Yamaguchi T, Kanatsuka A *et al.* (1995) Formation of islet amyloid fibrils in beta secretory granules of transgenic mice expressing islet amyloid polypeptide amylin. *Eur. J. Endocrinol.* **132**: 487–496.

Yki-Jarvinen H, Young A, Lamkin C, Foley JE. (1987) Kinetics of glucose disposal in whole body and across skeletal muscle in man. *J. Clin. Invest.* **79**: 1713–1719.

Yki-Jarvinen H, Sahlin K, Ren JM, Koivisto VA. (1990) Localization of rate limiting defect for glucose disposal in skeletal muscle of insulin resistant type 1 diabetic patients. *Diabetes* **39**: 157–167.

Zhang Y, Proenca R, Maffei M, Barone M, Leopold L, Friedman JM. (1994) Positional cloning of the mouse obese gene and its human homologue. *Nature* **372**: 425–432.

Zierath JR. (1995) In vitro studies of human skelatal muscle: hormonal and metabolic regulation of glucose transport. *Acta Physiol. Scand.* **155** (Suppl 626): 1–95

Zierath JR, Galuska D, Nolte LA, Thorne A, Smedgaard-Kristensen J, Wallberg-Henriksson H. (1994) Effects of glycaemia on glucose transport in isolated skeletal muscle from patients with NIDDM: in vitro reversal of muscular insulin resistance. *Diabetologia* **37**: 270–277.

6

PTHrP in calcium homeostasis

T. John Martin and Matthew T. Gillespie

6.1 Introduction

The discovery of parathyroid hormone-related protein (PTHrP) resulted from investigations of the mechanisms by which certain cancers cause hypercalcaemia without necessarily metastasizing to bone. This syndrome, the humoral hypercalcaemia of malignancy (HHM), had previously been ascribed to inappropriate production of parathyroid hormone (PTH) by cancers. However, when studies throughout the 1970s indicated that the tumour product differed from PTH immunochemically, but nevertheless was associated with clinical effects similar to those resulting from PTH excess, the existence of a previously uncharacterized factor seemed most likely. This suspicion led to a number of excellent clinical studies which put the question beyond doubt (Kukreja *et al.*, 1980; Stewart *et al.*, 1980). By that time, PTH assays had improved to the point that rapid, sensitive, robust methods were available for the identification of PTH-like activity. Extracts of tumours from hypercalcaemic animals and humans, and of medium from cultured tumours were found to contain PTH-like activity, assayed as adenylate cyclase responses in osteoblasts or kidney. This factor was shown to be a protein, actions of which were inhibited by specific peptide antagonists of PTH, but not by antisera capable of blocking PTH action (Broadus *et al.*, 1988; Martin and Suva, 1989).

6.2 Hypercalcaemia in cancer – the role of PTHrP

Hypercalcaemia is a very common complication of many cancers. Those patients with the syndrome of HHM constitute approximately 40% of all those with hypercalcaemia in malignant disease (Mundy and Martin, 1982). The cancers most commonly associated with this condition are squamous cell carcinomas especially of the lung, renal cortical carcinoma, other squamous cancers, neuroendocrine cancer and several miscellaneous types. The biochemical similarities between these patients and those with primary hyperparathyroidism have been alluded to

Molecular Endocrinology, edited by G. Rumsby and S.M. Farrow.

already, with the common features of high serum calcium, low serum phosphate, and increased nephrogenous cAMP. Confirmation of the aetiologic link between elevated levels of PTHrP and hypercalcaemia associated with malignancy has been achieved by measurement of circulating levels by radioimmunoassay (RIA) (Burtis *et al.*, 1990; Grill *et al.*, 1991). Despite a 60% homology over the first 13 amino acids, PTH and PTHrP are immunologically distinct. RIAs and two-site immunoradiometric assays (IRMAs) have documented circulating levels of PTHrP in a large number of subjects with malignancy-associated hypercalcaemia (Budayr *et al.*, 1989b; Burtis *et al.*, 1990; Fraser *et al.*, 1993; Grill *et al.*, 1991; Ikeda *et al.*, 1994; Ratcliffe *et al.*, 1991). Results so far suggest that PTHrP circulates at extremely low levels in healthy subjects, if at all, and may only rarely be detected by N-terminal RIAs.

Applying strict validation procedures to samples giving results near the assay detection limit (2 pmol l^{-1}) circulating PTHrP levels have been detected in up to 100% of patients with hypercalcaemia associated with malignancy without bone metastases (Grill *et al.*, 1991). There is little doubt that PTHrP is the major, if not the sole mediator, of hypercalcaemia in patients with the HHM syndrome. However, it is still possible that in some cases other bone-resorbing factors could contribute to the development of hypercalcaemia on a humoral basis. A number of tumour-derived factors have been identified to be potent resorbers of bone: interleukin-1(IL-1), tumour necrosis factor (TNF)-α and -β and transforming growth factor (TGF)-α. PTHrP and IL-1 can synergistically stimulate bone resorption *in vitro* and increase the serum calcium concentration in mice *in vivo* (Sato *et al.*, 1989a). Production of IL-1 has been identified in clonal cell lines established from squamous cell cancers associated with hypercalcaemia and leukocytosis (Sato *et al.*, 1988, 1989b). The significance of these factors and their possible interplay with PTHrP should become clear as our knowledge of the role of cytokines in bone metabolism is increased.

It should be noted, however, that there are some discrepancies between the features of HHM and hyperparathyroidism which may relate either to interactions with other tumour factors which may be co-secreted with PTHrP or possibly to actions mediated by regions of the PTHrP molecule beyond the first 34 amino acids. One example is the hypokalaemic alkalosis seen in hypercalcaemia of cancer, whereas mild hypochloraemic acidosis is more commonly noted in primary hyperparathyroidism. As a possible explanation for this, the renal handling of bicarbonate by the rat kidney perfused with PTHrP (1–141) differs from that in response to PTHrP (1–34) (Ellis *et al.*, 1990).

A second possible difference relates to vitamin D metabolism. As is the case with PTH, N-terminal fragments of PTHrP promote renal 1-α hydroxylation of vitamin D *in vivo*. Early clinical observations indicated that, in contrast to primary hyperparathyroidism, low levels of 1,25-dihydroxyvitamin D_3 ($1,25(OH)_2D_3$) are present in most patients with HHM (Stewart *et al.*, 1980). However, this is not true in all cases (Yamamoto *et al.*, 1987) and, in a recent study, serum $1,25(OH)_2D_3$ levels were not found to be generally suppressed in HHM, with a positive relationship between plasma PTHrP and serum $1,25(OH)_2D_3$ in the absence of demonstrable bone metastases (Schweitzer *et al.*, 1994). It is possible that, in some cases, factors other than PTHrP released by tumours causing hypercalcaemia may modify the capacity of the amino-terminal portion of PTHrP to stimulate the 1-α hydroxylase enzyme (Fukumoto *et al.*, 1989). Alternatively, the suppression of 1-α hydroxylase may be mediated by another region of the PTHrP molecule, or by the level of serum calcium, which is generally appreciably higher in the cancer patients. The controversy over vitamin D metabolism in HHM is yet to be resolved.

Finally, whereas in primary hyperparathyroidism there is an osteoblastic response to the excessive bone resorption, this does not seem to be the case in patients with HHM in whom osteoblasts have been noted to be quiescent, with no evidence of the response which couples bone formation to resorption. There are several possible explanations for this difference, including the co-production by the cancers of cytokines which depress the osteoblastic response, or due to the more severe hypercalcaemia, or simply the fact that these patients are usually much more seriously and chronically ill.

As development of a PTHrP assay progressed, it became clear that its involvement in the hypercalcaemia of cancer was not confined to those patients with HHM alone. We have analysed hypercalcaemic patients with solid tumours in two groups, according to presence or absence of bone metastases, as demonstrated by isotope scanning (Grill *et al.*, 1991). Breast cancer patients with hypercalcaemia were analysed separately. In patients with hypercalcaemia associated with solid tumours and no evidence of bone metastases, PTHrP was always detected in plasma. Elevated PTHrP levels were also found in 64% of patients with hypercalcaemia and metastatic malignancy to bone, from primary sources other than breast. Consistent with this is the finding that in the metastatic group, all squamous cell cancer patients had elevated PTHrP levels, as did one patient with a pancreatic neuroendocrine tumour in this group. Therefore, the presence or absence of bone metastases, a feature that was used to distinguish between humoral and osteolytic mechanisms of hypercalcaemia (Ralston *et al.*, 1984; Stewart *et al.*, 1980), can no longer be used to define the cancer syndromes.

A humoral mechanism for the hypercalcaemia may still exist with metastatic bone disease. Of particular interest is the finding of elevated plasma levels of PTHrP in patients with hypercalcaemia associated with breast cancer and bone metastases. The hypercalcaemia in this situation had not generally been considered to have a humoral basis (Mundy and Martin, 1982). However, patients with hypercalcaemia of breast cancer have been identified with biochemical features similar to those with HHM (Percival *et al.*, 1985). PTHrP has been purified from breast cancer tissue (Burtis *et al.*, 1987), and we have found evidence using immunohistology and *in situ* hybridization for the presence of PTHrP in 60% of cases from an unselected series of breast cancers (Southby *et al.*, 1990; Vargas *et al.*, 1992). We have also found using immunohistochemistry an increased incidence of positive localization of PTHrP in breast cancer metastases to bone compared with other sites (Powell *et al.*, 1991). A clinical study demonstrated a positive correlation between PTHrP expression in primary breast cancers and subsequent development of bone metastases (Kohno *et al.*, 1994). All of these observations focus on a likely role of PTHrP in malignant breast disease, and raise one possibility which is of particular interest, that PTHrP production might contribute to the ability of breast cancers to erode bone and establish there as metastases. There is strong evidence in support of this in a mouse model of bone metastases in which inoculation of a human cancer cell line into the left ventricle of the nude mouse reproducibly causes osteolytic metastases (Guise *et al.*, 1994). When the malignant cells were engineered to overexpress PTHrP (by transfection with the cDNA for preproPTHrP), an increase in the number of osteolytic metastases was observed (Guise *et al.*, 1994).

Elevated PTHrP levels have also been found in a proportion of patients with haematological malignancies and hypercalcaemia (Grill *et al.*, 1991; Ikeda *et al.*, 1994). For example, circulating levels of PTHrP of the order of those associated with HHM have been detected in cases of non-Hodgkin's lymphoma of B-cell lineage and in cases of multiple myeloma. Immunohistochemical staining demonstrated intracellular PTHrP in some of the neoplastic cells from a lymph node section in one of the cases. In another case of blastic transformation of chronic myeloid leukaemia, elevated plasma levels of PTHrP were temporally related to the development of hypercalcaemia, and a fall in PTHrP concentrations was associated with regression of disease following chemotherapy (Seymour *et al.*, 1993).

It is now well established that the retrovirus (human T-cell leukaemia virus, HTLV-1)-associated adult T-cell leukaemia/lymphoma often produces hypercalcaemia associated with increased urinary excretion of cAMP and low–normal PTH levels. Expression of PTHrP mRNA within HTLV-1-infected T cells in culture has been demonstrated (Motokura *et al.*, 1988). PTHrP has been found in lymph nodes involved with neoplastic tissue by immunohistochemistry (Moseley *et al.*, 1991a) and in serum from patients with HTLV-1-positive human adult T-cell leukaemia/lymphoma (Ikeda *et al.*, 1994).

These observations indicate that PTHrP-mediated hypercalcaemia not only occurs in association with solid tumours with or without skeletal metastases but also in association with haematological malignancies. In the latter case, PTHrP-mediated hypercalcaemia can extend not only to malignant processes involving the T-lymphocyte series but also to those involving the B-lymphocyte series and the myeloid series. It is possible that PTHrP has a role as local mediator of increased bone resorption in multiple myeloma, and that it is at times produced in sufficient quantities to reach the circulation and produce an endocrine effect. Such a process could contribute to the osteoporosis in multiple myeloma as well as to the hypercalcaemia.

6.3 Characterization of PTHrP

Molecular cloning revealed PTHrP to be substantially larger than PTH, the cDNA clones predicting a protein of 141 amino acids in length, with a prepropeptide of 36 amino acids (Suva *et al.*, 1987). The deduced primary amino acid sequence revealed that eight of the first 13 amino acids were identical to those of PTH with the remainder of the two sequences unique (*Figure 6.1*) (Suva *et al.*, 1987). Shortly after the original cloning, reports from other groups predicted additional proteins of 173 (Mangin *et al.*, 1988a, b, 1989) and 139 (Thiede *et al.*, 1988) amino acids in length. All PTHrP isoforms were identical up to amino acid 139; however, the 173-amino-acid isoform possessed a unique C-terminal sequence of an additional 35 amino acids (Mangin *et al.*, 1988a, 1989). The cDNAs encoding the 139- and 173-amino-

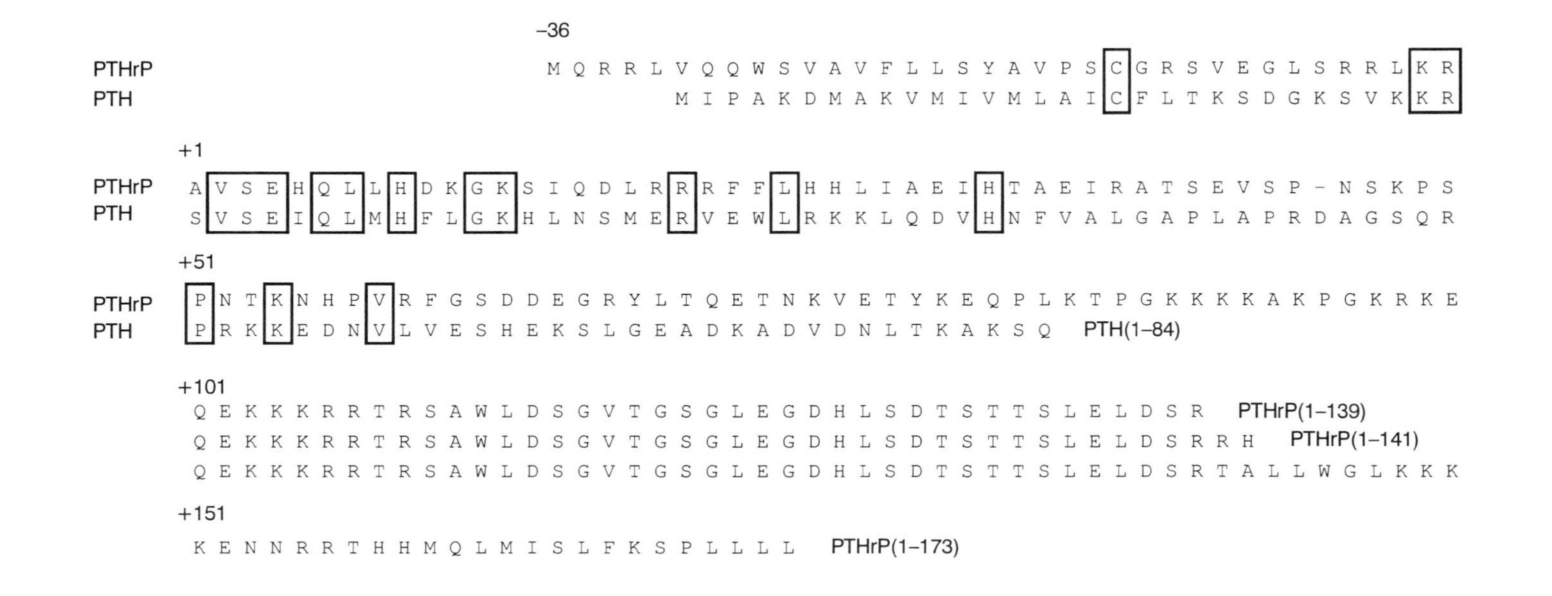

Figure 6.1. Amino acid sequence of human PTHrP and PTH. The sequences are aligned from amino acid +1, with conserved amino acids boxed. A space has been introduced into the PTHrP sequence in order to maximize sequence similarity. The different C-terminal sequences of the three isoforms of human PTHrP (139, 141 and 173 amino acids) are shown. The single letter code for amino acids is used.

acid isoforms have different 3′ untranslated regions (3-UTRs) to that of the 141-amino-acid isoform (Mangin *et al.*, 1988a, 1989).

The PTHrP sequence contains no cysteines, methionines or potential NH_2-linked glycosylation sites (Suva *et al.*, 1987), although O-linked glycosylation on serine and/or threonine residues was reported on PTHrP produced by keratinocytes (Wu *et al.*, 1991). The amino acid region between 86 and 106 is highly basic and contains three consensus nuclear localization signals which have been demonstrated to be operative in chondrocytes. In these cells, nuclear localization of PTHrP appears to reduce apoptosis, suggesting that intracellular targeting of PTHrP is important for cell viability (Henderson *et al.*, 1995). Juxtaposed to the nuclear localization signal is a potential phosphorylation site at threonine 85. The phosphorylation status of this amino acid may be important in the intracellular targeting of PTHrP, although it is of interest to note that the chicken protein contains a valine at this residue, which would not be phosphorylated (*Figure 6.2*). This suggests that the second and third nuclear localization sequences of PTHrP may be the important nuclear targeting sequences for PTHrP. Additionally in the experiments conducted by Henderson *et al.* (1995), the putative phosphorylation site at threonine 85 had been deleted from the construct and, by analogy with the SV40 nuclear localization signal, phosphorylation of this amino acid would promote nuclear exclusion of PTHrP and not nuclear import.

Rat, mouse and canine PTHrP cDNAs have been isolated and predict a single protein species with a 36-amino-acid prepro-leader sequence. Rat and canine PTHrP are predicted to be 141 amino acids in length (Rosol *et al.*, 1995; Thiede and Rodan, 1988; Yasuda *et al.*, 1989b), while mouse PTHrP is predicted to be 139 amino acids in length (Mangin *et al.*, 1990a). Mouse PTHrP is two amino acids shorter than the rat PTHrP protein as a result of a 6 bp deletion. There is marked conservation between the predicted rat, mouse, canine and human PTHrP amino acid sequences until the C-terminal region where the sequences diverge; however, the rat, mouse, human and canine PTHrP sequences are still very similar in this region. The first 111 amino acids of rat and mouse PTHrP are identical to human PTHrP, except for two substitutions, whilst canine PTHrP differs by one amino acid (*Figure 6.2*). Rat PTHrP also shares nine of the first 13 amino acids with rat PTH (Thiede and Rodan, 1988). Additionally, the cDNA sequences revealed greater than 90% conservation between the 3′ UTR of rat, mouse and canine PTHrP and the 3′-UTR of the 141-amino-acid isoform of human PTHrP (Rosol *et al.*, 1995; Thiede and Rodan, 1988; Yasuda *et al.*, 1989b).

The amino acid sequence of chicken PTHrP is also similar to human PTHrP; however, it too diverges from the human sequence in the C-terminal region (*Figure 6.2*). The first 111 share all but 18 amino acids with the rat and human PTHrP sequences (Schermer *et al.*, 1991). In contrast to rat, mouse and canine, there are two isoforms of chicken PTHrP that are 139 and 141 amino acids in length (reviewed in Gillespie and Martin, 1994).

The marked species conservation of the PTHrP sequence over these 111 amino acids indicates that important functions are likely to reside in this region. Indeed, this region contains the known biological activities of PTHrP which include PTH/PTHrP receptor binding, trans-placental calcium transport, renal bicarbonate excretion and *in vitro* osteoclast inhibition (Moseley and Gillespie, 1995; Orloff and Stewart, 1995).

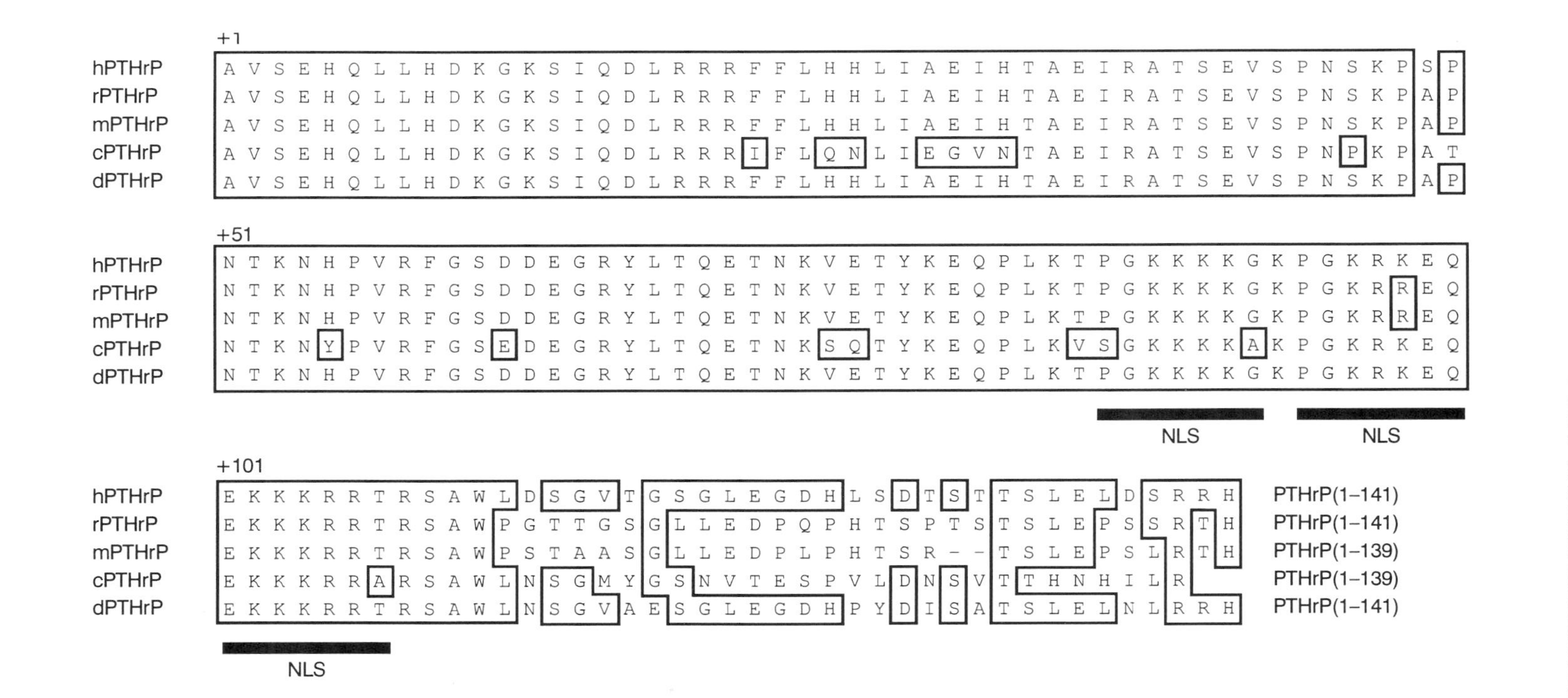

Figure 6.2. The amino acid sequences of rat (rPTHrP), mouse (mPTHrP), chicken (cPTHrP) and canine (dPTHrP) are aligned with the human (hPTHrP) sequence from amino acid +1, with conserved residues boxed. The two dashes in the mouse PTHrP sequence denote the two amino acids which are absent due to a 6 bp deletion in the mouse gene sequence. Only the sequence of the human 141-amino-acid isoform is given since the 173-amino-acid isoform of PTHrP has not been identified in any species other than human, and the 139-amino-acid isoform is identical except for the loss of the last two amino acids. The single letter code for amino acids is used. Potential nuclear localization signals (NLS) are indicated below the alignments.

6.4 The PTHrP gene

The existence of multiple PTHrP mRNA species as shown by Northern blot analysis and the demonstration of 5′ and 3′ divergent cDNA clones indicated that PTHrP transcripts may be alternatively spliced or that the transcripts result from a multigene family. Support for the former resulted from mapping a single copy gene to the short arm of chromosome 12 (chromosome 12p11–12) (Mangin *et al.*, 1988b; Suva *et al.*, 1989b). In the cells from which PTHrP was first purified and cloned, the gene for PTHrP exists on an amplicon with the oncogene *KRAS2* in approximately 14 copies per chromosome (Rudduck *et al.*, 1993). It is unknown whether the amplification contributed to the development of the original cancer from which the cell line was derived or was a consequence of cell culture.

The assignment of PTHrP to chromosome 12p is significant since chromosomes 11 and 12 are thought to have arisen through duplication of a single common ancestral chromosome (Comings, 1972). The related gene, PTH, maps to chromosome 11p supporting a common evolutionary relationship between PTHrP and PTH which may account for their shared biochemical characteristics.

The structure of the human PTHrP gene was characterized by three independent groups in several stages (Mangin *et al.*, 1989, 1990b; Suva *et al.*, 1989b; (Vasavada *et al.*, 1993; Yasuda *et al.*, 1989a) and, in comparison, is considerably more complex than the PTH gene (Vasicek *et al.*, 1983) (*Figure 6.3*). The PTHrP gene is composed of nine exons, may be alternatively spliced both at the 5′ and 3′ ends, and is under the control of three promoters (Mangin *et al.*, 1989, 1990b; Suva *et al.*, 1989a; Vasavada *et al.*, 1993; Yasuda *et al.*, 1989a; reviewed in Gillespie and Martin, 1994), while the PTH gene is composed of three exons and is under the control of a single promoter (Vasicek *et al.*, 1983). Exons I to IV of PTHrP encode 5′-untranslated sequences (Mangin *et al.*, 1988b; Suva *et al.*, 1987, 1989b; Thiede *et al.*, 1988; Yasuda *et al.*, 1989a), while exon V contains 22 bp of 5′-untranslated sequence and encodes the majority of the prepro-coding region. Exon VI encodes the final 7 bp of the prepro-leader sequence and the majority of the coding sequence, up to residue 139 where a donor splice site is located (Mangin *et al.*, 1989; Yasuda *et al.*, 1989a). Exons V and VI are invariant amongst PTHrP mRNA transcripts. Alternative 3′ splicing of exons VI, VII, and IX results in potential protein products of 139 (Thiede *et al.*, 1988; Yasuda *et al.*, 1989a), 173 (Mangin *et al.*, 1988a, 1989) and 141 (Mangin *et al.*, 1988b; Suva *et al.*, 1987) amino acids, respectively, with different 3′-UTRs (*Figure 6.3*).

Although there is no intronic sequence between human PTHrP exons VI and VII, we have designated this read-through exon as two separate exons based on homology with the PTHrP genes from other species. It could be argued that exons VI and VII comprise a single exon with a splice donor site at the junction of exons VI and VII. Other discrepancies in the literature relating to the organization of the PTHrP gene result from exons I, II and III being designated as exons 1a, 1b and 1c, respectively, and exons VII, VIII and IX being designated 4a, 4b and 4c, respectively (Mangin *et al.*, 1989, 1990b).

The human PTHrP gene is under the transcriptional control of three promoters (*Figure 6.3*). The promoters 5′ to exons I (P1) and IV (P3) both contain TATA elements (Mangin *et al.*, 1989, 1990b; Suva *et al.*, 1989a; Yasuda *et al.*, 1989a), while

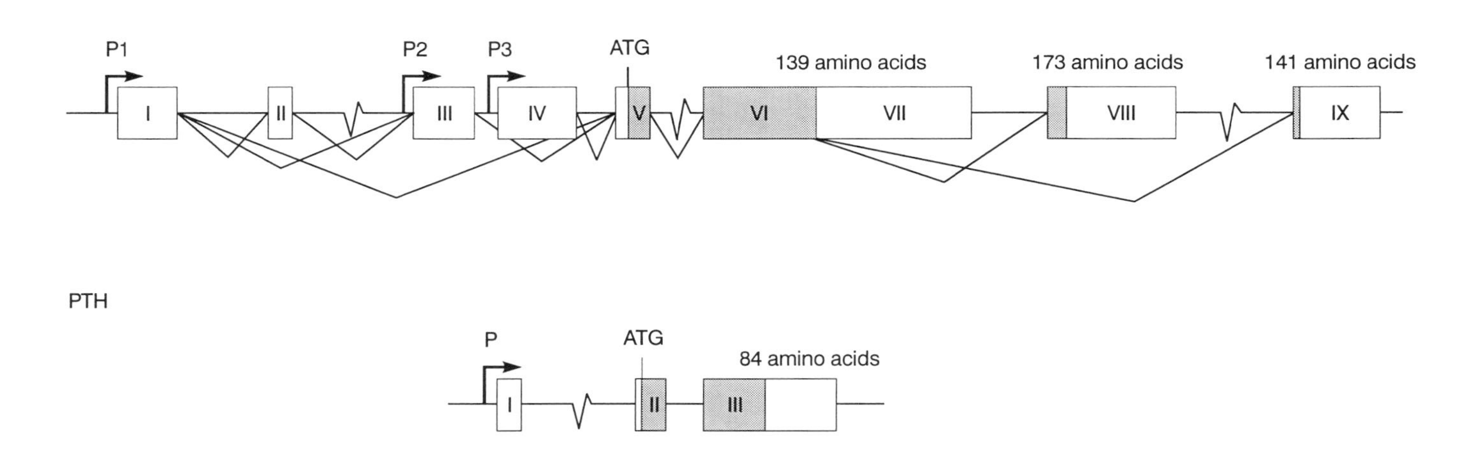

Figure 6.3. Diagrammatic representation of the structural organization of the human PTHrP and PTH genes. Exons are shown by boxes and introns by horizontal lines. The coding regions and the untranslated regions are shown by the shaded and open boxes, respectively. Indicated above the PTHrP gene are the positions of the three promoters, P1 (TATA), P2 (GC-rich) and P3 (TATA), initiating ATG and the predicted PTHrP protein isoforms produced by 3′ alternative splicing. Known PTHrP splicing events are shown below the gene. Indicated above the PTH gene is the promoter (P), initiating ATG and the size of the mature PTH protein. The PTH gene is aligned with the PTHrP gene at the common intron–exon boundary at the start of the exon encoding the mature protein (exon VI in the PTHrP gene and exon III in the PTH gene).

the promoter located 5′ to exon III (P2) is a GC-rich promoter (Suva *et al.*, 1989b; Vasavada *et al.*, 1993). Expression of the human PTHrP gene has been determined to be influenced by its methylation status. Specific demethylation of GC-rich sequences within the second intron results in transcriptional activation of the PTHrP gene (Holt *et al.*, 1993), thus indicating that methylation is an active mechanism for controlling PTHrP transcription.

The structural organization of the rat, mouse and chicken PTHrP genes has also been resolved (Karaplis *et al.*, 1990; Mangin *et al.*, 1990a; Schermer *et al.*, 1991). The rat and mouse PTHrP genes have a much simpler structure than the human gene and do not undergo alternative 3′ splicing, producing only one form of cDNA (*Figure 6.4*). The rodent genes consist of five exons corresponding to human PTHrP exons III, IV, V, VI and IX (Karaplis *et al.*, 1990; Mangin *et al.*, 1990a; Vasavada *et al.*, 1993). Mouse exon IV contains a 6 bp deletion resulting in a final PTHrP protein product of 139 amino acids, in comparison to the predicted 141-amino-acid PTHrP protein encoded by the rat gene. A TATA promoter equivalent to the P3 promoter of the human gene has been identified in both the rat and mouse genes (Karaplis *et al.*, 1990; Vasavada *et al.*, 1993). There is also substantial homology to the human gene over the GC-rich promoter (P2) region, inferring that this second promoter region exists in the rat and mouse genes (Vasavada *et al.*, 1993). There is no evidence to suggest that the rat or mouse PTHrP genes have exons equivalent to exons I or II of the human gene. PTHrP has been assigned to mouse chromosome 6 (Hendy *et al.*, 1990; Seldin *et al.*, 1992) and rat chromosome 4, where interspecies chromosome homology has been reported (Szpirer *et al.*, 1991). The chicken PTHrP gene consists of five exons corresponding to exons IV, V, VI, VII and IX of the human gene (*Figure 6.4*). Chicken PTHrP can encode two isoforms of 139 and 141 amino acids, since it has exons equivalent to human VII and IX. There is only one apparent promoter region in the chicken gene equivalent to the human P3 (reviewed in Gillespie and Martin, 1994). The structural organization of the canine PTHrP gene has not yet been resolved; however, canine PTHrP cDNAs have sequence homology with exon I of the human gene implying that the canine gene will have a human P1-like element (Rosol *et al.*, 1995). The conservation of the exons equivalent to human exons IV, V, VI and IX among the rat, mouse and chicken PTHrP genes implies that these exons constitute the minimum PTHrP gene structure. Moreover, the TATA promoter region 5′ to human exon IV appears to be the major promoter for the PTHrP gene because it is conserved across species. Similarly, the 141-amino-acid isoform encoded by human PTHrP exon IX appears to be the predominant molecular form of the protein.

Differential use of the different promoter regions in the human gene may result in tissue-specific and/or developmentally regulated expression of PTHrP, although a comprehensive analysis of this and 3′ alternative splicing has not been undertaken. These studies have been hindered by the lack of alternative splicing in rodent genes and promoter differences to their human counterpart which have limited these analyses to human tissues. Preliminary reports indicated the potential for tumour-specific promoter usage and alternative 3′ splicing; however, only limited numbers of tumours were examined and not all splicing outcomes were considered (Brandt *et al.*, 1994; Campos *et al.*, 1992, 1994). In the most extensive

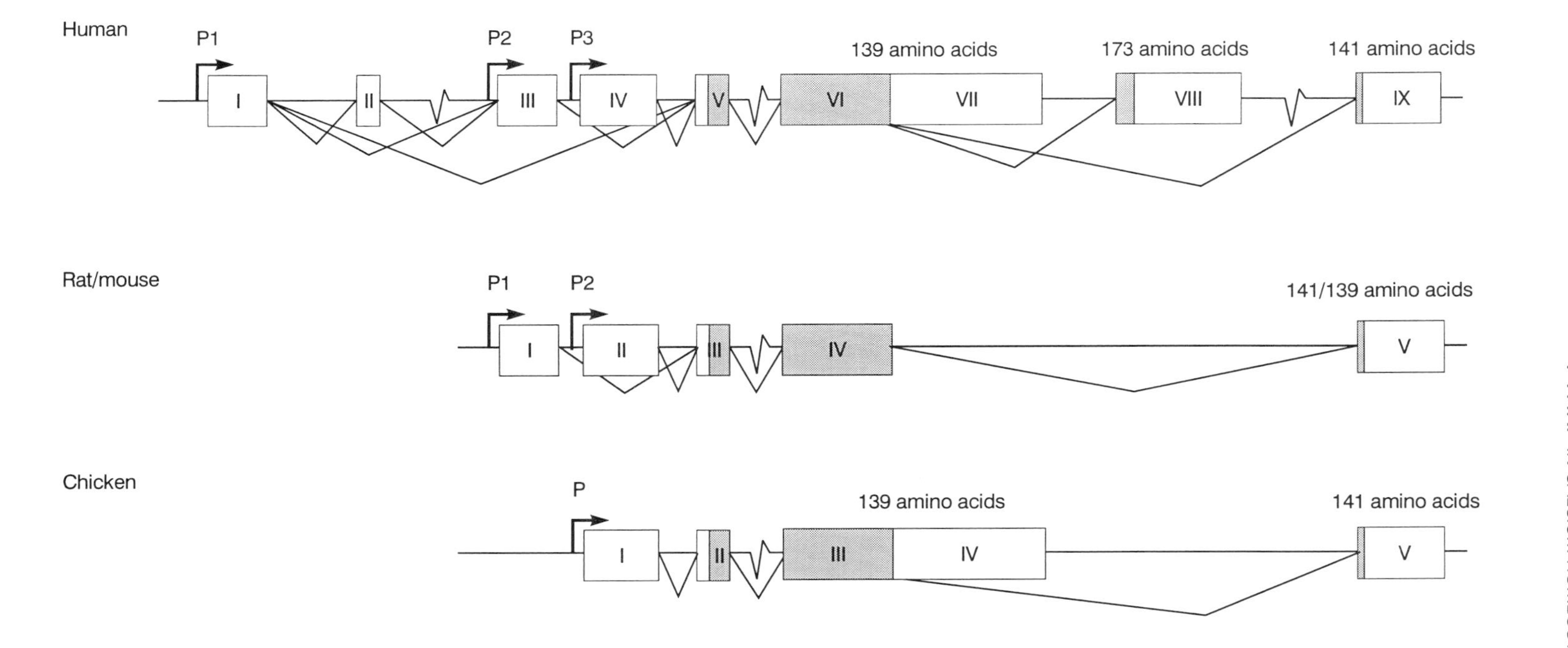

Figure 6.4. Diagrammatic representation of the structural organization of the human, rat, mouse and chicken PTHrP genes. Exons are shown by boxes and introns by horizontal lines. The coding regions and the untranslated regions are shown by the shaded and open boxes, respectively. The rat, mouse and chicken genes are aligned with the human PTHrP gene over regions of exon sequence homology. Indicated above each map are the promoters (P) and the PTHrP isoform(s) encoded by the gene. The rat PTHrP gene encodes a protein of 141 amino acids, while the mouse PTHrP gene encodes a protein of 139 amino acids. P1 and P3 of the human gene, P2 of the mouse gene and the single promoters of the rat and chicken genes are TATA promoters. P2 of the human gene and P1 of the mouse gene are GC-rich promoters. Known splicing events are shown below the gene.

analysis of human tissues and tumours (Southby *et al.*, 1995), different splicing patterns and promoter usage were detected; however, no tissue- or tumour-specific transcripts were detected. Future studies need to address the role of expression of PTHrP in a temporal and developmental manner to determine the relative contribution of each of the transcripts.

Apart from specifying different isoforms of PTHrP, the alternative transcripts may exhibit different properties (see Chapter 1). While the three isoforms of human PTHrP have different 3′-UTRs, they all contain copies of an AUUUA instability motif (Mangin *et al.*, 1988a, 1988b; Suva *et al.*, 1987; Thiede *et al.*, 1988; Yasuda *et al.*, 1989a), which is associated with the rapid turnover of mRNA from cytokines and oncogenes (Ikeda *et al.*, 1990; Shaw and Kamen, 1986). Exon VII and exon IX each contain four copies of this motif whilst exon VIII only contains two copies. Based on the presence of these instability motifs in exons VII and IX, it would be expected that mRNA species containing either exon VII or IX would be unstable. In a human keratinocyte cell line, HaCaT, we have determined that transcripts containing exon VII or VIII are stable in the presence of epidermal growth factor (EGF), whilst exon IX is unstable both in the presence and absence of EGF (Heath *et al.*, 1995). Examination of the sequences of these 3′-untranslated regions reveals that exon IX contains two copies of a recently described nonamer sequence (UUAUUUAUU) which is the key AU-rich sequence motif that mediates mRNA degradation, while neither exon VII nor exon VIII contains this consensus nonamer sequence nor the less effective octamer sequence UAUUUAUU or UUAUUUAU (Zubiaga *et al.*, 1995).

In addition to the potential involvement of mRNA instability motifs contained within the 3′-untranslated regions, it is not known to what extent the length of the poly(A) tail plays in determining PTHrP mRNA stability. It is clear from a number of Northern blot analyses using RNA extracted from different cell lines and tissues that PTHrP mRNAs migrate at different molecular weights. This suggests that polyadenylation may be a contributing factor for mRNA stability. Deadenylation of PTHrP transcripts has been attributed as a step for translational activation of the mRNA (Karaplis *et al.*, 1995).

Another feature shared by PTHrP with members of the early immediate response gene family is the presence of the heptamer sequence, TTTTGTA, or its inverted repeat, TACAAAA, in its 3′-untranslated sequences. The function of this sequence is not known, but it does not appear to function in polyadenylation, splicing or destabilization of transcripts. Since all immediate early genes are responsive to serum induction, the heptamer sequence, along with the serum response element (SRE), may be essential for trafficking of mRNA for protein synthesis.

The properties outlined above suggest that PTHrP should be considered as a cytokine and support the proposal that PTHrP is a member of the early response gene family. However, given the stability of exon VIII-containing transcripts, PTHrP may well have different functions such as that of an immediate early gene and an essential gene for cellular maintenance. Further work is required to determine how the different isoforms of PTHrP affect cellular functions and how these isoforms are processed to elaborate their biological activities. Evidence is now emerging to suggest that the isoforms are differentially processed by members of the prohormone convertase family of enzymes. Thus, consideration will have to be given to the isoforms of PTHrP produced by a tissue or cell type and to the complement of processing enzymes harboured by a cell type which may influence the final form of PTHrP.

6.5 Regulation of PTHrP expression

Production of PTHrP mRNA and protein is regulated by a diverse range of factors, the majority of which enhance PTHrP mRNA or protein levels (*Table 6.1*).

Most PTHrP regulators act primarily via transcriptional mechanisms; however, post-transcriptional mechanisms involving stabilization of PTHrP mRNA transcripts have also been reported following treatment with cycloheximide (Ikeda *et al.*, 1990), TGF-β (Kiriyama *et al.*, 1993; Merryman *et al.*, 1994), angiotensin II (Pirola *et al.*, 1993), IL-2 (Ikeda *et al.*, 1993b), and EGF (Heath *et al.*, 1995; Southby *et al.*, 1996). There have also been recent studies in human cancer cell lines indicating that EGF and TGF-β preferentially modulate the transcriptional activity of promoters P1 and P2, not P3 (Kiriyama *et al.*, 1993; Southby *et al.*, 1996). A study in HTLV-1-transformed cells also demonstrated that the *tax* gene product responsible for HTLV-1 transactivation of PTHrP acted preferentially on what is now designated as the P3 promoter (Dittmer *et al.*, 1993; Ejima *et al.*, 1993).

Studies on the effects of calcium in the regulation of PTHrP mRNA expression in various cell lines have produced ambiguous and conflicting results. Short-term cultures (< 96 h) of keratinocytes and BEN cells in raised calcium concentrations had no effect on PTHrP secretion (Emly *et al.*, 1994; Kremer *et al.*, 1991). An increase in extracellular calcium from 0.4 to 3 mM for 6–14 h was reported to enhance PTHrP secretion by cultured Leydig tumour cells by a post-transcriptional mechanism (Rizzoli and Bonjour, 1989). Similarly, an increased calcium concentration stimulated PTHrP secretion in long-term cultures of keratinocytes which was associataed with an increase in keratinocyte differentiation. However, under similar conditions other workers demonstrated a decrease in PTHrP secretion which was also associated with an increase in differentiation (Löwik *et al.*, 1992; Werkmeister *et al.*, 1993). Raised extracellular calcium also inhibited the release of PTHrP from a parathyroid cell line and from cytotrophoblast cells of the human placenta (Hellman *et al.*, 1992; Zajac *et al.*, 1989). The reason for this disparity is unknown but may relate to tissue-specific functions and may be complicated by the effects of calcium on cellular differentiation and on secretion of peptides.

6.6 PTH, PTHrP and calcium receptors

N-terminal PTH and PTHrP interact with a common receptor, the 'classical' PTH/PTHrP receptor (Jüppner *et al.*, 1988, 1991). The PTH/PTHrP receptor is most abundant in bone and kidney and, following activation by PTH or PTHrP, can stimulate the cAMP and intracellular calcium pathways. Cloning of the PTH/PTHrP receptor from rat osteosarcoma cells and opossum kidney cells

Table 6.1. Regulators of PTHrP expression

	References
Positive regulators	
Angiotensin II	Okano *et al.* (1995); Pirola *et al.* (1993)
cAMP	Deftos *et al.* (1989a); Ikeda *et al.* (1993a); Inoue *et al.* (1993); Rizzoli *et al.* (1994); Zajac *et al.* (1989)
Cycloheximide	Allinson and Drucker (1992); Ikeda *et al.* (1990); Rodan *et al.* (1989); Southby *et al.* (1996); Streutker and Drucker (1991)
1,25-dihydroxyvitamin D_3	Henderson *et al.* (1991); Kremer *et al.* (1991); Merryman *et al.* (1993); Streutker and Drucker (1991)
EGF	Allinson and Drucker (1992); Ferrari *et al.* (1994); Heath *et al.* (1995); Rodan *et al.* (1989); Southby *et al.* (1996)
Oestradiol	Casey *et al.* (1993); Gillespie *et al.* (1992); Suva *et al.* (1991); Thiede *et al.* (1991a)
IL-1β, -2, -6	Casey *et al.* (1993); Ikeda *et al.* (1993a,b); Inoue *et al.* (1993); Rizzoli *et al.* (1994)
Phorbol ester	Casey *et al.* (1993); Deftos *et al.* (1989a); Emly *et al.* (1994); Ferrari *et al.* (1994); Okano *et al.* (1995); Rodan *et al.* (1989)
Prolactin	Dvir *et al.* (1995); Thiede (1989)
Prostaglandin E_1	Ikeda *et al.* (1993a)
Retinoic acid	Chan *et al.* (1990); Merryman *et al.* (1993); van de Stolpe *et al.* (1993); Streutker and Drucker (1991)
Serum	Allinson and Drucker (1992); Casey *et al.* (1993); Hongo *et al.* (1991); Kremer *et al.* (1991); Okano *et al.* (1995)
Sodium butyrate	Streutker and Drucker (1991)
Stretch	Daifotis *et al.* (1992); Thiede *et al.* (1990); Yamamoto *et al.* (1992b)
tax gene product	Dittmer *et al.* (1993); Ejima *et al.* (1993); Watanabe *et al.* (1990)
TGF-β	Allinson and Drucker (1992); Casey *et al.* (1992); Kiriyama *et al.* (1993); Merryman *et al.* (1993, 1994); Southby *et al.* (1996)
TNF-α	Casey *et al.* (1993); Rizzoli *et al.* (1994)
Negative regulators	
Chromogranin A	Deftos *et al.* (1989b)
1,25-dihydroxyvitamin D_3	Allinson and Drucker (1992); Henderson *et al.* (1991); Ikeda *et al.* (1989); Inoue *et al.* (1993); Kremer *et al.* (1991); Liu *et al.* (1993)
Glucocorticoids	Glatz *et al.* (1994); Ikeda *et al.* (1989); Kasono *et al.* (1991); Liu *et al.* (1993); Löwik *et al.* (1992); Lu *et al.* (1989); Southby *et al.* (1996)
Retinoic acid	Suda *et al.* (1996)
Testosterone	Liu *et al.* (1993)

revealed a seven-transmembrane-domain protein-linked receptor (see Section 3.2.3). The PTH/PTHrP receptor has significant homology to the G-protein-linked secretin and calcitonin receptors but lacks homology with other G-protein-linked receptors. Renal and bone PTH/PTHrP receptors are identical (Schipani *et al.*, 1993), and slight differences occur between species. Recent studies have identified two different promoter regions in the PTH/PTHrP receptor gene that are tissue specific, providing the potential for tissue-specific regulation of receptor expression (McCuaig *et al.*, 1995). Mutation analysis of this receptor demonstrates that the cytoplasmic tail is not required for ligand binding or cAMP or calcium signalling, but is essential for proper cell surface localization and expression of the receptor.

Analogous to PTH, the first 34 amino acids of PTHrP are required for receptor binding and activation (Kemp *et al.*, 1987). N-terminal truncated analogues of PTH or PTHrP, to yield peptides with residues 3–34 or 7–34, act as partial receptor antagonists, demonstrating the importance of the first six residues for full bioactivity (Caulfield and Rosenblatt, 1990). This is consistent with the predicted structure of PTHrP (1–34), depicted as a compact structure with the N- and C-termini forming the active site (Barden and Kemp, 1989).

Many tissues other than bone and kidney also express transcripts for the PTH/PTHrP receptor, including skin, where multiple mRNA species were observed (Ureña *et al.*, 1993). However, in keratinocytes the signal transduction pathways stimulated by PTH or PTHrP appear to be mediated by protein kinase C (PKC) and intracellular calcium but not through cAMP (Orloff *et al.*, 1992, 1995; Whitfield *et al.*, 1992). This was not due to the lack of a functioning receptor–cyclase coupling mechanism because the cells could be stimulated to synthesize cAMP by isoproterenol. Further studies revealed that the mRNA from keratinocytes was not of the expected size and analysis with region-specific probes revealed that the receptor in keratinocytes only had partial homology with the classical PTH/PTHrP receptor (Orloff *et al.*, 1995). This infers that keratinocytes may express novel receptors for N-terminal PTHrP that are distinct from the PTH/PTHrP receptor present in bone and kidney. Orloff *et al.* (1995) have designated these PKC-linked receptors as 'type II PTH/PTHrP receptors' and, although they are yet to be cloned, receptors with similar properties also appear to be expressed in lymphocytes and islet cells (McCauley *et al.*, 1992; Gaich *et al.*, 1993).

Given the intensely conserved mid- and C-terminal regions of PTHrP which share no sequence homology with PTH, the biological actions in ovine placental calcium transport (see section 6.8) that are unique to mid-region PTHrP and the demonstration that mid-region PTHrP is an authentic secretory product (Soifer *et al.*, 1992), it is possible that novel receptors for these PTHrP species exist but are yet to be identified. Moreover, two unique PTH receptors were recently identified. The 'PTH-2 receptor' recognizes PTH (1–34) but not PTHrP (1–36) (Usdin *et al.*, 1995). In studying the biological actions of synthetic analogues of PTHrP through this receptor, Gardella *et al.* (1996) found that by changing residues 5 and 23 in PTHrP to those in PTH (histidine and tryptophan, respectively), they converted PTHrP to a potency equivalent to that of PTH on the PTH-2 receptor.

A 'C-PTH receptor' has been found to be specific for the C-terminus of PTH (1–84) (Inomata *et al.*, 1995). C-terminal fragments of PTHrP and PTHrP (107–111) and (107–139), have also been shown to have biological actions on isolated osteoclasts *in vitro* (Fenton *et al.*, 1991a,b) and can stimulate PKC activity in several cell types, implying that other cells including keratinocytes have receptors that can be activated by C-terminal peptides (Gagnon *et al.*, 1993; Whitfield *et al.*, 1994, 1996). These findings infer that there may be a number of unique PTH and/or PTHrP receptors.

The discovery of a calcium-sensing receptor is an exciting finding which is directly relevant to the control of PTH secretion. The latter is increased in response to a decline in extracellular fluid calcium. The change is detected by a calcium sensor which is a G-protein-linked cell surface receptor of the seven-transmembrane superfamily. Transfection of the calcium receptor cDNA into cells produces an increase in cytosolic calcium concentration by activation of the phospholipase C/inositol phosphate signal transduction pathway. mRNA for this receptor is most abundant in the parathyroids but it is also expressed at relatively high levels in the kidney, and also in thyroid, where it may also mediate the calcium-sensitive secretion of calcitonin from the thyroid C cells.

The effects of mutations in these receptors are discussed later in this chapter (see section 6.11).

6.7 Tissue distribution of PTHrP – its functions as a cytokine

Although PTHrP functions as a hormone in those cancers in which it is produced in excess, its circulating levels in postnatal mammals and man are very low, and no assays at the present time have convincingly measured PTHrP in the circulation of normal human subjects, except during pregnancy and lactation.

The widespread expression of PTHrP in the developing embryo, particularly in epithelia at many locations and in a number of adult tissues, has supported the hypothesis that PTHrP is a cellular cytokine. PTHrP mRNA or protein has been detected in the following human tissues: adrenal, bone, brain, heart, intestine, kidney, liver, lung, mammary gland, ovary, parathyroid, placenta, prostate, skeletal muscle, skin, spleen, stomach and smooth muscle (reviewed in Moseley and Gillespie, 1995). PTHrP is expressed in the keratinocyte layer of normal skin (Danks *et al.*, 1989), by keratinocytes in culture (Merendino *et al.*, 1986) and during normal mammalian fetal development (Campos *et al.*, 1991). Although its role is yet to be defined, there is a growing body of evidence supporting a role for PTHrP in epithelial growth and differentiation in the embryo and the adult (Kaiser *et al.*, 1992, 1994). It may function as an autocrine and/or paracrine factor contributing to keratinocyte growth and differentiation, but may also be part of a paracrine interaction between keratinocytes and fibroblasts important in the normal physiology or development of skin.

PTH-induced relaxation of both vascular and non-vascular smooth muscle has been recognized for some time (Mok *et al.*, 1989) beginning with reports of the hypotensive actions of PTH. Since the normal circulating levels of PTH would be insufficient to affect vascular relaxation, the local production of PTHrP makes it a prime candidate as a regulator of vascular tone. The realization that PTHrP is a local paracrine regulator of smooth muscle tone comes from the demonstration that PTHrP is expressed at sites adjacent to smooth muscle beds and that PTHrP (1–34) is a potent relaxant of cardiac and of smooth muscle from many locations. PTHrP can relax smooth muscle from many locations including blood vessels (Crass and Scarpace, 1993), uterus (Shew *et al.*, 1991), gastrointestinal tract (Mok *et al.*, 1989) and bladder.

In the myometrium it is proposed that PTHrP plays a role in expansion of the uterus to accommodate fetal growth (Daifotis *et al.*, 1992; Thiede *et al.*, 1990). Moreover, pre-treatment of rats with 17β-oestradiol increased the sensitivity of the uterus to the effects of exogenous PTHrP and induced PTHrP mRNA expression (Paspaliaris *et al.*, 1992; Thiede *et al.*, 1991a). Angiotensin II, a potent vascoconstrictor, has been shown to produce a marked induction of PTHrP mRNA (Pirola *et al.*, 1993) in vascular smooth muscle cells. The detection of PTHrP in the human amnion and amniotic fluid is also suggested to relate to the ability of PTHrP to act as a vasorelaxant (Brandt *et al.*, 1992; Dvir *et al.*, 1995; Ferguson *et al.*, 1992; Germain *et al.*, 1992). In the amnion, PTHrP is thought to modulate fetal vessel tone and thereby fetal–placental blood flow (Germain *et al.*, 1992). After rupture of the fetal membranes, PTHrP mRNA in the amnion decreased by 78%, suggesting that PTHrP may have a role in the onset of labour (Ferguson *et al.*, 1992). PTHrP may also have paracrine effects on the contractility of the nearby myometrium and on fetal development, but no direct evidence for these actions exists.

In the chicken, PTHrP mRNA expression was detected in smooth muscle of the oviduct as the egg moved through the oviduct (Thiede *et al.*, 1991b). PTHrP was also present in the vascular smooth muscle of vessels in the shell gland where it is thought that it mediates the increased blood flow to the shell gland during the calcification phase of egg laying (Thiede *et al.*, 1991b). Thus, PTHrP (1–34) is proposed to function as a relaxant of both ductal and vascular smooth muscle.

6.8 PTHrP and placental calcium transport

The concentration of ionized calcium in mammalian fetal plasma is greater than that in maternal plasma and is maintained by active placental calcium transport. These are properties necessary for fetal skeletal growth. Thyroparathyroidectomy of the fetal lamb abolishes the placental calcium gradient, but this can be restored by extracts of fetal parathyroid glands or partially purified PTHrP (Rodda *et al.*, 1988). The gradient could not however be restored by PTH infusion (Rodda *et al.*, 1988). This, and the demonstration that PTHrP was produced by fetal parathyroid glands (Abbas *et al.*, 1990; Loveridge *et al.*, 1988; MacIsaac *et al.*, 1991; Rodda *et al.*, 1988), implied that production of PTHrP by fetal parathyroids could contribute to the relative hypercalaemia of the fetal lamb (Rodda *et al.*, 1988).

PTHrP (1–141), (1–108) and (1–84) were all able to stimulate placental calcium transport in thyroparathyroidectomized lambs, but PTHrP (1–34) was without effect (Abbas *et al.*, 1989; Rodda *et al.*, 1988). This implies that the effect is not mediated by the PTH-like N-terminal region of PTHrP but by a mid-molecule region, possibly between amino acids 35 and 84. Some evidence from sheep indicates that the region of PTHrP responsible for stimulating placental calcium transport is contained within PTHrP (67–86) (Care *et al.*, 1990). The presence of PTHrP in ovine parathyroid glands was detected immunohistochemically and was shown by Western blot analysis to correspond in size approximately to that of PTHrP (1–84) (MacIsaac *et al.*, 1991). This supports the suggestion that PTHrP produced in the fetal parathyroid is acting via a unique, and as yet undefined, receptor in the placenta through regions of the molecule not homologous to PTH, to maintain the placental calcium gradient. The role of PTHrP in placental calcium transport awaits confirmation in humans but evidence supporting this role for PTHrP in another species comes from recent studies in PTHrP knockout mice (Kovacs *et al.*, 1995). This study demonstrated that fetuses homozygous for the PTHrP gene deletion have a lower blood calcium and a reduced fetal–maternal calcium gradient, thereby providing direct evidence that PTHrP is required for placental calcium transport.

Production of PTHrP by the placenta has been noted (Rodda *et al.*, 1988) and localized to the chorionic plate and chorionic villi within fibroblasts, epithelium, Hofbauer cells, syncytiotrophoblasts and cytotrophoblasts (Asa *et al.*, 1990; Hellman *et al.*, 1992; Ishikawa *et al.*, 1992; Kramer *et al.*, 1991). Production of PTHrP by cytotrophoblasts was regulated by extracellular calcium levels such that an increase in extracellular calcium inhibited the release of PTHrP from the cytotrophoblasts (Hellman *et al.*, 1992). PTHrP may have effects on the development of a fully functional placenta and placental blood flow. It has also been suggested that the placenta could be an alternative source of PTHrP involved in the early regulation of placental calcium transport until this role can be taken over by the fetal parathyroid glands. PTHrP is present in the fetal ovine parathyroids from 116 days of gestation up to 180 days post birth (MacIsaac *et al.*, 1991), and removal of these glands in late pregnancy abolishes the placental calcium gradient confirming these glands as the major site for PTHrP production in later stages of pregnancy. Expression of PTH only predominates after birth.

6.9 PTHrP in lactation

A physiological role for PTHrP during lactation has been proposed because the protein is expressed in lactating mammary tissue (Thiede and Rodan, 1988), is present in the circulation of lactating mothers (Grill *et al.*, 1992) and is present at high concentrations in the milk of many mammalian species (Budayr *et al.*, 1989a).

In the rat, suckling increased the expression of PTHrP mRNA in the lactating mammary gland (Thiede and Rodan, 1988), and was mediated by prolactin (Thiede, 1989). PTHrP has also been demonstrated in nests of epithelial cells in the mammary gland where it is thought that it may be important in the development of the mammary gland during pregnancy (Rakopoulos *et al.*, 1992). Overexpression of PTHrP in myoepithelial cells of mammary glands in transgenic mice impaired branching morphogenesis of the developing mammary duct system (Wysolmerski *et al.*, 1995).

Increased calcium mobilization and PTH- and vitamin D-independent bone loss have long been recognized in association with lactation, and might be due to a humoral factor (Halloran and De Luca, 1980) with PTHrP a likely candidate. In lactating women, it may mobilize maternal bone calcium for milk production. Evidence for this role comes from observations that significant bone loss is associated with hyperprolactinaemia, which correlates with elevated PTHrP levels in non-lactating subjects. In addition, increased maternal renal excretion of cAMP and phosphate in the rat in response to suckling (Yamamoto *et al.*, 1992a) indicates that PTHrP may reach the circulation to act subsequently on the kidney.

PTHrP also appears to be involved in mammary cell growth since it is readily detectable in mammary tissue, predominantly in myoepithelial cells (Khosla *et al.*, 1990; Rakopoulos *et al.*, 1992), and there is a temporal pattern of expression during pregnancy and lactation (Rakopoulos *et al.*, 1992). Hypercalcaemia and overexpression of PTHrP in myoepithelial cells have been observed in a case of benign breast hypertrophy, which resolved following mastectomy, confirming that breast PTHrP can reach the circulation (Khosla *et al.*, 1990). In another case, high circulating levels of PTHrP were detected in a pregnant woman with hypercalcaemia and fell to normal when the calcium levels returned to normal after lactation (Lepre *et al.*, 1990).

6.10 PTHrP in bone

PTH-like actions of PTHrP in bone were recognized in patients with HHM long before its isolation and cloning. While PTHrP has an endocrine effect on bone in cancer and perhaps also on maternal bone during lactation, most evidence now indicates a local paracrine role for PTHrP in the growth and differentiation of bone cells in the fetus and possibly in adult life.

Localization of PTHrP in normal fetal bone (Karmali *et al.*, 1992; Moseley *et al.*, 1991b), adult bone (Walsh *et al.*, 1994) and cartilage cells (Tsukazaki *et al.*, 1995), and the gross abnormalities of endochondral bone formation seen in mice homozygous for the disrupted PTHrP gene (Amizuka *et al.*, 1994; Karaplis *et al.*, 1994), now indicate the importance of PTHrP in bone development and potentially also in adult life. As a result of abnormalities in the growth and differentiation of chondrocytes in these animals, there is accelerated progression of differentiation resulting in shortened limbs (Amizuka *et al.*, 1994). In addition to their abnormalities of endochondral bone formation, mice homozygous for the disrupted PTHrP gene die at or soon after birth. It is of particular interest that this phenotype can be 'rescued' by producing transgenic animals in which PTHrP is driven by a type II collagen promoter, which would be expected to be expressed in chondrocytes (Philbrick *et al.*, 1996).

The heterozygous mice show bone loss (Amizuka *et al.*, 1996), indicative of a gene dose effect that will be amenable to further study aimed at delineating the role of local PTHrP production in bone. Consistent with these observations in mice with the PTHrP gene ablated, overexpression of the PTHrP gene in chondrocytes *in vivo* inhibited differentiation and delayed osteogenesis (Henderson *et al.*, 1995).

It is noteworthy that although the PTHrP knockout mice can survive several hours after birth (Karaplis *et al.*, 1994) the homozygous PTH/PTHrP receptor knockout mouse, which has a very similar phenotype, dies *in utero* (Lanske *et al.*, 1996). This observation alone indicates the possibility of other ligands for the PTH/PTHrP receptor, either PTH itself or other as yet unidentified PTH-like substances.

Both PTHrP and PTH/PTHrP receptors have been identified in developing chondrocytes and osteoblasts of the periosteum in young bones (Iwamoto *et al.*, 1994; Jongen *et al.*, 1995; Lee *et al.*, 1995; Tsukuzaki *et al.*, 1995), and there is some evidence that their levels of expression are regulated during differentiation of these cells and by growth factors (Lee *et al.*, 1995; Suda *et al.*, 1995; Tsukazaki *et al.*, 1995). For example, retinoic acid treatment of pre-osteoblast UMR 201 cells, which induces a more mature phenotype, also decreases expression of PTHrP (Suda *et al.*, 1996). Furthermore, both PTHrP and PTH/PTHrP receptors are regulated by TGF-β, which is highly expressed in fetal growth plate chondrocytes and induces their differentiation (Jongen *et al.*, 1995; Tsukazaki *et al.*, 1995).

All these data point to actions of PTHrP being mediated via PTH/PTHrP receptors in bone development, in osteoblasts and in particular in chondrocyte differentiation. However, it may be that there are other parts of the PTHrP molecule which have a role during bone development and for which there is redundancy in the knockout animals. Examples of these are actions of PTHrP as a cellular cytokine, possibly mediated by nucleolar binding sites in the region 86–97 (Henderson *et al.*, 1995), and the osteoclast inhibitory action localized to residues 107–111 (Fenton *et al.*, 1991a,b).

6.11 Receptor-related genetic disorders

There is much to be learned of normal physiology from the study of genetic disorders, and recent studies of a few patients with mutations of the PTH/PTHrP receptor and of the calcium-sensing receptor have conveyed valuable insights into normal function.

Jansen-type metaphyseal chondrodysplasia is a rare syndrome characterized by short-limbed dwarfism and marked hypercalcaemia and hypophosphataemia, with normal or undetectable amounts of PTH or PTHrP in plasma. Schipani *et al.* (1995) have identified a mutation in the PTH/PTH receptor gene in several patients with Jansen's metaphyseal chondrodysplasia, which changes the histidine at residue 233 to arginine. This alteration, at the junction between the first intracellular loop and second membrane-spanning helix, leads to persistent activation of the receptor. This explains the patients' hypercalcaemia and hypophosphataemia, resembling that which occurs in patients with HHM. The explanation for the short-limbed dwarfism reflects the important role emerging for PTHrP in bone growth and development (see Section 6.10). Thus, the accelerated differentiation taking place in mice homozygous for the deleted PTHrP gene results in a gross abnormality of endochondral bone formation and shortened limbs. Overexpression of the PTHrP gene in mice inhibits the differentiation process and delays osteogenesis (Henderson *et al.*, 1995). The overactive PTH/PTHrP receptor in Jansen's disease is more akin to the latter.

In the case of the calcium-sensing receptor, it was predicted that mutations in the receptor gene might lead to alterations in the 'set point'; that is, the calcium concentration at which PTH secretion is half maximally inhibited. Such alterations would be expected to lead to inappropriate secretion of PTH. Missense mutations of the calcium-sensing receptor gene have been found to cause three disorders. Autosomal dominant hypocalcaemia results from an inherited defect in the gene giving rise to persistently activated receptor. Inactivation of the receptor through a number of missense mutations leads to the dominantly inherited familial hypocalciuric hypercalcaemia (FHH), a clinical syndrome associated with mild elevations of serum calcium and relative hypocalciuria. The third of the inherited conditions is that occurring in individuals who are homozygous for

FHH, and who present with neonatal severe hyperparathyroidism. Elegant confirmation of these observations comes from the finding that mice heterozygous for the deleted calcium-sensing receptor gene express the phenotype of human FHH, while homozygous mice resemble human infants with neonatal severe hyperparathyroidism (Ho *et al.*, 1995). The combined clinical and experimental data strongly suggest that setting of the extracellular calcium levels is determined by the number of calcium-sensing receptors at the cell surface.

References

Abbas SK, Pickard DW, Rodda CP, Heath JA, Hammonds RG, Wood WI, Caple IW, Martin TJ, Care AD. (1989) Stimulation of ovine placental calcium transport by purified natural and recombinant parathyroid hormone-related protein (PTHrP) preparations. *Quart. J. Exp. Physiol.* **74:** 549–552.

Abbas SK, Pickard DW, Illingworth D *et al.* (1990) Measurement of parathyroid hormone-related protein in extracts of fetal parathyroid glands and placental membranes. *J. Endocrinol.* **124:** 319–325.

Allinson ET, Drucker DJ. (1992) Parathyroid hormone-like peptide shares features with members of the early response gene family: rapid induction by serum, growth factors, and cycloheximide. *Cancer Res.* **52:** 3103–3109.

Amizuka N, Warshawsky H, Henderson JE, Goltzman D, Karaplis AC. (1994) Parathyroid hormone-related peptide-depleted mice show abnormal epiphyseal cartilage development and altered endochondral bone formation. *J.Cell Biol.* **126:** 1611–1623.

Amizuka N, Karaplis AC, Henderson JE *et al.* (1996) Haploinsufficiency of parathyroid hormone-related peptide (PTHrP) results in abnormal postnatal bone. *Devel. Biol.* **175:** 166–176.

Asa SL, Henderson J, Goltzman D, Drucker DJ. (1990) Parathyroid hormone-like peptide in normal and neoplastic human endocrine tissues. *J. Clin. Endocrinol. Metab.* **71:** 1112–1118.

Barden JA, Kemp BE. (1989) NMR study of a 34-residue N-terminal fragment of the parathyroid-hormone-related protein secreted during humoral hypercalcemia of malignancy. *Eur. J. Biochem.* **184:** 379–394.

Brandt DW, Bruns ME, Bruns DE, Ferguson JE II, Burton DW, Deftos LJ. (1992) The parathyroid hormone-related protein (PTHrP) gene preferentially utilizes a GC-rich promoter and the PTHrP 1–139 coding pathway in normal human amnion. *Biochem. Biophys. Res. Commun.* **189:** 938–943.

Brandt, DW, Wachsman W, Deftos LJ. (1994) Parathyroid hormone-like protein: alternative messenger RNA splicing pathways in human cancer cell lines. *Cancer Res.* **54:** 850–853.

Broadus AE, Mangin M, Ikeda K, Insogna KL, Weir EC, Burtis WJ, Stewart AF. (1988) Humoral hypercalcemia of cancer. Identification of a novel parathyroid hormone-like peptide. *New Engl. J. Med.* **319:** 556–563.

Budayr AA, Halloran BP, King JC, Diep D, Nissenson RA, Strewler GJ. (1989a) High levels of parathyroid hormone-like peptide in milk. *Proc. Natl Acad. Sci. USA* **86:** 7183–7185.

Budayr AA, Nissenson RA, Klein RF, Pun KK, Clark OH, Diep D, Arnaud CD, Strewler GJ. (1989b) Increased levels of a parathyroid hormone-like protein in malignancy-associated hypercalcemia. *Ann. Intern. Med.* **111:** 807–812.

Burtis WJ, Wu J, Bunch CM, Wysolmerski JJ, Insogna K.I, Weir EC, Broadus AE, Stewart AF. (1987) Identification of a novel 17,000-dalton parathyroid hormone-like adenylate cyclase-stimulating protein from a tumor associated with humoral hypercalcemia of malignancy. *J. Biol. Chem.* **262:** 7151–7156.

Burtis WJ, Brady TG, Orloff JJ *et al.* (1990) Immunochemical characterization of circulating parathyroid hormone-related hormone in patients with humoral hypercalcemia of cancer. *New Engl. J. Med.* **322:** 1106–1112.

Campos RV, Asa SL, Drucker DJ. (1991) Immunocytochemical localization of parathyroid hormone-like peptide in the rat fetus. *Cancer Res.* **51:** 6351–6357.

Campos RV, Wang C, Drucker DJ. (1992) Regulation of parathyroid hormone-related peptide (PTHrP) gene transcription: cell- and tissue-specific promoter utilization mediated by multiple positive and negative *cis*-acting DNA elements. *Mol. Endocrinol.* **6:** 1642–1652.

Campos RV, Zhang L, Drucker DJ. (1994) Differential expression of RNA transcripts encoding unique carboxy-terminal sequences of human parathyroid hormone-related peptide. *Mol. Endocrinol.* **8:** 1656–1666.

Care AD, Abbas SK, Pickard DW, Barri M, Drinkhill M, Findlay JBC, While IR, Caple IW. (1990) Stimulation of ovine placental transport of calcium and magnesium by mid-molecule fragments of human parathyroid hormone-related protein. *Exp. Physiol.* **75:** 605–608.

Casey ML, Mibe M, Erk A, MacDonald PC. (1992) Transforming growth factor-β_1 stimulation of parathyroid hormone-related protein expression in human uterine cells in culture: mRNA levels and protein secretion. *Clin. Endocrinol. Metab.* **74:** 950–952.

Casey ML, Mibe M, MacDonald PC. (1993) Regulation of parathyroid hormone-related protein gene expression in human endometrial stromal cells in culture. *Clin. Endocrinol. Metab.* **77:** 188–194.

Caulfield MP, Rosenblatt, M. (1990) Parathyroid hormone–receptor interactions. *Trends Endocrinol. Metab.* **1:** 164–168.

Chan SDH, Strewler GJ, King KL, Nissenson RA. (1990) Expression of a parathyroid hormone-like protein and its receptor during differentiation of embryonal carcinoma cells. *Mol. Endocrinol.* **4:** 638–646.

Comings DE. (1972) Evidence for ancient tetraploidy and conservation of linkage groups in mammalian chromosomes. *Nature* **238:** 455–457.

Crass MF, Scarpace PJ. (1993) Vasoactive properties of a parathyroid hormone-related protein in the rat aorta. *Peptides* **14:** 179–183.

Daifotis AG, Weir EC, Dreyer BE, Broadus AE. (1992) Stretch-induced parathyroid hormone-related peptide gene expression in the rat uterus. *J. Biol. Chem.* **267:** 23455–23458.

Danks JA, Ebeling PR, Hayman JA, Chou ST, Moseley JM, Dunlop J, Kemp BE, Martin TJ. (1989) Parathyroid hormone-related protein: immunohistochemical localization in cancers and in normal skin. *J. Bone Min. Res.* **4:** 273–278.

Deftos LJ, Gazdar AF, Ikeda K, Broadus AE. (1989a) The parathyroid hormone-related protein associated with malignancy is secreted by neuroendocrine tumors. *Mol. Endocrinol.* **3:** 503–508.

Deftos LJ, Hogue-Angeletti R, Chalberg C, Tu S. (1989b) PTHrP secretion is stimulated by CT and inhibited by CgA peptides. *Endocrinology* **125:** 563–565.

Dittmer J, Gitlin SD, Reid RL, Brady JN. (1993) Transactivation of the P2 promoter of parathyroid hormone-related protein by human T-cell lymphotropic virus type I Tax_1: evidence for the involvement of transcription factor Ets1. *J. Virol.* **67:** 6087–6095.

Dittmer J, Gégonne A, Gitlin SD, Ghysdael J, Brady JN. (1994) Regulation of parathyroid hormone-related protein (PTHrP) gene expression. Sp1 binds through an inverted CACCC motif and regulates promoter activity in cooperation with Ets1. *J. Biol. Chem.* **269:** 21428–21434.

Dvir R, Golander A, Jaccard N, Yedwab G, Otremski I, Spirer Z, Weisman Y. (1995) Amniotic Fluid and plasma levels of parathyroid hormone-related protein and hormonal modulation of its secretion by amniotic fluid cells. *Eur. J. Endocrinol.* **133:** 277–282.

Ejima E, Rosenblatt JD, Massari M, Quan E, Stephens D, Rosen CA, Prager D. (1993) Cell-type-specific transactivation of the parathyroid hormone-related protein gene promoter by the human T-cell leukemia virus type I (HTLV-I) tax and HTLV-II proteins. *Blood* **81:** 1017–1024.

Ellis AG, Adam WR, Martin TJ. (1990) Comparison of the effects of parathyroid hormone (PTH) and recombinant PTH-related protein on bicarbonate excretion by the isolated perfused rat kidney. *J. Endocrinol.* **126:** 403–408.

Emly JF, Hughes S, Green E, Ratcliffe WA. (1994) Expression and secretion of parathyroid hormone-related protein by a human cancer cell line. *Biochim. Biophys. Acta* **1220:** 193–198.

Fenton AJ, Kemp BE, Hammonds RG Jr, Mitchelhill K, Moseley JM, Martin TJ, Nicholson GC. (1991a) A potent inhibitor of osteoclastic bone resorption within a highly conserved pentapeptide region of parathyroid hormone-related protein; PTHrP[107–111]. *Endocrinology* **129:** 3424–3426.

Fenton AJ, Kemp BE, Kent GN, Moseley JM, Zheng M-H, Rowe DJ, Britto JM, Martin TJ, Nicholson GC. (1991b) A carboxyl-terminal peptide from the parathyroid hormone-related protein inhibits bone resorption by osteoclasts. *Endocrinology* **129:** 1762–1768.

Ferguson JE, Gorman JV, Bruns DE, Weir EC, Burtis WJ, Martin TJ, Bruns ME. (1992) Abundant expression of parathyroid hormone-related protein in human amnion and its association with labor. *Proc. Natl Acad. Sci. USA* **89:** 8384–8388.

Ferrari SL, Rizzoli R, Bonjour J-P. (1994) Effects of epidermal growth factor on parathyroid hormone-related protein production by mammary epithelial cells. *J. Bone Min. Res.* **9:** 639–644.

Fraser WD, Robinson J, Lawton R, Durham B, Gallacher SJ, Boyle IT, Beastall GH, Logue FC. (1993) Clinical and laboratory studies of a new immunoradiometric assay of parathyroid hormone-related protein. *Clin. Chem.* **39:** 414–419.

Fukumoto S, Matsumoto T, Yamamoto H, Kawashima H, Ueyama Y, Tamaoki N, Oagata E. (1989) Suppression of serum 1,25 dihydroxyvitamin D in humoral hypercalcemia of malignancy is caused by elaboration of a factor that inhibits renal 1,25 dihydroxyvitamin D production. *Endocrinology* **124:** 2057–2062.

Gagnon L, Jouishomme H, Whitfield JF, Durkin JP, Maclean S, Neugebauer W, Willick G, Rixon RH, Chakravarthy B. (1993) Protein kinase C-activating domains of parathyroid hormone-related protein. *J. Bone Min. Res.* **8:** 497–503.

Gaich G, Orloff JJ, Atillasoy EJ, Burtis WJ, Ganz MB, Stewart AF. (1993) Amino-terminal parathyroid hormone-related protein: specific binding and cytosolic calcium responses in rat insulinoma cells. *Endocrinology* **132:** 1402–1409.

Gardella TJ, Jensen GS, Luck M, Usdin TB, Jüpper H. (1996) Converting parathyroid hormone-related peptide (PTHrP) into a potent PTH-2 receptor agonist. *J. Bone Min. Res.* **11** (Suppl 1): S114 (Abstract 77).

Germain AM, Attaroglu H, Macdonald PC, Casey ML. (1992) Parathyroid hormone-related protein mRNA in avascular human amnion. *J. Clin. Endocrinol Metab.* **75:** 1173–1175.

Gillespie MT, Martin TJ. (1994) The parathyroid hormone-related protein gene and its expression. *Mol. Cell. Endocrinol.* **100:** 143–147.

Gillespie MT, Kiriyama T, Glatz JA, Suva LJ, Fukumoto S, Heath JK, Moseley JM, Rodan GA, Martin TJ. (1992) Human PTHrP gene transcription is stimulated by estrogen, EGF and TGFβ. In: *Calcium regulating hormones and bone metabolism* (eds DV Cohn, C Gennari, AH Tashjian Jr). Excerpta Medica, New York, pp. 52–56.

Glatz JA, Heath JK, Southby J, O'Keeffe LM, Kiriyama T, Moseley JM, Martin TJ, Gillespie MT. (1994) Dexamethasone regulation of parathyroid hormone-related protein (PTHrP) expression in a squamous cancer cell line. *Mol. Cell. Endocrinol.* **101:** 295–306.

Grill V, Ho P, Body JJ, Johanson N, Lee SC, Kukreja SC, Moseley JM, Martin TJ. (1991) Parathyroid hormone-related protein: elevated levels in both humoral hypercalcemia of malignancy and hypercalcemia complicating metastatic breast cancer. *J. Clin. Endocrinol. Metab.* **73:** 1309–1315.

Grill V, Hillary J, Ho PMW, Law FMK, MacIsaac RJ, MacIsaac IA, Moseley JM, Martin TJ. (1992) Parathyroid hormone-related protein: a possible endocrine function in lactation. *Clin. Endocrinol.* **27:** 405–410.

Guise TA, Taylor SD, Yoneda T, Sasald A, Wright K, Boyce BF, Chirgwin JM, Mundy GR. (1994) Parathyroid hormone-related protein (PTHrP) expression by breast cancer cells enhance osteolytic bone metastases in vivo. *J. Bone Min. Res.* **9** (Suppl 1): S128 (Abstract).

Halloran BP, Deluca HF. (1980) Skeletal changes during pregnancy and lactation: the role of vitamin D. *Endocrinology* **107:** 1923–1929.

Heath JK, Southby J, Fukumoto S, O'Keeffe LM, Martin TJ, Gillespie MT. (1995) Epidermal growth factor-stimulated parathyroid hormone-related protein expression involves increased gene transcription and mRNA stability. *Biochemistry* **307:** 159–167.

Hellman P, Ridefelt P, Juhlin C, Åkerström G, Rastad J, Gylfe E. (1992) Parathyroid-like regulation of parathyroid-hormone-related protein release and cytoplasmic calcium in cytotrophoblast cells of human placenta. *Arch. Biochem. Biophys.* **293:** 174–180.

Henderson J, Sebag M, Rhim J, Goltzman D, Kremer R. (1991) Dysregulation of parathyroid hormone-like peptide expression and secretion in a keratinocyte model of tumor progression. *Cancer Res.* **51:** 6521–6528.

Henderson JE, Amizuka N, Warshawsky H, Biasotto D, Lanske BMK, Goltzman D, Karaplis AC. (1995) Nucleolar localization of parathyroid hormone-related peptide enhances survival of chondrocytes under conditions that promote apoptotic cell death. *Mol. Cell. Biol.* **15:** 4064–4075.

Hendy GN, Sakaguchi AY, Lalley PA, Martinez L, Yasuda T, Banville D, Goltzman D. (1990) Gene for parathyroid hormone-like peptide is on mouse chromosome 6. *Cytogenet. Cell Genet.* **53:** 80–82.

Ho C, Conner DA, Pollak MR, Ladd DJ, Kifor O, Warren HB, Brown EB, Seidman JG, Sidman CE. (1995) A mouse model of human familial hypocalciuric hypercalcemia and neonatal severe hyperparathyroidism. *Nature Genet.* **11:** 389–394.

Holt EH, Vasavada RC, Bander NH, Broadus AE, Philbrick WM. (1993) Region-specific methylation of the parathyroid hormone-related peptide gene determines its expression in human renal carcinoma cell lines. *J. Biol. Chem.* **268:** 20639–20645.

Hongo T, Kupfer J, Enomoto *et al.* (1991) Abundant expression of parathyroid hormone-related protein in primary rat aortic smooth muscle cells accompanies serum-induced proliferation. *J. Clin. Invest.* **88:** 1841–1847.

Ikeda K, Lu C, Weir EC, Mangin M, Broadus AE. (1989) Transcriptional regulation of the parathyroid hormone-related peptide gene by glucocorticoids and vitamin D in a human C-cell line. *J. Biol. Chem.* **264:** 15743–15746.

Ikeda K, Lu C, Weir EC, Mangin M, Broadus AE. (1990) Regulation of parathyroid hormone-related peptide gene expression by cycloheximide. *J. Biol. Chem.* **265:** 5398–5402.

Ikeda K, Okazaki R, Inoue D, Ogata E, Matsumoto T. (1993a) Transcription of the gene for parathyroid hormone-related peptide from the human is activated through a cAMP-dependent pathway by prostaglandin E1 in HTLV-I-infected T cells. *J. Biol. Chem.* **268:** 1174–1179.

Ikeda K, Okazaki R, Inoue D, Ohno H, Ogata E, Matsumoto T. (1993b) Interleukin-2 increases production and secretion of parathyroid hormone-related peptide by human T cell leukemia virus type I-infected T cells: possible role in hypercalcemia associated with adult T cell leukemia. *Endocrinology* **132:** 2551–2556.

Ikeda K, Ohno H, Hane M *et al.* (1994) Development of a sensitive two-site assay for parathyroid hormone-related peptide: evidence for elevated levels in plasma from patients with adult T-cell leukaemia/lymphoma and B-cell lymphoma. *J. Clin. Endocrinol. Metab.* **79:** 1322–1327.

Inomata N, Akiyama M, Kubota N, Jüppner H. (1995) Characterization of a novel parathyroid hormone (PTH) receptor with specificity for the carboxyl-terminal region of PTH-(1–84). *Endocrinology* **136:** 4732–4740.

Inoue D, Matsumoto T, Ogata E, Ikeda K. (1993) 22-Oxacalcitriol, a noncalcemic analogue of calcitriol, suppresses both cell proliferation and parathyroid hormone-related peptide gene expression in human T cell lymphotrophic virus, type I-infected T cells. *J. Biol. Chem.* **268:** 16730–16736.

Ishikawa E, Katakami H, Hidaka H, Ushiroda Y, Ikeda T, Ikenoue T, Matsukura S. (1992) Characterization of parathyroid hormone-related protein in the human term placenta. *Endocrinol. Jpn* **39:** 555–561.

Iwamoto M, Jikko A, Murakami H *et al.* (1994) Changes in parathyroid hormone receptors during chondrocyte cytodifferentiation. *J. Biol. Chem.* **269:** 17245–17251.

Jongen JWJM, Willemstein-Van Hove EC, Van Der Meer JM, Jüppner H, Segre GV, Abou-Samra AB, Feyen JHM, Herrmann-Erlee MPM. (1995) Down-regulation of the receptor for parathyroid hormone (PTH) and PTH-related peptide by transforming growth factor-β in primary fetal rat osteoblasts. *Endocrinology* **136:** 3260–3266.

Jüppner H. Abou-Samra A-B, Uneno S, Gu W-X, Potts JT Jr, Segre GV. (1988) The parathyroid hormone-like peptide associated with humoral hypercalcemia of malignancy and parathyroid hormone bind to the same receptor on the plasma membrane of ROS 17/2.8 cells. *J. Biol. Chem.* **263:** 8557–8560.

Jüppner H, Abou-Samra A-B, Freeman M *et al.* (1991) A G protein-linked receptor for parathyroid hormone and parathyroid hormone-related peptide. *Science* **254:** 1024–1026.

Kaiser SM, Laneuville P. Bernier SM, Rhim JS. Kremer R, Goltzman D. (1992) Enhanced growth of a human keratinocyte cell line induced by antisense RNA for parathyroid hormone-related peptide. *J. Biol. Chem.* **267:** 13623–13628.

Kaiser SM, Sebag M, Rhim JS, Kremer R, Goltzman D. (1994) Antisense-mediated inhibition of parathyroid hormone-related peptide production in a keratinocyte cell line impedes differentiation. *Mol. Endocrinol.* **8:** 39–147.

Karaplis AC, Yasuda T, Hendy GN, Goltzman D, Banville D. (1990) Gene-encoding parathyroid hormone-related peptide: nucleotide sequence of the rat gene and comparison with the human homologue. *Mol. Endocrinol.* **4:** 441–446.

Karaplis AC, Luz A, Glowacki J, Bronson RT, Tybulewicz VLJ, Kronenberg HM, Mulligan RC. (1994) Lethal skeletal dysplasia from targeted disruption of the parathyroid hormone-related peptide gene. *Genes Devel.* **8:** 277–289.

Karaplis AC, Hiou-Tim FFT, Henderson JE. (1995) Translational silencing of maternal PTHrP mRNA by deadenylation in murine primary oocytes. *J. Bone Min. Res.* **10** (Suppl 1): S142 (Abstract 14).

Karmali R, Schiffmann SN, Vanderurinden J-M, Nys-De Wolf N, Corvilain J, Bergmann P, Vanderhaegen J-J. (1992) Expression of mRNA of parathyroid hormone-related peptide in fetal bones of the rat. *Cell. Tiss. Res.* **270:** 597–600.

Kasono K, Isozaki O, Sato K, Shizume K, Ohsumi K, Demura H. (1991) Effects of glucocorticoids and calcitonin on parathyroid hormone-related protein (PTHrP) gene expression and PTHrP release in human cancer cells causing humoral hypercalcemia. *Jpn Cancer Res.* **82:** 1008–1014.

Kemp BE, Moseley JM, Rodda CP *et al.* (1987) Parathyroid hormone-related protein of malignancy: active synthetic fragments. *Science* **238:** 1566–1570.

Khosla S, Johansen KL, Ory SJ, O'Brien PC, Kao PC. (1990) Parathyroid hormone-related peptide in lactation and in umbilical cord blood. *Mayo Clin. Proc.* **65:** 1408–1414.

Kiriyama T, Gillespie MT, Glatz JA, Fukumoto S, Moseley JM, Martin TJ. (1993) Transforming growth factor β stimulation of parathyroid hormone-related protein (PTHrP): a paracrine regulator? *Mol. Cell. Endocrinol.* **92:** 55–62.

Kohno N, Kitzawa S, Fukaze M, Sakoda Y, Kanabara Y, Furuya Y, Ohashi O, Ishikawa Y, Saitoh Y. (1994) The expression of parathyroid hormone-related protein in human breast cancer with skeletal metastases. *Surg. Today* **24:** 215–220.

Kovacs CS, Lanske B, Karaplis A, Kronenberg HM. (1995) PTHrP-knockout mice have reduced ionized calcium, fetal-maternal calcium gradient and 45-calcium transport *in utero. J. Bone Min. Res.* **10** (Suppl. 1): S157 (Abstract).

Kramer S, Reynolds FH Jr, Castillo M, Valenzuela DM, Thorikay M, Sorvillo JM. (1991) Immunological identification and distribution of parathyroid hormone-like protein polypeptides in normal and malignant tissues. *Endocrinology* **128:** 1927–1937.

Kremer R, Karaplis AC, Henderson J, Gulliver W, Banville D, Hendy GN, Goltzman D. (1991) Regulation of parathyroid hormone-like peptide in cultured normal human keratinocytes. Effect of growth factors and 1,25 dihydroxyvitamin D_3 on gene expression and secretion. *J. Clin. Invest.* **87:** 884–893.

Kukreja SC, Shermerdiak WP, Lad TE, Johnson PA. (1980) Elevated nephrogenous cyclic AMP with normal serum parathyroid hormone levels in patients with lung cancer. *J. Clin. Endocrinol. Metab.* **51:** 167–169.

Lanske B, Karaplis AC, Luz A, Jüppner H, Bunghurst R, Abou-Samra A-B, McLaughlin J, Mulligan R, Kronenberg HM. (1996) Characterization of mice heterozygous for the parathyroid hormone (PTH)/PTH-related peptide (PTHrP) receptor gene knockout. *Science* **273:** 663–666.

Lee K, Deeds JD, Segre GV. (1995) Expression of parathyroid hormone-related peptide and its receptor messenger ribonucleic acids during fetal development in rats. *Endocrinology* **136:** 453–463.

Lepre F, Grill V, Danks JA, Ho P, Law FM, Moseley JM, Martin TJ. (1990) Hypercalcemia in pregnancy and lactation due to parathyroid hormone-related protein production. *Bone Min.* **10:** S317.

Liu B, Goltzman D, Rabbani SA. (1993) Regulation of parathyroid hormone-related peptide production *in vitro* by the rat hypercalcemic Leydig cell tumor H-500. *Endocrinology* **1332:** 1658–1664.

Loveridge N, Caple IW, Rodda C, Martin TJ, Care AD. (1988) Further evidence for a parathyroid hormone-related protein in fetal parathyroid glands of sheep. *Quart. Exp. Physiol.* **73:** 781–784.

Löwik CWGM, Hoekman K, Offringa R, Groot CG, Hendy GN, Papapoulos SE, Ponec M. (1992) Regulation of parathyroid hormonelike protein production in cultured normal and malignant keratinocytes. *Invest. Dermatol.* **98:** 198–203.

Lu C, Ikeda K, Deftos LJ, Gazdar AF, Mangin M, Broadus AE. (1989) Glucocorticoid regulation of parathyroid hormone-related peptide gene transcription in a human neuroendocrine cell line. *Mol. Endocrinol.* **3:** 2034–2040.

MacIsaac RJ, Caple IW, Danks JA, Diefenbach-Jagger H, Grill V, Moseley JM, Southby J, Martin TJ. (1991) Ontogeny of parathyroid hormone-related protein in the ovine parathyroid gland. *Endocrinology* **129:** 757–764.

Mangin M, Ikeda K, Dreyer BE, Milstone L, Broadus AE. (1988a) Two distinct tumor-derived, parathyroid hormone-like peptides result from alternative ribonucleic acid splicing. *Mol. Endocrinol.* **2:** 1049–1055.

Mangin M, Webb AC, Dreyer BE *et al.* (1988b) Identification of a cDNA encoding a parathyroid hormone-like peptide from a human tumor associated with humoral hypercalcemia of malignancy. *Proc. Natl Acad. Sci. USA* **85:** 597–601.

Mangin M, Ikeda K, Dreyer BE, Broadus AE. (1989) Isolation and characterization of the human parathyroid hormone-like peptide gene. *Proc. Natl Acad. Sci. USA* **86:** 2408–2412.

Mangin M, Ikeda K, Broadus AE. (1990a) Structure of the mouse gene encoding parathyroid hormone-related peptide. *Gene* **95:** 195–202.

Mangin M, Ikeda K, Dreyer BE, Broadus AE. (1990b) Identification of an up-stream promoter of the human parathyroid hormone-related peptide gene. *Mol. Endocrinol.* **4:** 851–858.

Martin TJ, Suva LJ. (1989) Parathyroid hormone-related protein in hypercalcemia of malignancy. *Clin. Endocrinol.* **31:** 631–647.

McCauley LK, Rosol TJ, Merryman JI, Capen CC. (1992) Parathyroid hormone-related protein binding to human T-cell lymphotrophic virus type I-infected lymphocytes. *Endocrinology* **130:** 300–306.

McCuaig KA, Lee HS, Clarke JC, Assar H, Horsford J, White JH. (1995) Parathyroid hormone/parathyroid hormone-related peptide receptor gene transcripts are expressed from tissue-specific and ubiquitous promoters. *Nucl. Acids Res.* **23:** 1948–1955.

Merendino JJ Jr, Insogna KL, Milstone LM, Broadus AE, Stewart AF. (1986) A parathyroid hormone-like protein from cultured human keratinocytes. *Science* **231:** 388–390.

Merryman JI, Capen CC, McCauley LK, Werkmeister JR, Suter MM, Rosol TJ. (1993) Regulation of parathyroid hormone-related protein production by a squamous carcinoma cell line *in vitro*. *Lab. Invest.* **69:** 347–354.

Merryman JI, DeWille JW, Werkmeister JR, Capen CC, Rosol TJ. (1994) Effects of transforming growth factor-β on parathyroid hormone-related protein production and ribonucleic acid expression by a squamous carcinoma cell line *in vitro*. *Endocrinology* **134:** 2424–2430.

Mok LLS, Ajiwe E, Martin TJ, Thompson JC, Cooper CW. (1989) Parathyroid hormone-related protein relaxes gastric smooth muscle and shows cross-desensitization with parathyroid hormone. *Bone Min. Res.* **4:** 433–439.

Moseley JM, Gillespie MT. (1995) Parathyroid hormone-related protein. *Crit. Rev. Clin. Lab. Sci.* **32:** 299–343.

Moseley JM, Danks JA, Grill V, Lister TA, Horton MA. (1991a) Immunohistochemical demonstration of PTHrP protein in neoplastic tissue of HTLV-1 positive human adult T cell leukemia/lymphoma: implications for the mechanism of hypercalcemia. *Br. J. Cancer* **64:** 745–748.

Moseley JM, Hayman JA, Danks JA, Alcorn D, Grill V, Southby J, Horton MA. (1991b) Immunohistochemical detection of parathyroid hormone-related protein in human fetal epithelia. *Clin. Endocrinol. Metab.* **73:** 478–484.

Motokura T, Fukumoto S, Takahashi S, Watanabe T, Matsumoto T, Igarashi T, Ogata E. (1988) Expression of parathyroid hormone-related protein in a human T cell lymphotrophic virus type 1-infected T cell line. *Biochem. Biophys. Res. Commun.* **154:** 1182–1188.

Mundy GR Martin TJ. (1982) The hypercalcemia of malignancy: pathogenesis and management. *Metabolism* **31:** 1247–1267.

Okano K, Pirola CJ, Wang H, Forrester JS, Fagin JA, Clemens TL. (1995) Involvement of cell cycle and mitogen-activated pathways in induction of parathyroid hormone-related protein gene expression in rat aortic smooth muscle cells. *Endocrinology* **136:** 1782–1789.

Orloff JJ, Stewart AF. (1995) Parathyroid hormone-related protein as a prohormone: posttranslational processing and receptor interactions: update 1995. *Endocrine Rev.* 4: 207–210.

Orloff JJ, Ganz MB, Ribaudo AE, Burtis WJ, Reiss M, Milstone LM, Stewart AF. (1992) Analysis of PTHRP binding and signal transduction mechanisms in benign and maligant squamous cells. *Am. Physiol.* **262:** E599-E607.

Orloff JJ, Kats Y, Urena P *et al.* (1995) Further evidence for a novel receptor for amino-terminal parathyroid hormone-related protein on keratinocytes and squamous carcinoma cells lines. *Endocrinology* **136:** 3016–3023.

Paspaliaris V, Vargas SJ, Gillespie MT *et al.* (1992) Oestrogen enhancement of the myometrial response to exogenous parathyroid hormone-related hormone (PTHrP), and tissue localization of endogenous PTHrP and its mRNA in the virgin rat uterus. *Endocrinology* **134:** 415–425.

Percival K, Yates AJP, Gray RES, Galloway J, Rogers K, Neal FE, Kanis JA. (1985) Mechanisms of malignant hypercalcemia in carcinoma of the breast. *Br. Med. J.* **291:** 776–779.

Philbrick WM, Weir EC, Karaplis AC, Dreyer BE, Broadus AE. (1996) Rescue of the PTHrP-null mouse reveals multiple developmental defects. *J. Bone Min. Res.* **11** (Suppl 1): S157 (Abstract P268).

Pirola CJ, Wang H, Kamyar A, Wu S, Enomoto H, Sharifi B, Forrester JS, Clemens TL, Fagin JA. (1993) Angiotensin II regulates parathyroid hormone-related protein expression in cultured rat aortic smooth muscle cells through transcriptional and post-transcriptional mechanisms. *J. Biol. Chem.* **268:** 1987–1994.

Powell GJ, Southby J, Danks JA, Stillwell RG, Hayman JA, Henderson MA, Bennett RC, Martin TJ. (1991) Localization of parathyroid hormone-related protein in breast cancer metastases: increased incidence in bone compared with other sites. *Cancer Res.* **51:** 3059–3061.

Rakopoulos M, Vargas SJ, Gillespie MT *et al.* (1992) Production of parathyroid hormone-related protein by the rat mammary gland in pregnancy and lactation. *J. Am. Physiol.* **263:** E1077-E1085.

Ralston SH, Fogelman I, Gardiner MD, Boyle IT. (1984) Relative contribution of humoral and metastatic factors to the pathogenesis of hypercalcaemia in malignancy. *Br. Med. J.* **288:** 1405–1408.

Ratcliffe WA, Norbury S, Heath DA, Ratcliffe JG. (1991) Development and validation of an immunoradiometric assay of parathyrin-related protein in unextracted plasma. *Clin. Chem.* **37:** 678–685.

Rizzoli R, Bonjour J-P. (1989) High extracellular calcium increases the production of a parathyroid hormone-like activity by cultured Leydig tumor cells associated with humoral hypercalcemia. *Bone Min. Res.* **4:** 839–844.

Rizzoli R, Aubert ML, Sappino AP, Bonjour J-P. (1994) Cyclic AMP increases the release of parathyroid hormone-related protein from a lung-cancer cell line. *Cancer* **56:** 422–426.

Rodan SB, Wesolowski G, Ianacone J, Thiede MA, Rodan GA. (1989) Production of parathyroid hormone-like peptide in a human osteosarcoma cell line: stimulation by phorbol esters and epidermal growth factor. *Endocrinology* **122:** 219–227.

Rodda CP, Kubota M, Heath JA, Ebeling PR, Moseley JM, Care AD, Caple IW, Martin TJ. (1988) Evidence for a novel parathyroid hormone-related protein in fetal lamb parathyroid glands and sheep placenta: comparisons with a similar protein implicated in humoral hypercalcemia of malignancy. *Endocrinology* **117:** 261–271.

Rosol TJ, Steinmeyer CL, McCauley LK, Gröne A, DeWille JW, Capen CC. (1995) Sequences of the cDNAs encoding canine parathyroid hormone-related protein and parathyroid hormone. *Gene* **160:** 241–243.

Rudduck C, Duncan L, Cerner R, Garson OM. (1993) Co-amplification of the gene for parathyroid hormone-related protein (PTHRP) and *KRAS2* in a human lung cancer cell line. *Genes Chrom. Cancer* **7:** 213–218.

Sato K, Fujii Y, Kasono K, Tsishima T, Shizume K. (1988) Production of interleukin-1 alpha and parathyroid hormone-like factor by a squamous cell carcinoma of the esophagus (EC-GI) derived from a patient with hypercalcemia. *J. Clin. Endocrinol. Metab.* **67:** 592–601.

Sato K, Fujii YT, Kasono, K, Ozawa M, Imamura H, Kanaji Y, Kurosawa H, Tsushima,T, Shizume K. (1989a) Parathyroid hormone-related protein and interleukin-1 alpha synergistically stimulate bone resorption in vitro and increase serum calcium concentration in mice in vivo. *Endocrinology* **124:** 2172–2178.

Sato K, Fujii Y, Kakiuchi T *et al.* (1989b) Paraneoplastic syndrome of hypercalcemia and leukocytosis caused by squamous carcinoma cells (T3M-1) producing parathyroid hormone-related protein, interleukin 1 alpha, and granulocyte colony-stimulating factor. *Cancer Res.* **49:** 4740–4746.

Schermer DT, Chan SDH, Bruce R, Nissenson RA, Wood WI, Strewler GJ. (1991) Chicken parathyroid hormone-related protein and its expression during embryologic development. *J. Bone Min. Res.* **6:** 149–155.

Schipani E, Karga H, Karaplis AC, Potts JT Jr, Kronenberg HM, Segre GV, Abou-Samra A-B, Jüppner H. (1993) Identical complementary deoxyribonucleic acids encode a human renal and bone parathyroid hormone (PTH)/PTH-related peptide receptor. *Endocrinology* **132:** 2157–2165.

Schipani E, Kruse K, Jüppner H. (1995) A constitutively active mutant PTH-PTHrP receptor in Jansen-type metaphysical chondroplasia. *Science* **268:** 98–100.

Schweitzer DH, Hamdy NA, Frohlich M, Zwinderman AH, Papapoulos SE. (1994) Malignancy-associated hypercalcaemia: resolution of controversies over vitamin D metabolism by a pathophysiological approach to the syndrome. *Clin. Endocrinol.* **41:** 251–256.

Seldin MF, Mattei M-G, Hendy GN. (1992) Localization of mouse parathyroid hormone-like peptide gene (*Pthlp*) to distal chromosome 6 using interspecific backcross mice and in situ hybridization. *Cytogenet. Cell Genet.* **60:** 252–254.

Seymour J, Grill V, Lee N, Martin TJ, Firkin F. (1993) Hypercalcemia in the blastic phase of chronic myeloid leukemia associated with parathyroid hormone related protein. *Leukemia* **10:** 1672–1675.

Shaw G, Kamen R. (1986) A conserved AU sequence from the 3′ untranslated region of GM-CSF mRNA mediates selective mRNA degradation. *Cell* **46:** 659–667.

Shew RL, Yee JA, Kliewer DB, Keflemariam YJ, McNeill DL. (1991) Parathyroid hormone-related protein inhibits stimulated uterine contraction in vitro. *J. Bone Min. Res.* **6:** 955–959.

Soifer NE. Dee KE, Insogna KL *et al.* (1992) Parathyroid hormone-related protein. Evidence for secretion of a novel mid-region fragment by three different cell types. *J. Biol. Chem.* **267:** 18236–18243.

Southby J, Kissin MW, Danks JA, Hayman JA., Moseley JM., Henderson MA., Bennett RC, Martin TJ. (1990) Immunohistochemical localization of parathyroid hormone-related protein in human breast cancer. *Cancer Res.* **50:** 7710–7716.

Southby J, O'Keeffe LM, Martin TJ, Gillespie MT. (1995) Alternative promoter usage and mRNA splicing pathways for parathyroid hormone-related protein in normal tissues and tumors. *Br. J. Cancer* **72:** 702–707.

Southby J, Murphy LM, Martin TJ, Gillespie MT. (1996) Cell-specific and regulator-induced promoter usage and messenger ribonucleic acid splicing for parathyroid hormone-related protein. *Endocrinology* **137:** 1349–1357.

Stewart AF, Horst R, Deftos LJ, Cadman EC, Lang R, Broadus AE. (1980) Biochemical evaluation of patients with cancer-associated hypercalcemia. Evidence for humoral and non-humoral groups. *New Engl. J. Med.* **303:** 1377–1381.

van de Stolpe A, Karperien M, Löwik CWGM, Jüppner H, Segre GV, Abou-Samra A-B, de Laat SW, Defize LHK. (1993) Parathyroid hormone-related peptide as an endogenous inducer of parietal endoderm differentiation. *J.Cell. Biol.* **120:** 235–243.

Streutker C, Drucker DJ. (1991) Rapid induction of parathyroid hormone-like peptide gene expression by sodium butyrate in a rat islet cell line. *Mol. Endocrinol.* **5:** 703–708.

Suda N, Gillespie MT, Traianedes K, Zhou H, Ho PWM, Hards DK, Allan EH, Martin TJ, Moseley JM. (1996) Expression of parathyroid hormone-related protein in cells of osteoblast lineage. *J. Cell. Physiol.* **166:** 94–104.

Suva LJ, Winslow GA, Wettenhall REH *et al.* (1987) A parathyroid hormone-related protein implicated in malignant hypercalcemia: cloning and expression. *Science* **237:** 893–896.

Suva LJ, Gillespie MT, Wood WI, Martin TJ, Hudson PJ. (1989a) Multiple tissue-specific promoters and alternative splicing of the human parathyroid hormone-related protein gene. *J. Bone Min. Res.* **4** (Suppl.1): S134 (Abstract)

Suva LJ, Mather KA, Gillespie MT, Webb GC, Ng KW, Winslow GA, Wood WI, Martin TJ, Hudson PJ. (1989b) Structure of the 5′ flanking region of the gene encoding human parathyroid-hormone-related protein (PTHrP). *Gene* **77:** 95–105.

Suva LJ, Gillespie MT, Center RJ, Gardner RM, Rodan GA, Martin TJ, Thiede MA. (1991) A sequence in the human *PTHrP* gene promoter responsive to estrogen. *J. Bone Min. Res.* **6** (Suppl 1): S196 (Abstract).

Szpirer C. Riviére M, Szpirer J, Hanson C, Levan G, Hendy GN. (1991) Assignment of the rat parathyroid hormone-like peptide gene (PTHLH) to chromosome 4: evidence for conserved synteny between human chromosome 12, mouse chromosome 6, and rat chromosome 4. *Cytogenet. Cell Genet.* **56:** 193–195.

Thiede MA. (1989) The mRNA encoding a parathyroid hormone-like peptide is produced in mammary tissue in response to elevations in serum prolactin. *Mol. Endocrinol.* **3:** 1443–1447.

Thiede MA, Rodan GA. (1988) Expression of a calcium-mobilizing parathyroid hormone-like peptide in lactating mammary tissue. *Science* **242:** 278–280.

Thiede MA, Strewler GJ, Nissenson RA, Rosenblatt M, Rodan, GA. (1988) Human renal carcinoma expresses two messages encoding a parathyroid hormone-like peptide: evidence for the alternative splicing of a single copy-gene. *Proc. Natl Acad. Sci. USA* **85:** 4605–4609.

Thiede MA, Daifotis AG, Weir EC, Brines ML, Burtis WJ, Ikeda K. Dreyer BE, Garfield RE, Broadus AE. (1990) Intrauterine occupancy controls expression of the parathyroid hormone-related peptide gene in preterm rat myometrium. *Proc. Natl Acad. Sci. USA* **87:** 6969–6973.

Thiede MA, Harm SC, Hasson DM, Gardner RM. (1991a) *In vivo* regulation of parathyroid hormone-related peptide messenger ribonucleic acid in the rat uterus by 17β-estradiol. *Endocrinology* **128:** 2317–2323.

Thiede MA, Harm SC, McKee RL, Grasser WA, Duong LT, Leach RM Jr. (1991b) Expression of the parathyroid hormone-related protein gene in the avian oviduct: potential role as a local modulator of vascular smooth muscle tension and shell gland motility during the egg-laying cycle. *Endocrinology* **129:** 1958–1966.

Tsukazaki T, Ohtsuru A, Enomoto H, Yano H, Motomura K, Ito M, Namba H, Iwasaki K, Yamashita S. (1995) Expression of parathyroid hormone-related protein in rat articular cartilage. *Calc. Tissue Int.* **57:** 196–200.

Ureña P, Kong X-F, Abou-Samra A-B, Jüppner H, Kronenberg HM, Potts JT Jr, Segre GV. (1993) Parathyroid hormone (PTH)/PTH-related peptide receptor messenger ribonucleic acids are widely distributed in rat tissues. *Endocrinology* **133:** 617–623.

Usdin TB, Gruber C, Bonner TI. (1995) Identification and functional expression of a receptor selectively recognizing parathyroid hormone, the PTH2 receptor. *J. Biol. Chem.* **270:** 15455–15458.

Vargas SJ, Gillespie MT, Powell GJ, Southby J, Danks JA, Moseley JM, Martin TJ. (1992) Localization of parathyroid hormone-related protein mRNA expression in breast cancer and metastatic lesions by in situ hybridization. *J. Bone Min. Res.* **7:** 971–979.

Vasavada RC, Wysolmerski JJ, Broadus AE, Philbrick WM. (1993) Identification and characterization of a GC-rich promoter of the human parathyroid hormone-related peptide gene. *Mol. Endocrinol.* **7**: 273–282.

Vasicek TJ, McDevitt BE, Freeman MW, Fennick BJ, Hendy GN, Potts JT Jr, Rich A, Kronenberg HM. (1983) Nucleotide sequence of the human parathyroid hormone gene. *Proc. Natl Acad. Sci. USA* **80:** 2127–2131.

Walsh CA, Birch MA, Fraser WD, Robinson J, Lawton R, Dorgan J, Klenerman L, Gallagher JA. (1994) Primary cultures of human bone-derived cells produce parathyroid hormone related protein: a study of 40 patients of varying age and pathology. *Bone Min.* **27:** 43–50.

Watanabe T, Yamaguchi K, Takatsuki K, Osame M, Yoshida M. (1990) Constitutive expression of parathyroid hormone-related protein gene in human T cell leukemia virus type 1 (HTLV-1) carriers and adult T cell leukemia patients that can be *trans*-activated by HTLV-1 *tax* gene. *J. Exp. Med.* **172:** 759–765.

Werkmeister JR, Merryman JI, McCauley,LK, Horton JE, Capen CC, Rosol TJ. (1993) Parathyroid hormone-related protein production by normal human keratinocytes *in vitro*. *J. Exp. Cell Res.* **208:** 68–74.

Whitfield JF, Chakravarthy BR, Durkin JP, Isaacs RJ, Jouishomme H, Sikorska M, Williams RE, Rixon RH. (1992) Parathyroid hormone stimulates protein kinase C but not adenylate cyclase in mouse epidermal keratinocytes. *J. Cell. Physiol.* **150:** 299–303.

Whitfield JF, Isaacs RJ, Chakravarthy BR, Durkin JP, Morley P, Neugebauer W, Williams RE, Willick G, Rixon RH. (1994) C-terminal fragments of parathyroid hormone-related protein, PTHrP-(107–111) and (107–139), and the N-terminal PTHrP-(1–40) fragment stimulate membrane associated protein kinase C activity in rat spleen lymphocytes. *J. Cell. Physiol.***158:** 518–522.

Whitfield JF, Isaacs RJ, Jouishomme H, Maclean S, Chakravarthy BR, Morley P, Barisoni D, Regalia E, Armato U. (1996) C-terminal fragment of parathyroid hormone-related protein, PTHrP-(107–111), stimulates membrane associated protein kinase C activity and modulates the proliferation of human and murine skin keratinocytes. *J.Cell. Physiol.* **166:** 1–11.

Wu TL, Soifer NE, Burtis WJ, Milstone LM, Stewart AF. (1991) Glycosylation of parathyroid hormone-related peptide secreted by human epidermal keratinocytes. *Clin. Endocrinol. Metab.* **73:** 1002–1007.

Wysolmerski JJ, McCaughern-Carucci JF, Daifotis AG, Broadus AE, Philbrick WM. (1995) Overexpression of parathyroid hormone-related protein or parathyroid hormone in transgenic mice impairs branching morphogenesis during mammary gland development. *Development* **121:** 3539–3547.

Yamamoto I, Kitamura N, Aoki J, Kawamura J, Dokoh S, Morita R, Torizuka K. (1987) Circulating 1,25-dihydroxyvitamin D concentrations in patients with renal cell carcinoma-associated hypercalcemia are rarely suppressed. *J. Clin. Endocrinol. Metab.* **64:** 175–179.

Yamamoto M, Fisher JE, Thiede MA, Caulfield MP, Rosenblatt M, Duong LT. (1992a) Concentrations of parathyroid hormone-related protein in rat milk change with duration of lactation and interval from previous suckling, but not with milk calcium. *Endocrinology* **130:** 741–747.

Yamamoto M, Harm SC, Grasser WA, Thiede MA. (1992b) Parathyroid hormone-related protein in the rat urinary bladder: a smooth muscle relaxant produced locally in response to mechanical stretch. *Proc. Natl Acad. Sci. USA* **89:** 5326–5330.

Yasuda T, Banville D, Hendy GN, Goltzman, D. (1989a) Characterization of the human parathyroid hormone-like peptide gene. Functional and evolutionary aspects. *J. Biol. Chem.* **264:** 7720–7725.

Yasuda T, Banville D, Rabbani SA, Hendy GN, Goltzman D. (1989b) Rat parathyroid hormone-like peptide: comparison with the human homologue and expression in malignant and normal tissue. *Mol. Endocrinol.* **3:** 518–525.

Zajac JD, Callaghan J, Eldridge C, Diefenbach-Jagger H, Suva LJ, Hudson P, Moseley JM, Michelangeli VP, Pasquini G. (1989) Production of parathyroid hormone-related protein by a rat parathyroid cell line. *Mol. Cell. Endocrinol.* **67**: 107–112.

Zubiaga AM, Belasco JG, Greenberg ME. (1995) The nonamer UUAUUUAUU is the key AU-rich sequence motif that mediates mRNA degradation. *Mol. Cell. Biol.* **15**: 2219–2230.

7

Molecular biology of growth

W.R. Baumbach, B. Bingham and J.R. Zysk

7.1 The growth hormone axis

The growth of vertebrate animals from neonatal to adult size depends upon the proper functioning of the hypothalamic/pituitary/end organ system known as the growth hormone (GH) axis. Since the realization early in this century that GH deficiency leads to dwarfism and overabundance of GH leads to gigantism, a major goal of researchers has been to elucidate the regulation, production and mechanism of action of GH, also known as somatotropin. Investigation of the GH axis has led to the recognition of additional players: the regulatory factors growth hormone-releasing factor (GRF) and somatostatin (or somatotropin release-inhibiting factor, SRIF); and the major effector of GH-induced growth, insulin-like growth factor I (IGF-I). Each of these factors (including GH), which are all peptide hormones, rely upon cell surface receptors to transmit their signals. In addition, the actions of GH and IGF-I are further regulated via their interaction with soluble serum-binding proteins. Thus, the molecular biology of growth comprises not only the regulation of the genes of the four peptide hormones (SRIF, GRF, GH and IGF-I) and their receptors and binding proteins, but also the regulation of hormone secretion and the consequences of hormone–receptor and hormone–serum-binding protein interactions. Intertwined with this basic system is a variety of increasingly complex phenomena that fall outside the scope of this review: interactions between the GH axis and non-peptide hormones such as sex steroids and glucocorticoids, and interaction with physiological processes such as nutrition and insulin-regulated metabolism. This chapter addresses the structure and regulation of the major genes and proteins of the GH axis, while only hinting at the vast literature underlying our constantly evolving knowledge of this complex system.

7.2 The hypothalamic hormones GRF and SRIF and their receptors

GH, produced in the anterior pituitary, falls under the control of two peptidergic hypophysiotrophic hormones originating in specific neurons of the hypothalamus.

Molecular Endocrinology, edited by G. Rumsby and S.M. Farrow.

GRF is the premier positive regulator of GH, and SRIF the main negative regulator. Having only a single receptor that is expressed mainly in pituitary somatotrophs (i.e. GH-producing cells), the major endocrine effects of GRF appear to be limited to the pituitary. SRIF, in contrast, is more pleiotropic in nature, interacting with at least five different receptor subtypes on tissues throughout the body. GH release from the pituitary is episodic or pulsatile in nature, following an ultradian (i.e., less than a 24 h period) rhythm first described in the rat (Cronin *et al.*, 1986). GRF and SRIF are fundamental to the control of GH release and interact in at least two locations. The first is the hypothalamus, where secretion of GRF from neurons is influenced by various stimuli, including local GH and IGF-I concentration (Sato and Frohman, 1993), metabolic signals, central nervous system (CNS) input, glucocorticoids, stress, exercise and sleep (Malven, 1993). These GRF neurons are located in the arcuate nucleus, and the co-localization of SRIF receptor expression suggests regulation of GRF neurosecretion by SRIF as well (Bertherat *et al.*, 1992; McCarthy *et al.*, 1992). This level of control is believed to be important in regulating the pulsatile release of GH. Secondly, GRF and SRIF regulate GH at the pituitary somatotroph, where these hormones bind to specific cell surface receptors in a classic Yin–Yang system that influences GH synthesis and secretion. To understand the complex interplay between these two factors, it will be necessary to review each hormone and its attendant receptor/signal transduction system.

7.2.1 GRF and its receptor

GRF is a member of the glucagon/secretin family of peptide hormones. In humans, GRF is produced as a 108-amino-acid prohormone from a single copy gene encompassing five exons and 10 kb of genomic DNA on chromosome 20 (Mayo *et al.*, 1985a). The prohormone is expressed, synthesized and processed in neurons of the arcuate nucleus that secrete the active form of the hormone from axonal terminals in the median eminence. These sites of secretion are in proximity to capillary loops of the pituitary portal system that carry the hormone to target somatotrophs in the adenohypophysis. From the prohormone, GRF is processed to its active form of 44 amino acids, which is amidated at the C-terminus (*Figure 7.1a*). In the rat, the GRF gene also includes five exons, spanning almost 10 kb (Mayo *et al.*, 1985b) but encoding only a 104-amino-acid prohormone that is processed to a non-amidated, 43-amino-acid GRF, in addition to the recently discovered 30-amino-acid C-terminal peptide (GRF-related peptide) (Breyer *et al.*, 1996). It is the processed, 44- and 43-amino-acid forms in humans and rats that are found in serum and which interact with high affinity GRF receptors on the somatotroph cell surface.

GRF was originally isolated from ectopic tumours of two acromegalic patients. One of these tumours expressed the 44-amino-acid hormone, illustrated in *Figure 7.1a* (Guillemin *et al.*, 1982), while the other expressed an equally active GRF of 40 amino acids, lacking the four C-terminal residues (Rivier *et al.*, 1982). Structure/activity studies indicate that only the first 29 amino acids of GRF are required for biological activity. Human GRF is similar in structure to other mammalian GRF molecules, showing 93%, 89% and 67% amino acid identity with the

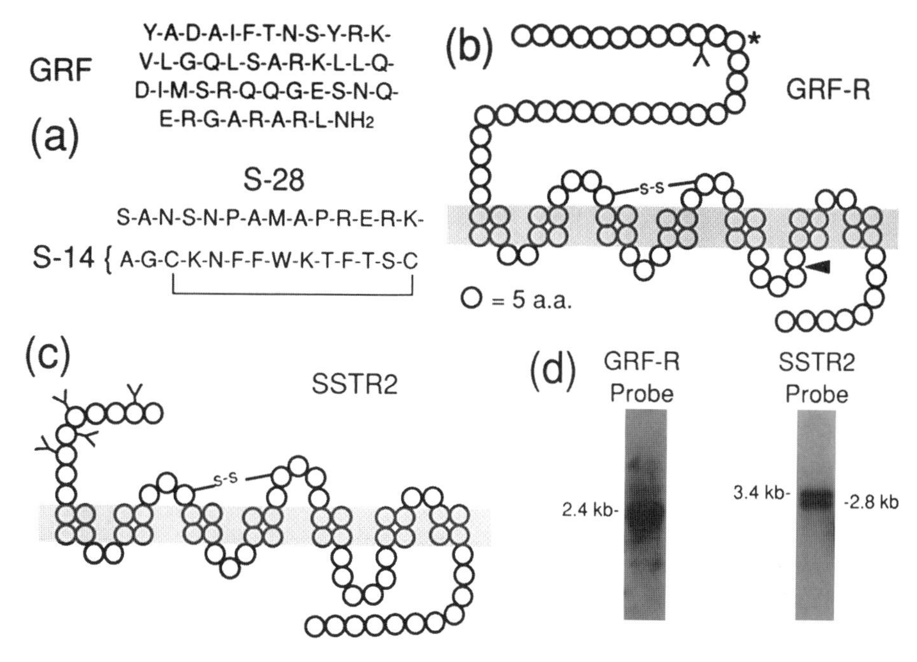

Figure 7.1. GRF and SRIF receptors and their ligands. (a) The amino acid sequence of the mature human GRF polypeptide and the two forms of SRIF are shown; S-14 is a cleavage product of S-28. The line indicates a disulphide linkage. (b) Schematic of the GRF receptor (GRF-R), each circle representing five amino acids, showing its traversal of the plasma membrane (shaded bar). In the extracellular domain (top), the symbol (Y) represents the site of glycosylation, and the asterisk (*) shows the location of D60. The symbol (-s-s-) represents a putative disulphide linkage between C202 and C272 (Dohlman *et al.*, 1990). The arrowhead indicates the location of an alternative 41-amino-acid insertion in the rat. (c) The SRIF receptor subtype 2 (SSTR2), using the same scale and symbols as part (b). The disulphide linkage is shown between C115 and C193. (d) Northern analysis of rat pituitary RNA. Poly(A) RNA (3 μg) was fractionated and probed with radiolabelled cDNAs from rat GRF-R (Mayo, 1992) and rat SSTR2 (Strnad *et al.*, 1993). Sizes of mRNAs are indicated.

hormones of pig, sheep and rat, respectively. These homologues are most similar among the first 29 amino acids, are active across species and are pharmacologically similar *in vitro*. Amino acids essential to the biological activity of GRF include tyrosine[1] (histidine in the rat) and alanine[2]. The replacement of alanine[2] with D-alanine results in a molecule with superagonist properties, while substitution with D-arginine at this position produces an antagonist (Coy *et al.*, 1986).

The GRF receptor (*Figure 7.1b*), a member of the non-neurokinin class of G-protein-linked, seven-transmembrane-domain receptors (Section 3.2.3), has been recently cloned from several mammalian species (Gaylinn *et al.*, 1993; Mayo, 1992; Lin *et al.*, 1992). This family of serpentine receptors includes secretin, glucagon, vasoactive intestinal peptide and pituitary adenylate cyclase-activating peptide. The human GRF receptor gene is located on chromosome 7 (Gaylinn *et al.*, 1994b) and encodes a 423-amino-acid molecule. The GRF receptor gene in mouse, located on chromosome 6, consists of 13 exons and remains to be fully characterized, but recent work on its 5′ flanking region reveals several putative regulatory elements, including binding sites for activating protein-1 (AP-1), cAMP response element (CRE), and pituitary-specific transcription factor (Pit-1) (Takahashi *et al.*, 1996). Expression of the receptor is mainly limited to the anterior pituitary (*Figure 7.1d*), although some minor expression of receptor mRNA has been detected in the hypothalamus and other tissues by the sensitive reverse transcriptase–polymerase chain reaction (RT-PCR) technique (Matsubara *et al.*, 1995). Only one form of the human GRF receptor protein has been characterized, but splicing variants have been identified in normal human pituitary and GH-secreting pituitary tumours (Tang *et al.*, 1995). Their functional relevance, however, remains to be determined. Alternative splicing of the rat GRF receptor RNA has also been described, with the alternative inclusion of a 123-base (41-amino-acid) insert in the third intracellular loop sequence (*Figure 7.1b*).

The GRF receptor contains a single N-linked glycosylation site (N50), resulting in a 10 kDa increase over its predicted molecular weight of 45 kDa (Gaylinn *et al.*, 1994a). Binding studies reveal that the formation of the receptor–ligand complex is sensitive to guanine nucleotides and that the receptor is rapidly downregulated. High affinity binding of GRF to its receptor ($K_D \sim 0.1$ nM) induces both GH release and synthesis. Activation of adenylate cyclase via coupling to G_s induces intracellular cAMP which, in turn, is instrumental in GH transcription and synthesis. Simultaneous activation of voltage-dependent calcium channels following GRF binding leads to an influx of extracellular calcium, resulting in depolarization of the cell membrane potential and secretion of GH (Chen, C. *et al.*, 1994). Thus, GH synthesis and secretion is interdependent upon cAMP and calcium.

Much evidence exists that GRF serves as a mitogenic factor at the pituitary level and that this, in fact, may be its primary role. Pituitary hyperplasia is evident in patients with GRF-secreting tumours and in transgenic mice overproducing GRF (Burton *et al.*, 1991). GRF has also been found to enhance somatotroph proliferation in primary cell culture. A genetic dwarfism, known as the *little* mouse (lit/lit), results from the substitution of glycine for aspartate at codon 60 in the GRF receptor (Lin *et al.*, 1993). These mice respond to GH, but not GRF, since the mutation prevents high affinity binding of GRF to its receptor in the pituitary. Perhaps more importantly, these mice suffer pituitary hypoplasia. Another dwarf mouse, the Snell dwarf (dw/dw), results from mutations in the gene encoding the tissue-specific transcription factor Pit-1, or GHF-1/Pit-1 (Lin *et al.*, 1992). This factor is important in the embryological development of the somatotroph and other pituitary cell types, and the mutation of this gene results in a proliferative defect in the anterior pituitary. Pit-1 is necessary for the development of somatotrophs, and thus the GRF receptor, and is also essential for appropriate expression of the GH gene (see *Figure 7.2*).

7.2.2 SRIF and its receptors

SRIF, a 14-amino-acid peptide originating in the hypothalamus, was discovered during early attempts to find a GRF (Brazeau *et al.*, 1973; Krulich *et al.*, 1968). In addition to affecting GH expression, SRIF is found widely throughout the CNS and other tissues. It affects a diversity of physiological processes both centrally

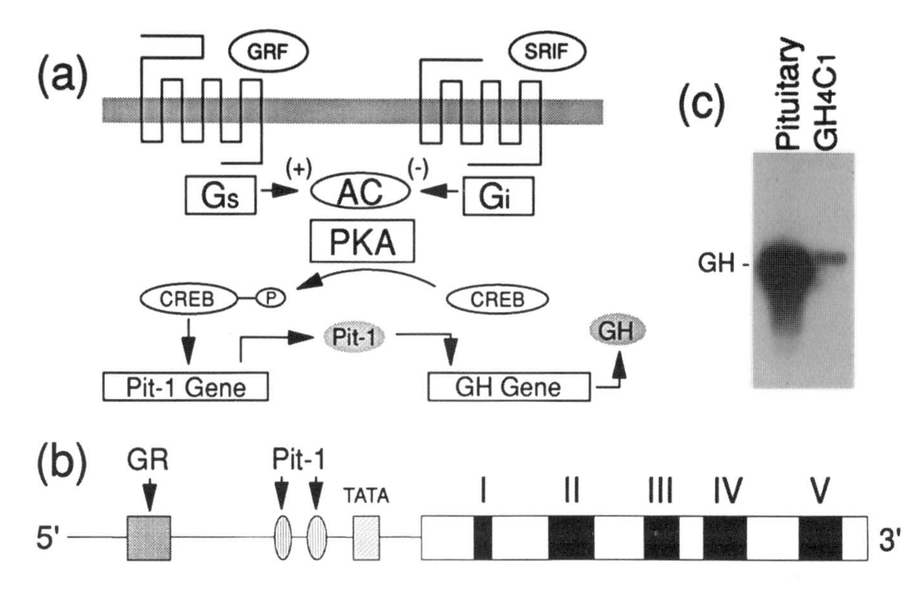

Figure 7.2. GH gene activation. (a) Positive and negative regulators GRF and SRIF, arising from hypothalamic neurons, bind to cell surface receptors of the somatotroph. Signal transduction through G_s or G_i controls cAMP production and the activation of protein kinase A (PKA). PKA acts upon CREB and other substrates to enhance the transcription factors responsible for GH gene activation. AC, adenylate cyclase. (b) Schematic of the GH gene. The coding region of exons I-V are denoted by filled boxes; introns and untranslated regions are clear boxes. The 5′-flanking region (single line) contains several *cis*-acting elements (shaded symbols) that enhance GH gene transcription. GR, glucocorticoid response element. (c) Northern analysis of RNA from rat pituitary (1 μg) and GH_4C_1 (rat pituitary adenoma) cells (3 μg), hybridized to a rat GH probe.

and peripherally, including locomotion, cognition, and insulin and glucagon release (Reichlin, 1982). Owing to their inhibitory effects on GH secretion, SRIF and its functional analogues (most notably, octreotide) have been used in the treatment of pituitary tumours. Although effective at reducing pituitary GH output, such treatment is, not surprisingly, often accompanied by wide-ranging side effects (Lamberts, 1988). In addition, SRIF and its analogues have been used in the treatment of some gastrointestinal disorders (Reichlin, 1982) and SRIF deficiency in the brain has been associated with many neuropsychiatric disorders (Rubinow, 1986). Unlike GRF, SRIF antagonists have only recently been reported, and have not been fully characterized (Wilkinson *et al.*, 1996).

The somatostatin gene has been best characterized in rats and humans, and its basic structure consists of two exons separated by a short intron. The upstream elements of the gene include a CRE sandwiched between CAAT and TATA sequences (Section 1.3.1). In humans, the primary translation product of the somatostatin gene is the 116-amino-acid preprosomatostatin molecule that includes a 24-amino-acid signal sequence at the N-terminus and the biologically active 14-amino-acid fragment (S-14) at its C-terminus. Alternative post-translational modification of preprosomatostatin also produces the 28-amino-acid

C-terminal fragment, somatostatin-28 (S-28), which is biologically active and pharmacologically distinguishable from S-14 (*Figure 7.1a*).

The diverse effects of SRIF are mediated by at least five cell surface receptor subtypes, termed somatostatin receptor (SSTR) 1–5 (Hoyer *et al.*, 1994). All of these molecules are seven-transmembrane, G-protein-linked receptors of the neurokinin class. Each is encoded in humans by a distinct genetic locus, and their sequences are sufficiently similar, particularly in the transmembrane domains, such that low stringency hybridization and PCR-based appproaches have been successfully employed in their cloning. SSTR1 was cloned from a set of 'orphan' G-protein-linked receptors isolated from pancreas islet RNA by RT–PCR (Yamada *et al.*, 1992), and SSTR2 (*Figure 7.1c*) was cloned by homology with SSTR1. Simultaneously, SSTR2 was purified from GH_4C_1 rat pituitary cells and identified by partial amino acid sequencing (Eppler *et al.*, 1992; Hulmes *et al.*, 1992), demonstrating that SSTR2 is the dominant subtype in the pituitary (*Figure 7.1d*). Based on structural similarities, two sets of receptor subtypes can be assigned, with SSTR1 and SSTR4 composing one group, and SSTR2, -3, and -5 the other (Hoyer *et al.*, 1994). Some pharmacological data tend to reinforce this pattern of classification, particularly the binding of the synthetic SRIF agonist, MK678, which shows high affinity for SSTR2, -3, and -5, but little affinity for SSTR1 and -4. However, other synthetic ligands can distinguish among subtypes, even within these structural subdivisions (reviewed in Reisine and Bell, 1995). The receptor subtypes differ little in their affinity for the two naturally occurring ligands; only SSTR5 discriminates well between S-14 and S-28. The affinity of SSTR1–4 for each ligand is approximately 0.2 nM, as is the affinity of SSTR5 for S-28. SSTR5, however, shows a 10–50-fold lower affinity for S-14. Mutation of a phenylalanine residue in the sixth transmembrane domain (TM6) of SSTR5 to tyrosine (found in the other four receptor subtypes) restores high affinity binding of S-14 (Ozenberger and Hadcock, 1995).

Northern analysis, RNase protection analysis, RT–PCR and *in situ* hybridization have been used to determine the tissue distribution of the transcripts of the five SRIF receptor subtypes (Bell and Reisine, 1995). All are present in brain, with each receptor subtype being expressed in a variety of regions, albeit at varying levels. For example, all are present in the hypothalamus, but only SSTR3 is found in the cerebellum, and SSTR2 is the predominant subtype in the arcuate nucleus. Peripherally, SSTR4 is the predominant subtype in heart and SSTR3 in spleen. SSTR2 is the most highly expressed subtype in the pituitary, where it mediates the GH-inhibitory properties of SRIF agonists (Raynor *et al.*, 1993). The primary pituitary action of SRIF on GH production is the inhibition of GRF-stimulated adenylate cyclase activity (*Figure 7.2*). SRIF also inhibits GH secretion by hyperpolarization of the somatotroph membrane through activation of potassium channels, causing the blockage of calcium flow through voltage-gated channels (Chen, C. *et al.*, 1994).

7.3 GH

GH deficiency has long been recognized as a common, but treatable, cause of dwarfism. The species specificity of GH meant that cadavers served as the sole source of GH prior to the development of recombinant DNA technology. This practice was not only expensive but prone to the spread of certain diseases. Recombinant human GH is now readily available and is currently being tested for a variety of indications other than dwarfism, including bone loss, wasting associated with AIDS, reduction of muscle mass in the aged and trauma-induced catabolism. In addition, bovine GH is currently used commercially for increasing milk production in dairy cows, and porcine GH has been shown to reduce fat and increase lean production in swine. Thus, the metabolic actions of GH rather than its effect upon growth *per se* may have value in livestock production as well as for human health.

7.3.1 Structure of GH

Human GH is a 22 kDa protein that is secreted by a single class of cells, the somatotrophs, found in the anterior pituitary (reviewed in Strobl and Thomas, 1994). In humans, two GH genes exist in a cluster with three placental lactogen (or chorionic somatomammotropin) genes in a 67 kb segment of DNA on the long arm of chromosome 17. All of these closely related genes are thought to have originated from a common ancestor. The first GH gene in this stretch (GH-N) is expressed in the pituitary and is essential for normal growth. The most severe form of growth deficiency, called isolated GH deficiency, type IA, is caused by mutation or deletion of GH-N. GH-N consists of five exons (as do all of the GH-like genes) which are spliced to form a mature mRNA encoding a 26-amino-acid leader, 191 amino acids for the mature hormone, and 5′ and 3′-untranslated regions (UTRs). Alternate splicing of the second exon produces a deletion in amino acids 32–46, resulting in a 20 kDa GH variant. This 20 kDa molecule contributes 5–15% of circulating GH, and can induce growth in transgenic mice (Stewart *et al.*, 1992), but binds to liver GH receptors (GHRs) with lower affinity than the normal 22 kDa form. The remaining four genes are expressed in the placenta, but are not essential for normal pregnancy or growth. One of these, GH-V, encodes a GH variant with 13 substituted amino acids that is sometimes glycosylated, producing 22, 25 and 26 kDa species, none of which have growth-promoting activity.

The mature GH polypeptide has a secondary and tertiary structure characterized by two disulphide-linked bridges and an antiparallel array of four α-helices. Crystallization of porcine GH (Abdel-Meguid *et al.*, 1987), co-crystallization of human GH with the GHR (De Vos *et al.*, 1992), and extensive scanning mutagenesis studies (Cunningham and Wells, 1989) have resulted in a detailed knowledge of the hormone structure and binding interactions. From this and other work (Cunningham *et al.*, 1991), it is now known that GH contains two binding sites of differing affinities (sites 1 and 2), that sequentially bind two GHRs thus allowing a signal to be transduced into the cell.

7.3.2 GH regulation and secretion

GH regulation is both simple and complex. Beneath the simplicity of GRF and SRIF up- and downregulating GH expression is the complex network of regulatory elements that mediate this process, as illustrated in *Figure 7.2*. cAMP plays a major role in the action of GRF upon GH gene regulation, affecting both synthesis and mitogenesis at the post-receptor level. Indeed, cAMP mimics, such as 8-bromo cAMP, which bypass the GRF receptor, activate the GH gene in a GRF-like manner. Transgenic mice in which somatotrophs are targeted by a cholera toxin transgene, leading to overproduction of cAMP, exhibit gigantism and somatotroph tumour development (Burton *et al.*, 1991). In the rat, a cAMP-responsive region has been identified in the GH gene (Copp and Samuels, 1989), and the cAMP-responsive transcription factor CREB has been found to be constitutively activated in human somatotroph adenomas (Bertherat *et al.*, 1995). Other upstream elements associated with the GH gene include a T_3 response element (in rat), a Pit-1 site specific for the homodimer form of Pit-1, a strongly responsive glucocorticoid element in the first intron and a more weakly responsive element in the 5′-UTR (*Figure 7.2b*). The general response to GRF probably involves the cAMP-dependent phosphorylation of CREB or possibly a co-activator which stimulates the transcription of Pit-1, GH and the oncogene *c-fos*. Thus, the cAMP-mediated action of GRF responsible for enhancing GH synthesis is also essential for its mitogenic action. Pit-1, in a positive regulatory role, further enhances transcription of GH. In the rat, Pit-1 and T_3 activate the GH promoter in a manner dependent upon protein kinase A and C (Gutierrez-Hartmann, 1994). The result of these interactions at the GH promoter is the extraordinarily high level expression of GH RNA in the pituitary, as compared with RNA from a GH-producing cell line (*Figure 7.2c*).

Our understanding of the GH secretory machinery is further complicated by consideration of the GH-releasing peptides (GHRPs), small synthetic molecules that act as GH secretagogues (Bowers *et al.*, 1984). Whether they do so by engaging specific hypothalamic or pituitary receptors is an issue that is only now coming into focus (Robinson, 1996). Other topics of current investigation include the identity and structure of the GHRP receptor (Pong *et al.*, 1996), and the nature (or existence) of a naturally occurring ligand that mimics GHRPs. *In vivo*, GHRPs synergize with GRF in stimulating GH release (Bowers *et al.*, 1990), and are being actively tested as a clinical alternative to GH.

7.4 GH receptors and binding proteins

The specific interaction of GH with cell surface receptors has been recognized for several decades. The liver, which is rich in GHRs, provided the material to produce specific antibodies which, along with GH itself, were effectively used to purify the rabbit GHR (Spencer *et al.*, 1988). During the past decade, the GHR has been one of the best studied cell surface receptors, work that was largely motivated by the important commercial market for recombinant GH. Thus, a partial amino acid sequence obtained from purified rabbit GHR led to the cloning of cDNAs encoding first rabbit and human (Leung *et al.*, 1987), and later rat, mouse, pig, sheep, cow and chicken GHRs [see Oldham *et al.* (1993) for references]. Additionally, mutated GHR genes have long been suspected to be the underlying defect in the GH-resistant dwarfism known as Laron syndrome, and studies of the GHR genomic structure have confirmed this for both humans and for GH-resistant dwarf chickens (Burnside *et al.*, 1991; Godowski *et al.*, 1989). Finally, through the production of large amounts of normal and mutant recombinant GHRs, a detailed three-dimensional structure of the GH–GHR interaction has been elucidated (De Vos *et al.*, 1992), and signalling has been shown to occur via dimerization of two GHRs by a single GH molecule (Cunningham *et al.*, 1991). Many of these recent structural and biochemical studies have been performed using an engineered soluble form of the GHR, mimicking the naturally occurring GH-binding protein (GHBP), which is described below. GH receptors are now known to be found in practically all tissues (Bingham *et al.*, 1994; Waters *et al.*, 1990) and the myriad growth and metabolic effects of GH are thus a result of interactions at the GHR both in the liver and at a variety of peripheral sites. The molecular basis of these effects is now just beginning to be understood in some detail.

7.4.1 GHR and GHBP structure

The GHR is a member of the cytokine/haematopoietin superfamily of single-transmembrane-domain cell surface receptors (Bazan, 1990). It is a 620-amino-acid polypeptide after processing of the signal sequence that is cleaved during transport to the plasma membrane. Its predicted molecular weight is 70 kDa, but is closer to 120 kDa *in vivo* due to post-translational modifications such as glycosylation and possibly ubiquitination (Leung *et al.*, 1987). A hydrophobic membrane-spanning domain divides the extracellular, ligand-binding domain and the intracellular, signalling domain (*Figure 7.3a*).

Members of this family feature characteristic structural motifs in their extracellular domains. The secondary and tertiary configuration conferred by seven antiparallel β-strands is reminiscent of the structure of fibronectin (Bazan, 1990). For GH (and prolactin) receptors, three disulphide-linked loops are anchored by highly conserved cysteine residues (De Vos *et al.*, 1992), while a seventh, unpaired, cysteine is found near the transmembrane domain. Much has been deduced regarding specific amino acids of GH and its receptor that are involved in binding (Bass *et al.*, 1991; Cunningham and Wells, 1989). A few key residues (e.g. W104 and W169) lend the majority of binding free energy to this interaction, while the rest of the binding surface is made up of many residues each contributing relatively little binding free energy (Clackson and Wells, 1995). Single amino acid changes in the receptor-binding domain can also alter its binding characteristics for different GH molecules (Souza *et al.*, 1995).

Intracellular signalling for the GHR has only recently been characterized in detail, beginning with the discovery that physical and biochemical interactions occur between the GHR intracellular domain and the tyrosine kinase, JAK2 (Janus kinase family), in direct response to GH binding (Argetsinger *et al.*, 1993). It is this interaction which appears to be central to GHR signalling; JAK2 is likely to be responsible for activation (via tyrosine phosphorylation of itself and the GHR) of several pathways leading to both cellular effects and transcriptional regulatory events (reviewed by Carter-Su *et al.*, 1996). For instance, activation of insulin receptor substrate (IRS) leads to regulation of glucose transport; activation of mitogen-activated protein (MAP) kinases, via the signalling cascade consisting of Shc, Grb2, Sos, Ras, Raf and MEK, results in gene activation at, for example, the serum response element (SRE) leading to cell growth and differentiation; and activation of STATs (signal transducers and activators of transcription) 1, 3 and 5 leads directly to transcriptional regulation in the nucleus. Other effects, possibly unrelated to JAK2 activation, include increases in intracellular calcium (Schwartz *et al.*, 1992) and activation of protein kinase C. Globally, the full range of GH actions (metabolic, developmental, endocrine and mitogenic) are mediated by these, or similar, pathways even though many of its actions are long term (i.e. months or years). Perhaps the most dramatic effect of GHR signalling with regard to growth is the induction of IGF-I synthesis and secretion. This occurs within minutes of GH injection *in vivo* (Bichell *et al.*, 1992), and has long been thought to be a major mediator of GH action (Daughaday, 1989). Many of these activities have been correlated with conserved structural features of the intracellular domain that have been shown to be important for signalling, such as 'box 1 and box 2,' and a series of tyrosine residues [*Figure 7.3a*, reviewed by Carter-Su *et al.* (1996)].

The soluble GHBP is also a product of the GHR gene, having an identical amino acid sequence to the extracellular domain of the GHR (approximately 290 amino acids, *Figure 7.3a*). However, the GHBP is produced in different species by two entirely distinct mechanisms: in humans, rabbits, pigs, and probably sheep and chickens, the GHBP is produced by the proteolytic processing of GHR proteins, which releases the soluble extracellular domain (Bingham *et al.*, 1994; Sotiropoulos *et al.*, 1993); in rats and mice, however, it is produced by alternative splicing of the primary GHR mRNA (Baumbach *et al.*, 1989; Sadeghi *et al.*, 1990; Smith *et al.*, 1989). This is illustrated in *Figure 7.3b*, which shows Northern blots of liver RNA from rat, pig and sheep, probed with the homologous GHR probes.

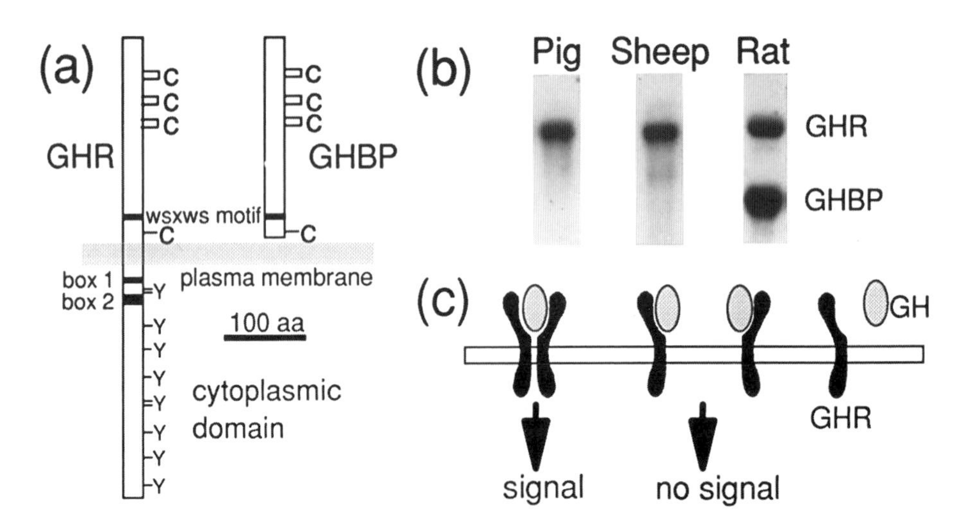

Figure 7.3. GHR and GHBP. (a) Scale diagram showing structural features of the rat proteins. Disulphide-linked cysteine residues in extracellular domain and tyrosine residues in the cytoplasmic domain are indicated by the letters C and Y, respectively. (b) Northern analysis of liver RNA from pig, sheep and rat. Each lane was probed with the homologous GHR extracellular domain. Only rat expresses a separate RNA for GHBP. (c) Cartoon indicating that dimerization of the GHR by a single GH molecule activates signalling. In the presence of excess GH (or a ligand that cannot bind two receptors), dimerization and signalling are prevented.

The lower band in the rat lane encodes the GHBP, while the larger band for each species encodes the GHR. As illustrated in *Figure 7.3c*, signalling by GHRs involves dimerization of two receptors by a single GH molecule. Thus, GHR signalling can be antagonized by either high concentrations of hormone, which prevent dimerization by binding a single receptor per GH molecule, or by mutant GH molecules that are incapable of binding two receptors simultaneously. Such a mutant GH has been recently designed whereby a single mutation in the third α-helix of GH (G119R in mouse; G120R in human) results in the loss of site 2 binding and effective antagonism at the GH receptor (Chen, W.Y. *et al.*, 1994).

7.4.2 GHR regulation

The structure of the GHR gene has not yet been fully elucidated. In the human, it spans more than 80 kilobases on chromosome 5 (Godowski *et al.*, 1989). The full length mRNA results from the transcription of 10 exons; exons 2–10 encompass the coding region and 3′-UTR. Exon 1 is non-coding and highly variable (see below). Exon 3, which encodes 22 amino acids near the N-terminus of the fully processed receptor, is also variable. In humans, two forms of GHR RNA are detected, either containing or lacking exon 3. In human placenta, only the isoform lacking exon 3 is found (Urbanek *et al.*, 1992) whereas, in other tissues, expression of both isoforms varies widely among individuals (Wickelgren *et al.*, 1995). In the pig, exon 3 can be found as a single copy, as a tandem duplication or absent entirely, all in the same tissue (Bingham *et al.*, 1994; Knapp and Kopchick, 1996), while in the chicken, exon 3 has never been detected (Burnside *et al.*, 1991). Thus far, no specific function for exon 3 has been discovered, as GHR function appears to be equal among variants (Knapp and Kopchick, 1996). Exon 10, which includes the 3′-UTR, can also vary due to alternative polyadenylation (Huang *et al.*, 1993). This does not affect the protein structure, but may lead to differences in RNA stability or translation efficiency. Another form of alternative polyadenylation has been observed in avians whereby a conserved polyadenylation site is embedded in the extracellular GHR coding region, resulting in the alternative production of a severely truncated, putative GHBP with no known function (Oldham *et al.*, 1993).

The 5′-UTR of GHR mRNA is encoded by exon 1, except for 11 bp arising from exon 2. Early cloning efforts revealed that this region varies due to alternative splicing (Baumbach *et al.*, 1989; Leung *et al.*, 1987). Later, PCR experiments using liver cDNA as template revealed as many as eight different variants of exon 1 in humans and at least six in rats (Pekhletsky *et al.*, 1992). To date, there is no evidence that additional exons exist upstream of exon 1. As a consequence, in genomic DNA, each alternative exon 1 must be preceded by a unique set of promoter/transcriptional control sequences, all of which must be upstream of the GHR coding sequence. *Figure 7.4a* shows diagramatically how this might be arranged. Since the presence of one or another exon 1 identifies any single RNA as originating under the transcriptional control of a unique promoter region, GHR RNA can be classified and named as shown in *Figure 7.4a*.

Since the various exons 1 are distinguishable by methods such as Northern analysis, RNAse protection assay and PCR, their regulation can be studied *in vivo*.

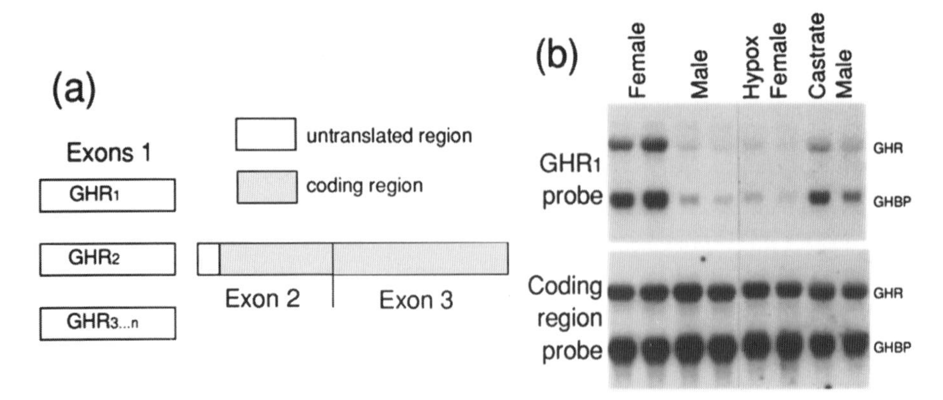

Figure 7.4. Alternatively spliced exon 1 in the GHR gene. (a) Diagram illustrating that several different 5′ ends of the GHR (and GHBP in rats) mRNA are alternatively spliced to the same coding exons. Each exon 1 is likely to be controlled by distinct promoter elements. (b) Northern analysis of rat liver mRNA. Although little difference is apparent among treatments for total GHR and GHBP mRNA (identified with a coding region probe), the class of mRNA containing the GHR_1 region does vary. Males and hypophysectomized (Hypox) females both lack, while male castration partially induces, the typical female pattern of GH secretion.

Thus far, the most thoroughly studied alternative exon 1 has been the rat GHR_1 (Baumbach and Bingham, 1995). Its transcription is tissue specific, sex specific and developmentally regulated, being found only in post-adolescent female liver. These studies revealed that the GHR_1 class of RNA is regulated in a similar manner to liver- and female-specific genes (e.g. the P450 gene 2C-12) involved in sex steroid biosynthesis. In fact, as was deduced more than a decade ago, the regulatory sex determining factor that exerts control over the transcription of these genes is GH itself, specifically the sexually dimorphic, episodic pattern of GH secretory pulses (Mode *et al.*, 1983; Norstedt and Palmiter, 1984). In the Northern analyses shown in *Figure 7.4b*, a GHR_1-specific probe is used to illustrate how GHR_1 RNA, encoding both GHR and GHBP in the rat, is regulated by several factors affecting the GH secretory pattern. Removal of GH (by hypophysectomy) in females reduces GHR_1 transcription, while castration of male rats or prolonged GH infusion into males (not shown) induces GHR_1 RNA. These latter treatments mimic the female pattern of GH release (characterized by a high baseline GH level and small, irregular pulses). Treatment of males with oestrogen also increases GHR_1 RNA (Gabrielsson *et al.*, 1995).

A final instance of GHR regulation that involves alternative splicing is the production of GHBP in rodents. Although GHBP production in humans does not appear to be regulated (Ip *et al.*, 1996), Cramer *et al.* (1992) have shown changes in the ratio of GHR to GHBP RNA during pregnancy in the mouse that are mirrored by changes in protein levels. How such regulation could be achieved in mice is not yet known, but is currently under study (Zhou *et al.*, 1996). Proposed functions of the GHBP are the regulation of GH availability to tissues (Lim *et al.*, 1990), the prolongation of GH half-life in the serum (Baumann *et al.*, 1987) and the modulation of the amplitude of GH secretory pulses. None of these functions has been demonstrated unambiguously, thus rendering the precise role of the GHBP a mystery. Not only is its regulation controversial, but GHBP has also been found to be attached to the cell surface (Goodman *et al.*, 1994), localized within the nucleus (Lobie *et al.*, 1994) and in pituitary secretory granules (Harvey *et al.*, 1993). Even the activity of GHBPs in serum seems enigmatic, as periodic trough levels of serum GH in male rats reach practically zero in spite of high GHBP levels. Additionally, other unknown serum factors may also be involved (Leung *et al.*, 1996).

7.5 IGF-I, its receptor and binding proteins

The most dramatic result of GH treatment in animals is the induction of IGF-I production (Rotwein *et al.*, 1994). In fact, the somatomedin hypothesis (Daughaday, 1989) is the assertion that all GH effects are mediated by IGF-I, which was first known as somatomedin C. Though not entirely correct, the somatomedin hypothesis remains instructive in drawing attention to the causal link between the high levels of hepatic expression of GHR and IGF-I, the interaction of GH and IGF-I RNA (*Figure 7.5a*), and the widespread growth promoting effects of IGF-I in peripheral tissues. Because the IGFs and their receptors and binding proteins have been the subject of several recent reviews (Cohick and Clemmons, 1993; Jones and Clemmons, 1995; LeRoith, 1991; LeRoith *et al.*, 1995), the following discussion incorporates only a small fraction of the published research in these areas.

7.5.1 IGF-I

IGF-I has strong mitogenic activity in many cell types and, in some cells, promotes differentiation. Among the general metabolic responses elicited by IGF-I are the synthesis of DNA, RNA and protein and the uptake of glucose and amino acids. In addition, many tissues display specific metabolic responses to IGF-I. One example is the stimulation by IGF-I of the incorporation of sulphate into cartilage (Salmon Jr and Daughaday, 1957). It was this effect that caused IGF-I to be named sulphation factor before being assigned the name somatomedin.

The mature IGF-I polypeptide is a single-chain, 70-amino-acid molecule containing B and A domains that show strong structural resemblance to the equivalent domains of insulin and IGF-II. B and A are separated by a short C domain, and C-terminal to A is the extension peptide, termed the D domain. This mature peptide is cleaved from a propeptide that includes a variable signal sequence upstream and some additional extension sequence (the E domain) downstream.

The IGF-I gene is composed of six exons, spanning more than 70 kb of genomic sequence in rat. As illustrated in *Figure 7.5b*, transcription of IGF-I involves a variety of alternative RNA processing events, some of which vary among mammalian species. Exons 3 and 4 encode the mature IGF-I peptide, and are found in all of the IGF-I RNAs. Variability in IGF-I RNA comes about as a result of alternative splicing in exons 1, 2 and 5 (inclusion of exon 5 encodes a frame shift, thus

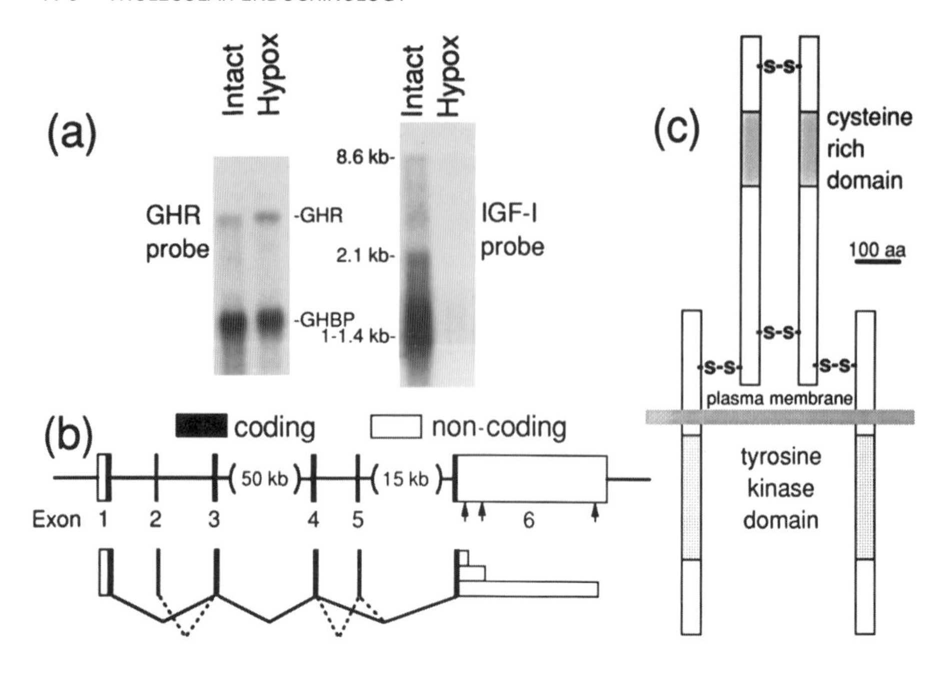

Figure 7.5. IGF-I and the IGF-I receptor. (a) Northern analysis of liver mRNA from pituitary intact vs. hypophysectomized (Hypox) rats. Lanes probed with a GHR coding region probe show little difference, whereas an IGF-I probe reveals reduced expression in Hypox animals, lacking GH. (b) Schematic diagram of the genomic structure of the rat IGF-I gene. Alternative splicing in the coding region leads to a variety of IGF-I peptide variants, and alternative polyadenylation in the 3′-non-coding region results in the variety of mRNAs seen in part (a). (c) Schematic diagram of the IGF-I receptor. The cysteine-rich (binding) domain and tyrosine kinase (signalling) domains are shown, as are the disulphide bridges among the monomeric units of the heterotetrameric structure.

altering the amino acid sequence encoded by exon 6). Alternative polyadenylation occurs in the 3′-UTR and accounts for essentially all of the considerable variation in size seen in IGF-I transcripts (*Figure 7.5a, b*). The variability in the composition of the 3′-UTR is believed to affect the stability and translatability of the alternative IGF-I messages (Hepler *et al.*, 1990).

In rats, IGF-I mRNA is present in a vast array of tissues, but is most abundant and most demonstrably GH dependent in the liver (Mathews *et al.*, 1986). Liver is believed to be the predominant source of IGF-I in circulation, and IGF-I of hepatic origin is most often characterized as the endocrine effector of GH. One prediction of the somatomedin hypothesis is that the *in vivo* effects of exogenous GH can be duplicated by exogenous IGF-I. IGF-I does stimulate somatic growth in GH-deficient animals. However, the growth promoting effects of IGF-I on specific organs are distinguishable from those of GH, probably as a result of the direct action of GH on some end organs. In addition, the effects of IGF-I and GH constitutively overexpressed in transgenic mice are also distinguishable. IGF-I transgenic mice achieve approximately 130% of the weight of their non-transgenic littermates, whereas GH transgenic mice are approximately twice the weight of control mice. Again, there are some organ-specific differences as well. In Laron-type dwarves, in whom GH insensitivity results from the absence of GHR signalling, increased growth velocity can be stimulated with the injection of IGF-I, although hypoglycaemia is a common side effect.

In addition to the endocrine action of IGF-I, the autocrine/paracrine role of IGF-I has been studied in many cell types, and a well-characterized paradigm for this role is found in cancer. Many tumour cells express high levels of IGF-I receptor. In small cell lung cancer (Macaulay *et al.*, 1990), the tumour cells also express IGF-I, with consequent autocrine stimulation of cell growth. In contrast, breast epithelial tumour cells have no IGF-I transcription, but surrounding stromal cells do express IGF-I, providing stimulation of tumour growth in a paracrine manner (Yee *et al.*, 1989). Disruption of the IGF-I–IGF-I receptor interaction has been proposed as a cancer therapy, but its potential utility remains speculative (Yee, 1994).

7.5.2 IGF-I receptor

The majority of the biological effects of IGF-I are mediated by the type I IGF receptor (IGF-I receptor). However, IGF-I is also known to interact with at least six soluble serum-binding proteins (IGFBPs) as well as the IGF-II/mannose-6-phosphate receptor, the insulin receptor and the IGF-I receptor–insulin receptor hybrid.

Similar in structure to the insulin receptor, the IGF-I receptor is a heterotetrameric membrane-spanning complex, composed of two α- and two β-subunits (*Figure 7.5c*). The extracellular α-subunits each include a cysteine-rich ligand-binding site, and the membrane-spanning β-subunits each include a tyrosine kinase signalling domain. Two disulphide bridges attach the two α-subunits to each other, and one disulphide bridge links each α-subunit to a β-subunit. The IGF-I receptor gene encodes the IGF-I proreceptor, a single precursor molecule composed of a signal sequence and an α- and a β-subunit separated by a proteolytic cleavage site. Expression of IGF-I receptor mRNA is widespread across tissue types, although not ubiquitous. In adult rats, highest levels are found in brain, but no IGF-I receptor RNA is found in the liver, the major site of IGF-I production. Furthermore, most tissues show a developmentally regulated decrease in mRNA abundance from the perinatal period through early adulthood (Werner *et al.*, 1989).

Ligand–receptor cross-talk exists between insulin, IGF-I and IGF-II, with all three ligands showing some affinity for all three receptors. The binding affinities of each ligand to its own receptor are all of the same order of magnitude ($K_D \sim 1$ nM), whereas the non-specific ligand–receptor interactions range from a 2–10-fold reduction in affinity for IGF-II binding to the IGF-I receptor to a greater than 100-fold reduction for insulin binding to either the IGF-I or the IGF-II/mannose-6-phosphate receptor. Moreover, the IGF-I receptor and the insulin receptor are sufficiently similar in structure that in cells expressing both receptors, hybrids exist, consisting of one $\alpha\beta$ complex of each receptor type. These hybrids bind well to IGF-I but poorly to insulin. *In vivo*, IGF-I half receptors pair with insulin half receptors no less readily than with other IGF-I half receptors. Cells expressing near equal amounts of both receptors have thus been shown to produce as many receptor hybrids as homologous receptors. However, the precise role of receptor hybrids *in vivo* remains under investigation.

Targeted disruption of the IGF-I receptor in mice has severe developmental effects. IGF-I receptor null mutants die at birth, having only 45% normal birthweight and suffering general organ hypoplasia. In contrast, targeted disruption of the IGF-I gene reported in the same study is not as severe, resulting in neonates that are approximately 60% normal birthweight. While their postnatal survival is reduced, some of these IGF-I null mutants survive to adulthood, albeit achieving only 30% of normal adult weight. The greater severity of the receptor null mutation can be attributed to the fact that embryonically the IGF-I receptor also serves as the primary binding site of IGF-II (Baker *et al.*, 1993; Liu *et al.*, 1993).

7.5.3 IGF-binding proteins

Six IGF-binding proteins (IGFBPs) have been described. Although bearing much structural similarity to one another, especially in a highly conserved array of cysteine residues, each IGFBP in humans is encoded by a separate genetic locus. IGFBP-3 is the most abundant IGFBP in circulation and is the only IGFBP positively regulated by GH. Some 95% of circulating IGF-I and IGF-II is bound to IGFBP-3, and much of it (~75%) is further bound in the form of a ternary complex of IGF, IGFBP-3 and the acid-labile subunit. The ternary complex is believed not to leave the vascular space, whereas unbound IGFs and the IGFs bound only to an IGFBP can easily diffuse through capillary walls. The ternary complex has a greatly increased biological half-life (12–15 h) compared to any of the IGF–IGFBP binary complexes (<2 h) or free IGF-I (minutes) (Baxter, 1996). IGFBP-1, -2 and -4 are also present in the circulation at substantial concentrations; levels of IGFBP-5 and -6 are very much lower.

Given the complexity of the system, it is not unexpected that reports of both the potentiation and the inhibition of IGFs by IGFBPs have been published (reviewed by Jones and Clemmons, 1995). In cell culture experiments, excess IGFBP-1, -2, -3, -4 and -5 have each been reported to inhibit IGF action. Under different conditions, however, IGFBP-1, -2, -3 and -5 have been shown to potentiate IGF action. Typically, the potentiation comes at lower concentrations of the IGFBP and under conditions that allow the IGFBP to form an association with the cell surface. In some cases, the effects of IGFBPs (especially with IGFBP-1) is dependent on the state of post-translational modification. IGFBP-4 is unique in that it appears to be an unambiguous inhibitor of IGF action and shows no association with the cell surface. IGFBP-6 is the least studied of the IGFBPs, and its biological role remains largely to be determined. Finally, IGFBP-3 has been shown to have some direct, ligand-independent, cell growth-inhibitory properties.

7.6 Concluding remarks

While it might be reasonable to think that the major molecular players in animal growth have been identified, it would be short-sighted to conclude that a full understanding of this system has been achieved. It is recognized that GRF and SRIF are released from the hypothalamus and regulate GH production and release by the pituitary. GH engages receptors in the liver and elsewhere, giving rise both to direct end organ growth effects and to production of circulating IGF-I of hepatic origin. IGF-I, in turn, also promotes cellular growth by engaging cell surface receptors in peripheral tissues. However, in spite of more than a quarter of a century of intensive investigation, only the simplest of the interactions in this endocrine cascade, such as the binding of ligands to their receptors, are well understood. Others, such as negative feedback loops within the cascade, are recognized but only poorly understood, and no doubt further subtleties of the GH axis have so far escaped our attention entirely, for example, the putative role of endogenous GHRPs.

The complexity of the hypothalamic/pituitary growth axis can be further appreciated by consideration of the variety of growth dynamics among species. At one extreme are animals such as rats and pigs, which continue somatic growth (i.e., longitudinal bone growth) indefinitely into adulthood. At the other extreme are most bird species, which achieve adult stature very early in life relative to sexual maturity. Humans offer an intermediate condition, in which somatic growth ceases not long after sexual maturity. Moreover, some species display enormous differences in stature between males and females, while others show very little sexual dimorphism. Sexual status and gonadal steroids have not only direct somatic effects but also affect growth via their interaction with the GH axis. In particular, the effect of sex steroids on the pulsatility of GH release is a phenomenon affecting not only growth, but also the expression of genes that are responsive to the pattern of GH release.

Much of the *in vivo* work on the molecular biology of growth comes from laboratory experiments in which high quality nutrition has been maintained. That nutrition affects growth is an ancient observation but, at the molecular level, the results are frequently contradictory or confusing. However, when considering human disease states or the growth of animals reared in the wild, nutritional status is a variable that cannot be ignored.

Although animal growth is a complex polygenic process, we can reasonably expect that, as the tools of molecular biology continue to expand, so too will our knowledge of the genes and proteins affecting growth. For example, PCR-based techniques have already contributed to the characterization of several naturally occurring mutant alleles of the human GH and GHR genes (Amselem *et al.*, 1991; Cogan *et al.*, 1995). In addition, an ever increasing number of animal model systems, both naturally occurring and transgenic, have facilitated the *in vivo* study of animal growth. Deeper questions, such as the origin of size variation among species, will also have molecular explanations, but these explanations are likely to involve many more, and possibly entirely different, genetic loci than have been discussed here.

Acknowledgements

The authors wish to thank Deborah Chaleff for actively fostering novel research on the GH axis. We also thank Kathleen Heaney for her support, and Mark Eppler and Catherine Thrash-Bingham for kindly reviewing the manuscript.

References

Abdel-Meguid SS, Shieh H, Smith WW, Dayringer HE, Violand BN, Bentle LA. (1987) Three-dimensional structure of a genetically engineered variant of porcine growth hormone. *Proc. Natl Acad. Sci. USA* **84:** 6434–6437.

Amselem S, Sobrier ML, Duquesnoy P, Rappaport R, Postel-Vinay MC, Gourmelen M, Dallapiccola B, Goossens M. (1991) Recurrent nonsense mutations in the growth hormone receptor from patients with Laron dwarfism. *J. Clin. Invest.* **87:** 1098–1102.

Argetsinger LS, Campbell GS, Yang X, Witthugn BA, Silvennoinen O, Ihle JN, Carter-Su C. (1993) Identification of JAK2 as a growth hormone receptor-associated tyrosine kinase. *Cell* 74: 237–244.

Baker J, Liu J, Robertson EJ, Efstratiadis A. (1993) Role of insulin-like growth factors in embryonic and postnatal growth. *Cell* **75:** 73–82.

Bass SH, Mulkerrin MG, Wells JA. (1991) A systematic mutational analysis of hormone-binding determinants in the human growth hormone receptor. *Proc. Natl Acad. Sci. USA* **88:** 4498–4502.

Baumann G, Amburn KD, Buchanan TA. (1987) The effect of circulating growth hormone binding protein on metabolic clearance, distribution and degradation of human growth hormone. *J. Clin. Endocrinol. Metab.* **64:** 657–660.

Baumbach WR, Bingham B. (1995) One class of growth hormone (GH) receptor and binding protein messenger ribonucleic acid in rat liver, GHR_1, is sexually dimorphic and regulated by GH. *Endocrinology* **136:** 749–760.

Baumbach WR, Horner D, Logan JS. (1989) The growth hormone-binding protein in rat serum is an alternatively spliced form of the rat growth hormone receptor. *Genes Devel.* **3:** 1199–1205.

Baxter RC. (1996) Regulation of IGF bioavailability – a complex problem. In: *Program of the 10th International Congress of Endocrinology*, San Francisco, CA, Abstract L6.

Bazan JF. (1990) Structural design and molecular evolution of a cytokine receptor superfamily. *Proc. Natl Acad. Sci. USA* **87:** 6934–6938.

Bell GI, Reisine T. (1995) Molecular biology of somatostatin receptors. *Trends Neurosci.* 16: 34–38.

Bertherat J, Dournaud P, Berod A, Normand E, Bloch B, Rostene W, Kordon C, Epelbaum J. (1992) Growth hormone-releasing hormone neurons are a subpopulation of somatostatin receptor-labelled cells in the arcuate nucleus: a combined *in situ* hybridization and receptor light-microscopic radiographic study. *Neuroendocrinology* **56:** 25–31.

Bertherat J, Chanson P, Montminy M. (1995) The cyclic adenosine 3′,5′-monophosphate-responsive factor CREB is constitutively activated in human somatotroph adenomas. *Mol. Endocrinol.* 9: 777–783.

Bichell DP, Kikuchi K, Rotwein P. (1992) Growth hormone rapidly activates insulin-like growth factor I gene transcription *in vivo*. *Mol. Endocrinol.* **6:** 1899–1908.

Bingham B, Oldham ER, Baumbach WR. (1994) Regulation of growth hormone receptor and binding protein expression in domestic species. *Proc. Soc. Exp. Biol. Med.* **206:** 195–199.

Bowers CY, Momany FA, Reynolds GA, Hong A. (1984) On the *in vitro* and *in vivo* activity of a new synthetic hexapeptide that acts on the pituitary to specifically release growth hormone. *Endocrinology* **114:** 1537–1545.

Bowers CY, Reynolds GA, Durham D, Barrera CM, Pezzoli SS, Thorner MO. (1990) Growth hormone (GH)-releasing peptide stimulates GH release in normal men and acts synergistically on GH-releasing hormone. *J. Clin. Endocrinol. Metab.* **70:** 975–982.

Brazeau P, Vale W, Burgus R, Ling N, Butcher M, Rivier J, Guillemin R. (1973) Hypothalamic peptide that inhibits the secretion of immunoreactive pituitary growth hormone. *Science* **257:** 77–79.

Breyer PR, Rothrock JK, Beaudry N, Pescovitz OH. (1996) A novel peptide from the growth hormone releasing hormone gene stimulates Sertoli cell activity. *Endocrinology* **137:** 2159–2162.

Burnside J, Liou SS, Cogburn LA. (1991) Molecular cloning of the chicken growth hormone receptor complementary deoxyribonucleic acid: mutation of the gene in sex-linked dwarf chickens. *Endocrinology* **128:** 3183–3192.

Burton F, Hasel K, Bloom F, Sutcliffe J. (1991) Pituitary hyperplasia and gigantism in mice caused by a cholera toxin transgene. *Nature* **350:** 77.

Carter-Su C, Schwartz J, Smit LS. (1996) Molecular mechanisms of growth hormone action. *Ann. Rev. Physiol.* **58:** 187–207.

Chen C, Vincent J, Clarke IJ. (1994) Ion channels and the signal transduction pathways in the regulation of growth hormone secretion. *Trends Endocrinol. Metab.* 5: 227–233.

Chen WY, Chen NY, Yun J, Wagner TE, Kopchick JJ. (1994) *In vitro* and *in vivo* studies of antagonistic effects of human growth hormone analogs. *J. Biol. Chem.* **269:** 15892–15897.

Clackson T, Wells JA. (1995) A hot spot of binding energy in a hormone-receptor interface. *Science* **267:** 383–386.

Cogan JD, Ramel B, Lehto M, Phillips J, Prince M, Blizzard RM, deRavel TJL, Brammert M, Groop L. (1995). A recurring dominant negative mutation causes autosomal dominant growth hormone deficiency – a clinical research center study. *J. Clin. Endocrinol. Metab.* **80:** 3591–3595.

Cohick WS, Clemmons DR. (1993) The insulin-like growth factors. *Ann. Rev. Physiol.* **55:** 131–153.

Copp RP, Samuels HH. (1989) Identification of an adenosine 3′,5′-monophosphate (cAMP)-responsive region in the rat growth hormone gene: evidence for independent and synergistic effects of cAMP and thyroid hormone on gene expression. *Mol. Endocrinol.* **3:** 790–796.

Coy DH, Murphy WA, Lance VA, Heiman ML. (1986) Strategies in the design of synthetic agonists and antagonists of growth hormone releasing factor. *Peptides* **7** (Suppl 1): 49–52.

Cramer SD, Barnard R, Engbers C, Ogren L, Talamantes F. (1992) Expression of the growth hormone receptor and growth hormone-binding protein during pregnancy in the mouse. *Endocrinology* **131:** 876–882.

Cronin MJ, Zysk JR, Baertschi AJ. (1986) Protein kinase C potentiates corticotropin releasing factor stimulated cyclic AMP in pituitary. *Peptides* **7:** 935–938.

Cunningham BC, Wells JA. (1989) High-resolution epitope mapping of hGH-receptor interactions by alanine-scanning mutagenesis. *Science* **244:** 1081–1085.

Cunningham BC, Ultsch M, De Vos AM, Mulkerrin MG, Clauser KR, Wells JA. (1991) Dimerization of the extracellular domain of the human growth hormone receptor by a single hormone molecule. *Science* **254:** 81–82.

Daughaday WH. (1989) A personal history of the somatomedin hypothesis and recent challenges to its validity. *Perspect. Biol. Med.* **32:** 194–211.

De Vos AM, Ultsch M, Kossiakoff AA. (1992) Human growth hormone and extracellular domain of its receptor: crystal structure of the complex. *Science* **255:** 306–312.

Dohlman H, Caron M, DeBlasi A, Frielle T, Lefkowitz R. (1990) Role of extracellular disulfide-bonded cysteines in the ligand binding function of the beta$_2$-adrenergic receptor. *Biochemistry* **29:** 2335–2342.

Eppler CM, Zysk JR, Corbett M, Shieh H-M. (1992) Purification of a pituitary receptor for somatostatin. *J. Biol. Chem.* **267:** 15603–15612.

Gabrielsson BG, Carmignac DF, Robinson ICAF. (1995) Steroid regulation of growth hormone receptor (GHR) and GH binding protein (GHBP) messenger RNAs in the rat. *Endocrinology* **136:** 209–217.

Gaylinn BD, Harrison JK, Zysk JR, Lyons CE, Lynch KR, Thorner MO. (1993) Molecular cloning and expression of a human pituitary receptor for growth hormone-releasing hormone. *Mol. Endocrinol.* **7:** 77–84.

Gaylinn BD, Lyons CE, Zysk JR, Clarke IJ, Thorner MO. (1994a) Photoaffinity cross-linking to the pituitary receptor for growth hormone-releasing factor. *Endocrinology* **135:** 950–955.

Gaylinn BD, von Kap-Herr C, Golden WL, Thorner MO. (1994b) Assignment of the growth hormone-releasing hormone receptor gene (GHRHR) to 7p14 by *in situ* hybridization. *Genomics* **19:** 193–195.

Godowski PJ, Leung DW, Meacham LR *et al.* (1989) Characterization of the human growth hormone receptor gene and demonstration of a partial gene deletion in two patients with Laron-type dwarfism. *Proc. Natl Acad. Sci. USA* **86:** 8083–8087.

Goodman HM, Frick GP, Tai L. (1994) The significance of the short form of the growth hormone receptor in rat adipocytes. *Proc. Soc. Exp. Biol. Med.* **206:** 304–308.

Guillemin R, Brazeau P, Bohlen P, Esch F, Ling N, Wehrenberg WB. (1982) Growth hormone-releasing factor from a human pancreatic tumor that caused acromegaly. *Science* **218:** 585–587.

Gutierrez-Hartmann A. (1994) Insight: Pit-1/GHF-1: a pituitary-specific transcription factor linking general signaling pathways to cell-specific gene expression. *Mol. Endocrinol.* **8:** 1447–1449.

Harvey S, Baumbach WR, Sadeghi H, Sanders EJ. (1993) Ultrastructural colocalization of growth hormone binding protein and pituitary hormones in adenohypophyseal cells of the rat. *Endocrinology* **133:** 1125–1130.

Hepler JE, Van Wyk JJ, Lund PK. (1990) Different half-lives of insulin-like growth factor I mRNAs that differ in length of 3′ untranslated sequence. *Endocrinology* **127:** 1550–1552.

Hoyer D, Bell GI, Berelowitz M *et al.* (1994) Classification and nomenclature of somatostatin receptors. *Trends Pharmacol. Sci.* **16:** 86–88.

Huang N, Cogburn LA, Agarwal SK, Marks HL, Burnside J. (1993) Overexpression of a truncated growth hormone receptor in the sex-linked dwarf chicken: evidence for a splice mutation. *Mol. Endocrinol.* **7:** 1391–1398.

Hulmes JD, Corbett M, Zysk JR, Bohlen P, Eppler CM. (1992) Partial amino acid sequence of a somatostatin receptor isolated from GH_4C_1 pituitary cells. *Biochem. Biophys. Res. Commun.* **184:** 131–136.

Ip TP, Hoffman DM, Hussain M, O'Sullivan AJ, Leung KC, Ho KKY. (1996) Neither growth hormone (GH) nor IGF-I regulates serum GH binding protein level in man. In: *10th International Congress of Endocrinology*, Abstract P1–497 Q11.

Jones JI, Clemmons DR. (1995) Insulin-like growth factors and their binding proteins: biological actions. *Endocr. Rev.* **16:** 3–34.

Knapp JR, Kopchick JJ. (1996) Ligand binding kinetics of porcine growth hormone receptor exon 3 variants. In: *Program of the 10th International Congress of Endocrinology*, San Francisco, CA, Abstract P2-350.

Krulich L, Dhariwal APS, McCann SM. (1968) Stimulatory and inhibitory effects of purified hypothalamic extracts on growth hormone release from rat pituitary *in vitro*. *Endocrinology* **83:** 783–790.

Lamberts SWJ. (1988) The role of somatostatin in the regulation of anterior pituitary hormone secretion and the use of its analog in the treatment of human pituitary tumors. *Endocr. Rev.* **9:** 417–436.

LeRoith D. (1991) *Insulin-like Growth Factors: Molecular and Cellular Aspects.* CRC Press, Boca Raton, FL.

LeRoith D, Werner H, Beitner-Johnson D, Roberts CT Jr. (1995) Molecular and cellular aspects of the insulin-like growth factor I receptor. *Endocr. Rev.* **16:** 143–163.

Leung DW, Spencer SA, Cachianes G, Hammonds RG, Collins C, Henzel WJ, Barnard R, Waters MJ, Wood WI. (1987) Growth hormone receptor and serum binding protein: purification, cloning and expression. *Nature* **330:** 537–543.

Leung KC, Markus I, Peters E, Baumbach WR, Ho KKY. (1996) Modulation of binding properties of rat GH binding protein by a high-molecular-weight factor in rat serum. In: *Program of the 10th International Congress of Endocrinology*, San Francisco, CA, Abstract P2-339.

Lim L, Spencer SA, McKay P, Waters MJ. (1990) Regulation of growth hormone (GH) bioactivity by a recombinant human GH-binding protein. *Endocrinology* **127:** 1287–1291.

Lin C, Lin S-C, Chang C-P, Rosenfeld MG. (1992) Pit-1-dependent expression of the receptor for growth hormone releasing factor mediates pituitary cell growth. *Nature* **360:** 765–768.

Lin S-C, Lin CR, Gukovsky I, Lusis AJ, Sawchenko PE, Rosenfeld MG. (1993) Molecular basis of the little mouse phenotype and implications for cell type-specific growth. *Nature* **364:** 208–213.

Liu J, Baker J, Perkins AS, Robertson EJ, Efstratiadis A. (1993) Mice carrying null mutations of the genes encoding insulin-like growth factor I (IGF-I) and type 1 IGF receptor. *Cell* **75:** 59–72.

Lobie PE, Wood TJ, Chen CM, Waters MJ, Norstedt G. (1994) Nuclear translocation and anchorage of the growth hormone receptor. *J. Biol. Chem.* **269:** 31735–31746.

Macaulay VM, Everard MJ, Teale JD, Trott PA, Van Wyk JJ, Smith IE, Millar JL. (1990) Autocrine function for insulin-like growth factor I in human small cell lung cancer cell lines and fresh tumor cells. *Cancer Res.* **50:** 2511–2517.

Malven P. (1993) *Mammalian Neuroendocrinology* CRC Press, Boca Raton, FL.

Mathews LS, Norstedt G, Palmiter RD. (1986) Regulation of insulin-like growth factor I gene expression by growth hormone. *Proc. Natl Acad. Sci. USA* **83:** 9343–9347.

Matsubara S, Sato M, Mizobuchi M, Niimi M, Takahara J. (1995) Differential gene expression of growth hormone (GH)-releasing hormone (GRH) and GRH receptor on various rat tissues. *Endocrinology* **136:** 4147–4150.

Mayo KE. (1992) Molecular cloning and expression of a pituitary-specific receptor for growth hormone-releasing hormone. *Mol. Endocrinol.* **6:** 1734–1744.

Mayo KE, Cerelli GM, Lebo RV, Bruce BD, Rosenfeld MG, Evans RM. (1985a) Gene encoding human growth hormone-releasing factor: structure, sequence, and chromosomal assignment. *Proc. Natl Acad. Sci. USA* **82:** 63–67.

Mayo KE, Cerelli GM, Rosenfeld MG, Evans RM. (1985b) Characterization of cDNA and genomic clones encoding the precursor of rat hypothalamic growth hormone-releasing factor. *Nature* **314:** 464–467.

McCarthy GF, Beaudet A, Tannenbaum GS. (1992) Colocalization of somatostatin receptors and growth hormone-releasing factor immunoreactivity in neurons of the rat arcuate nucleus. *Neuroendocrinology* **56:** 18–24.

Mode A, Norstedt G, Eneroth P, Gustafsson J-A. (1983) Purification of liver feminizing factor from rat pituitaries and demonstration of its identity with growth hormone. *Endocrinology* **113:** 1250–1260.

Norstedt G, Palmiter R. (1984) Secretory rhythm of growth hormone regulates sexual differentiation of mouse liver. *Cell* **36:** 805–812.

Oldham ER, Bingham B, Baumbach WR. (1993) A functional polyadenylation signal is embedded in the coding region of chicken growth hormone receptor RNA. *Mol. Endocrinol.* **7:** 1379–1390.

Ozenberger BA, Hadcock JR. (1995) A single amino acid substitution in somatostatin receptor subtype 5 increases affinity for somatostatin-14. *Mol. Pharmacol.* **47:** 82–87.

Pekhletsky RI, Chernov BK, Rubtsov PM. (1992) Variants of the 5′-untranslated sequence of human growth hormone receptor mRNA. *Mol. Cell. Endocrinol.* **90:** 103–109.

Pong S, Chaung L, Dean D, Nargund R, Patchett A, Smith R. (1996) Identification of a new G protein-linked receptor for growth hormone secretagogues. *Mol. Endocrinol.* **10:** 57–61.

Raynor K, Murphy WA, Coy DH, Taylor JE, Morear JP, Yasuda K, Bell GI, Reisine T. (1993) Cloned somatostatin receptors: identification of subtype-selective peptides and demonstration of high affinity binding of linear peptides. *Mol. Pharmacol.* **43:** 838–844.

Reichlin S. (1982) Somatostatin, parts 1 and 2. *New Engl. J. Med.* **309:** 1495–1501; 1556–1563.

Reisine T, Bell GI. (1995) Molecular biology of somatostatin receptors. *Endocr. Rev.* **16:** 427–442.

Rivier J, Spiess J, Thorner MO, Vale W. (1982) Characterization of a growth hormone-releasing factor from a human pancreatic islet tumour. *Nature* **300:** 276–278.

Robinson ICAF. (1996) Mechanism of action of GHRPs *in vivo*. In: *Program of the 10th International Congress of Endocrinology*, San Francisco, CA, Abstract S10-2.

Rotwein P, Gronowski AM, Thomas MJ. (1994) Rapid nuclear actions of growth hormone. *Hormone Res.* **42:** 170–175.

Rubinow DR. (1986) Cerebrospinal fluid somatostatin and psychiatric illness. *Biol. Psychiatry* **21:** 341–365.

Sadeghi HB, Wang B, Lumanglas AL, Logan JS, Baumbach WR. (1990) Identification of the origin of the growth hormone-binding protein in rat serum. *Mol. Endocrinol.* **4:** 1799–1805.

Salmon WD Jr, Daughaday WH. (1957) A hormonally controlled serum factor which stimulates sulfate incorporation by cartilage in vitro. *J. Lab. Clin. Med.* **49:** 825–836.

Sato M, Frohman LA. (1993) Differential effects of central and peripheral administration of growth hormone (GH) and insulin-like growth factor on hypothalamic GH-releasing hormone and somatostatin gene expression in GH-deficient dwarf rats. *Endocrinology* **133:** 793–799.

Schwartz Y, Yamaguchi H, Goodman HM. (1992) Growth hormone increases intracellular free calcium in rat adipocytes; correlation with actions on carbohydrate metabolism. *Endocrinology* **131:** 772–778.

Smith WC, Kuniyoshi J, Talamantes F. (1989) Mouse serum growth hormone (GH) binding protein has GH receptor extracellular and substituted transmembrane domains. *Mol. Endocrinol.* **3:** 984–990.

Sotiropoulos A, Goujon L, Simonin G, Kelly PA, Postel-Vinay M-C, Finidori J. (1993) Evidence for generation of the growth hormone-binding protein through proteolysis of the growth hormone membrane receptor. *Endocrinology* **132:** 1863–1865.

Souza SC, Frick GP, Wang X, Kopchick JJ, Lobo RB, Goodman HM. (1995) A single arginine residue determines species specificity of the human growth hormone receptor. *Proc. Natl Acad. Sci. USA* **92:** 959–963.

Spencer SA, Hammonds RG, Henzel WJ, Rodriguez H, Waters MJ, Wood WI. (1988) Rabbit liver growth hormone receptor and serum binding protein purification, characterization and sequence. *J. Biol. Chem.* **263:** 7862–7867.

Stewart TA, Clift S, Pitts-Meek S, Martin L, Terrell TG, Liggitt D, Oakley H. (1992) An evaluation of the functions of the 22-kilodalton (kDa), the 20-kDa, and the N-terminal polypeptide forms of human growth hormone using transgenic mice. *Endocrinology* **130:** 405–414.

Strnad J, Eppler CM, Corbett M, Hadcock JR. (1993) The rat SSTR2 somatostatin receptor subtype is coupled to inhibition of cyclic AMP accumulation. *Biochem. Biophys. Res. Commun.* 191: 968–976.

Strobl JS, Thomas MJ. (1994) Human growth hormone. *Pharmacol. Rev.* **46:** 1–23.

Takahashi T, Okimura Y, Mizuno I, Kaji H, Abe H, Chihara K. (1996) Cloning of 5′ flanking region of human growth hormone-releasing hormone receptor gene. In: *Program of the 10th International Congress of Endocrinology*, San Francisco, CA, Abstract P1-621.

Tang J, Lagace G, Castagne J, Collu R. (1995) Identification of human growth hormone-releasing hormone receptor splicing variants. *J. Clin. Endocrinol. Metab.* **80:** 2381–2387.

Urbanek M, MacLeod JN, Cooke NE, Liebhaber SA. (1992) Expression of a human growth hormone (hGH) receptor isoform is predicted by tissue-specific alternative splicing of exon 3 of the hGH receptor gene transcript. *Mol. Endocrinol.* **6:** 279–287.

Waters MJ, Barnard RT, Lobie PE, Lim L, Hamlin G, Spencer SA, Hammonds RG, Leung DW, Wood WI. (1990) Growth hormone receptors – their structure, location and role. *Acta Paediatr. Scand.* **366:** 60–72.

Werner H, Woloschak M, Adamo M, Shen-Orr Z, Roberts CT Jr, LeRoith D. (1989) Developmental regulation of the rat insulin-like growth factor I receptor gene. *Proc. Natl Acad. Sci. USA* **86:** 7451–7455.

Wickelgren RB, Landin KL, Ohlsson C, Carlsson LMS. (1995) Expression of exon 3-containing and exon 3-excluding isoforms of the human growth hormone receptor is regulated in an interindividual, rather than a tissue-specific, manner. *J. Clin. Endocrinol. Metab.* **80:** 2154–2157.

Wilkinson GF, Thurlow RJ, Sellers LA, Coote JE, Feniuk W, Humphrey PPA. (1996) Potent antagonism by BIM-23056 at the human recombinant somatostatin sst_5 receptor. *Br. J. Pharmacol.* **118:** 445–447 Q13.

Yamada Y, Post SR, Wang K, Tager HS, Bell GI, Seino S. (1992) Cloning and functional characterization of a family of human and mouse somatostatin receptors expressed in brain, gastrointestinal tract, and kidney. *Proc. Natl Acad. Sci. USA* **89:** 251–255.

Yee D. (1994) The insulin-like growth factor system as a target in breast cancer. *Breast Cancer Res. Treat.* **32:** 85–95.

Yee D, Paik S, Lebovic GS, Marcus RR, Favoni RE, Cullen DJ, Lippman ME, Rosen N. (1989) Analysis of insulin-like growth factor I gene expression in malignancy: evidence for a paracrine role in human breast cancer. *Mol. Endocrinol.* **3:** 509–517.

Zhou Y, He L, Baumann G, Kopchick JJ. (1996) Deletion of the mouse growth hormone binding protein (mGHBP) poly-adenylation signal(s) does not inhibit mGHBP production. In: *Program of the 10th International Congress of Endocrinology*, San Francisco, CA, Abstract P3-212.

8

Molecular biology of steroid biosynthesis

Gill Rumsby

8.1 Introduction

The biosynthesis of steroid hormones is the result of sequential hydroxylation and isomerization reactions which convert cholesterol to the physiologically active mineralocorticoids, glucocorticoids, androgens and oestrogens. The importance of these hormones in development and homeostasis is illustrated by the spectrum of symptoms arising from inherited disorders of steroidogenesis and encompass renal salt loss and impaired gluconeogenesis to inappropriate development of sexual phenotype.

The major sites of steroidogenesis are the ovaries, testes, adrenal glands and placenta although other organs such as the brain, liver and adipose tissue can play a significant role. Steroidogenesis is stimulated by the peptide hormones adrenocorticotrophic hormone (ACTH) in the adrenal gland, luteinizing hormone (LH) and follicle-stimulating hormone (FSH) in the gonads and human chorionic gonadotrophin (hCG) in placenta in a process mediated by cAMP and protein kinase A. Activators of the protein kinase C pathway can have an inhibitory effect on steroidogenesis (Black *et al.*, 1993; Moore *et al.*, 1990; Reyland, 1993) and the relative activities of these two systems may determine the steroid secretory pattern of a particular tissue.

The aim of this chapter is to familiarize the reader with the molecular genetics of the different enzymes involved in adrenal and gonadal steroidogenesis and to review the current literature on the ontogeny and regulation of these genes.

8.2 The enzymes

The steroid biosynthetic pathway is illustrated in *Figure 8.1* and it can be seen from this figure that some of the enzymes are common to both adrenal and gonadal tissues while the others are restricted to a specific tissue. Six of the enzymes are members of the cytochrome P450 group of mixed function oxidases. The nomenclature used for these enzymes and their genes in this chapter follows that proposed by

Molecular Endocrinology, edited by G. Rumsby and S.M. Farrow.

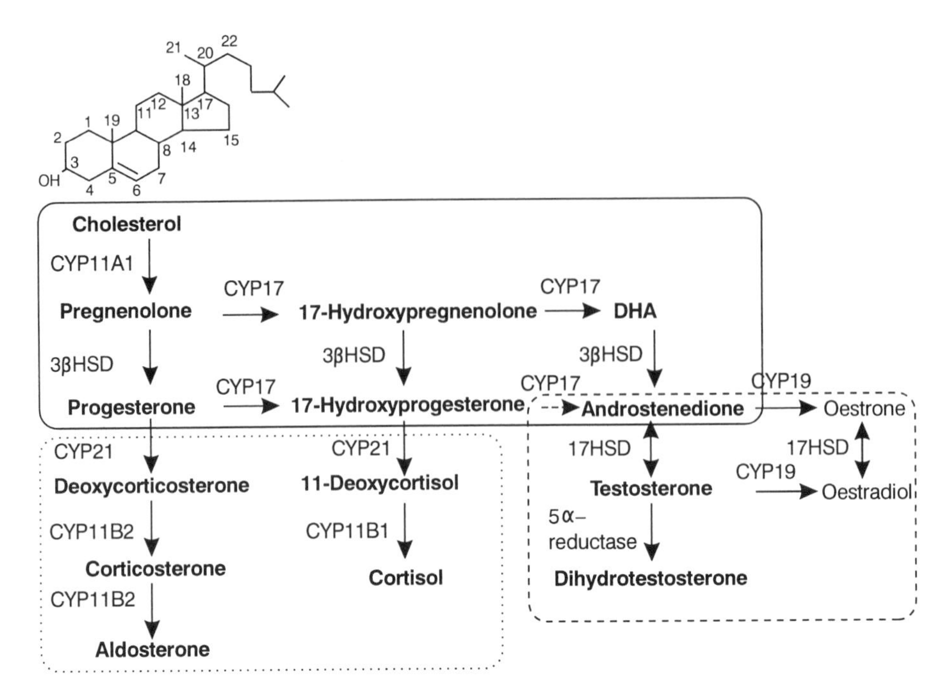

Figure 8.1. Adrenal and gonadal steroid biosynthesis. Pathways common to both organs are bordered by a solid line, those restricted to the adrenal glands are bordered by a dotted line and those which occur essentially in the gonads are bordered by a dashed line. The structure of cholesterol is shown with carbon atoms numbered. See text for nomenclature.

Nelson *et al*., (1993) in which the root CYP is used to denote human cytochrome P450 followed by arabic numerals for the family, for example CYP21 for 21-hydroxylase, the root *CYP* in italics denotes the gene. Cholesterol side-chain cleavage enzyme (CYP11A1), 11β-hydroxylase (CYP11B1) and aldosterone synthase (CYP11B2) are mitochondrial enzymes which accept electrons from the electron transport proteins ferredoxin and ferredoxin reductase (also known as adrenodoxin and adrenodoxin reductase, respectively). The microsomal enzymes 17α-hydroxylase (CYP17), 21-hydroxylase (CYP21) and aromatase (CYP19) obtain reducing equivalents from NADPH via the ubiquitous steroid P450 reductase and cytochrome b_5. The non-P450 enzymes are 3β-hydroxysteroid dehydrogenase (3βHSD), 17β-hydroxysteroid dehydrogenase (17HSD) and 5α-reductase.

8.2.1 Cholesterol side-chain cleavage enzyme (CYP11A1)

Cholesterol is converted to pregnenolone in three sequential steps, 20α-hydroxylation, 22-hydroxylation and C20,22 bond scission, catalysed by CYP11A1 on a single active site (Lambeth and Pember, 1983). This reaction is the rate-limiting step in steroidogenesis but the availability of cholesterol within the mitochondria is actually the limiting factor *in vivo* (Black *et al*., 1993). The human enzyme is synthesized as a peptide of 521 amino acids which includes an N-terminal mitochondrial targeting sequence. Once inside the mitochondria the enzyme is tightly bound to the inner mitochondrial membrane where it is in close contact with ferredoxin and ferredoxin reductase which are loosely associated with the inner mitochondrial membrane.

Ferredoxin functions as a mobile electron carrier from NADPH adrenodoxin reductase to CYP11A1 (Lambeth and Pember, 1983). Structural analysis of ferredoxin has identified the amino acids D76 and D79 as important for interaction of the protein with ferredoxin reductase and D72, E73, D76 and D79 for interaction with CYP11A1 indicating that the binding sites for the two proteins overlap but are not identical (Coghlan and Vickery, 1991). The interacting region of the CYP11A1 protein contains a number of lysine residues with those at codons 73, 109, 110, 126 and 148 probably the most important since these are conserved between the human and bovine protein sequences (Adamovich *et al.*, 1989).

The single human gene was mapped by somatic cell hybrids to the long arm of chromosome 15 (Chung *et al.*, 1986) at 15q23–24 (Sparkes *et al.*,1991). The adrenal *CYP11A1* cDNA is about 1850 bp in length and identical to that from the testis (Chung *et al.*, 1986). The nucleotide and amino acid sequences share 82% and 72% homology with the bovine cDNA and protein, respectively, the nucleotide differences clustering in a non-random fashion (Chung *et al.*, 1986). The gene is expressed in testis, placenta, adrenal, ovary (Voutilainen and Miller, 1986) and brain (Mellon and Deschepper, 1993) but expression is under the control of different *trans*-acting factors in the various tissues (see Section 8.3.3).

8.2.2 3β-Hydroxysteroid dehydrogenase/isomerase (3βHSD)

This family of enzymes catalyses the transformation of all 5-ene-3β-hydroxysteroids (that is with a double bond between carbons 5 and 6, the Δ^5 steroids) into 4-ene-3-oxosteroids (double bond between carbons 4 and 5, the Δ^4). The human enzymes prefer the co-factor NAD^+ to $NADP^+$ and are bifunctional, the NADH inducing a conformational change in the enzyme molecule which mediates the sequential reaction of dehydrogenation followed by isomerization (Thomas *et al.*, 1995), the two events possibly occurring at separate sites within the protein molecule (Luu-The *et al.*, 1991). The enzyme from bovine adrenal gland appeared to form a complex with CYP11A1, antibodies to either enzyme co-precipitating the other (Cherradi *et al.*, 1995).

Two isoenzymes have been identified in the human and at least four in the rat. The 3βHSDI enzyme is found in placenta and skin (Lorence *et al.*, 1990a) and 3βHSDII in the adrenals and gonads (Lachance *et al.*, 1991). The nomenclature as it now stands is rather confusing as the same enzyme in different species may have a different number (for a review see Mason, 1993).

Placental 3βHSD (type I) was the first of the human enzymes to be isolated and sequenced (Lorence *et al.*, 1990a; Luu-The *et al.*, 1989a). The 1.7 kb cDNA was used to screen a human EMBL3 genomic library and a genomic clone of approximately 8 kb was isolated (Lorence *et al.*, 1990a). This gene was subsequently mapped to chromosome 1p13.1 by *in situ* hybridization (Berube *et al.*, 1989) where it lies in tandem with the type II gene isolated from adrenal and gonadal tissue (Lachance *et al.*, 1991). The two genes share significant homology with one another and are located within a

0.29 megabase *Sac*II DNA fragment, closely linked to the D1S514 polymorphic marker (Morissette *et al.*, 1995). Analysis of genomic DNA by hybridization with exon-specific probes has identified several other sequences in the genome which may reflect the presence of other isoenzymes or pseudogenes (McBride *et al.*, 1995) and three of the latter have been found on chromosome 1. Both the type I and II genes have an identical intron–exon arrangement of four exons, of which the first one and part of the second are non-coding (*Figure 8.2*). Minor differences in the nucleotide sequence around the transcription start site of the type II gene lead to the shift of the ATG translational start codon of the type II gene and the loss of an amino acid; hence, the protein is 371 amino acids long compared to 372 amino acids in the type I protein but overall the proteins display 93.5% homology. In expression studies the type I enzyme possessed higher 3βHSD/Δ4–5 isomerase activity with K_m values for pregnenolone of 0.24 vs. 1.2 μM (Rheaume *et al.*, 1991). Thus, the type I enzyme can efficiently convert the lower concentrations of substrates present in non-steroidal tissues. Both enzymes were able to catalyse the conversion of pregnenolone, 17-hydroxypregnenolone and dehydroepiandrosterone but not the reverse reaction presumably because the isomerization reaction is irreversible. The enzymes could also catalyse the conversion of 5α-dihydrotestosterone to 5α-androstan-3β,17β-diol (Lorence *et al.*, 1990b; Rheaume *et al.*, 1991) in the presence of excess NADH and also, by reversal of the dehydrogenation reaction, of 5α-androstan-3β,17β-diol to 5α-dihydrotestosterone (Lorence *et al.*, 1990b).

Confirmation of the role of the type II enzyme in adrenal and gonadal steroidogenesis came from the identification of mutations in this gene from patients with congenital adrenal hyperplasia due to 3β-hydroxysteroid dehydrogenase deficiency (Rheaume *et al.*, 1992, Simard *et al.*, 1993a) while no mutation was observed in the type I gene (Simard *et al.*, 1993a). These patients typically have salt losing crises and variable ambiguity of the external genitalia. While an elevated ratio of Δ^5 to Δ^4 steroids is generally considered to be the best basis for diagnosis of this

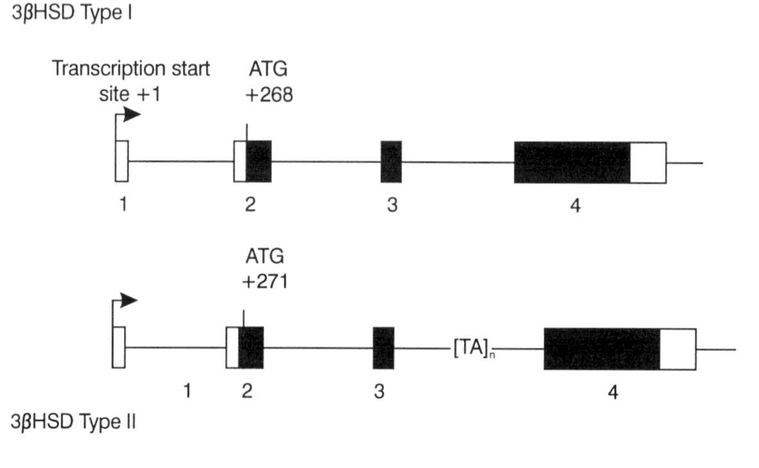

Figure 8.2. Comparison of gene structure of the type I and II 3βHSD genes. Arrow head denotes transcription start site. Solid rectangles represent coding and open rectangles non-coding exons. $[TA]_n$ denotes TA-rich sequence in third intron of the type II gene.

disorder, there is a paradoxical increase in some of the Δ^4 steroids such as 17-hydroxyprogesterone; presumably the result of conversion of adrenal or gonadal steroids in peripheral tissues by 3βHSD type I. The study of naturally occurring mutations has demonstrated the importance of E142, Y253 (Simard *et al.*,1993a) and Y254 (Sanchez *et al.*, 1994) for catalytic activity. The three amino acids are conserved in all the 3βHSD isoenzymes studied to date (Simard *et al.*, 1993a) and are associated with the substrate-binding site (Thomas *et al.*, 1993).

The 3βHSD type II has a TA-rich region of 179 nucleotides in the third intron which is absent in the type I gene. This type of sequence is prone to the formation of hairpin-like structures which may be involved in transcription termination (Ryan and Chamberlin, 1983) hence explaining the generation of multiple products from reverse transcriptase–polymerase chain reaction (RT–PCR) ending at different positions in this region (Lachance *et al.*, 1991) although the significance of such terminations *in vivo* is unclear.

8.2.3 17α-Hydroxylase (CYP17)

CYP17 catalyses 17α-hydroxylation and side-chain cleavage (17,20-lyase activity) at carbon 17 of the steroid molecule and is essential for both glucocorticoid and sex steroid production but not for mineralocorticoid biosynthesis. The same gene is expressed in adrenals and gonads (Chung *et al.*, 1987) but expression in these tissues is restricted to the testicular Leydig cells, ovarian theca and adrenal zona fasciculata and reticularis; the mRNA has not been found in the adrenal zona glomerulosa, ovarian granulosa and placenta (Voutilainen *et al.*, 1986). The preferred pathway of human adrenal steroid biosynthesis is via the Δ^5 steroid 17-hydroxypregnenolone to 17-hydroxyprogesterone and cortisol or to dehydroepiandrosterone, reflecting the lower intrinsic 3βHSD activity in human adrenal cells (Hornsby and Aldern, 1984) and also the preference of the human CYP17 enzyme for the Δ^5 steroids. The rat CYP17, which is able to use 17-hydroxyprogesterone, a Δ^4 steroid, as a substrate has a single residue, F343, which when mutated to the corresponding residue found in the bovine or human protein destroyed the ability of the enzyme to utilize Δ^4 substrates (Koh *et al.*, 1993).

A number of factors determine whether a steroid molecule will remain on the single active site of CYP17 and undergo cleavage of the 17,20 bond to produce androgens (*Figure 8.1*). These include the availability of reducing equivalents and the amount of P450 reductase (Lin *et al.*, 1993a) and cytochrome b_5 (Fisher *et al.*, 1992). There is also a suggestion that phosphorylation of serine and threonine residues may play a part in the augmentation of lyase activity. However, the study of naturally occurring mutations has failed to find one which affects lyase activity alone, although site-directed mutagenesis studies of the human enzyme found that mutation of arginine 347 could create a protein with 79% of hydroxylase activity but only 7% lyase activity (Kitamura *et al.*, 1991). The amount of CYP21 present does not appear to be a factor as removal of CYP21 by binding to an antibody made no difference to the amount of lyase activity observed (Yanagibashi and Hall, 1986). Some other more subtle factors must play a role as lyase activity is low in the prepubertal child but increases with adrenarche. The levels of dehydroepiandrosterone synthesized (and presumably of lyase activity) fall with stress, a finding which cannot be accounted for solely by the increase in cortisol production.

The gene for CYP17, *CYP17*, has been mapped to chromosome 10q24–25 (Matteson *et al.*, 1986; Sparkes *et al.*, 1991) and contains eight exons spanning 6.6 kb. The 1.7 kb mRNA encodes an enzyme of 508 amino acids and has a molecular weight of approximately 57 kDa. The cysteine residue involved in haem binding, a feature of the CYP family of enzymes, is located at codon 441.

Few polymorphisms have been identified in the gene although one in the 5′ promoter region introduced a putative Spl transcription factor-binding site (Carey *et al.*, 1994). This polymorphism has not been tested for its effect on gene transcription *in vitro* but did not appear to have an effect in terms of androgen production *in vivo* (Techatraisak *et al.*, 1996).

8.2.4 21-Hydroxylase (CYP21)

CYP21 is a member of the same gene family as *CYP17* sharing a similar intron/exon structure and has 10 exons spanning approximately 3.2 kb (White *et al.*, 1986) on chromosome 6p21.3 within the HLA class III region. Unlike *CYP17*, expression of CYP21 is restricted to the adrenal gland but is found in both zona glomerulosa and fasciculata (Voutilainen and Miller, 1986) where it catalyses the conversion of progesterone and 17-hydroxyprogesterone, respectively. The gene is located adjacent to the 3′ end of the gene for complement 4B (Carroll *et al.*, 1985) and some 30 kb from the pseudogene *CYP21P*, a gene arrangement which may have arisen as the result of an ancestral duplication. The pseudogene shares approximately 98% sequence homology with *CYP21* (Higashi *et al.*, 1986; Rodrigues *et al.*, 1987; White *et al.*, 1986) but has several point mutations and a minor deletion which render *CYP21P* inactive in terms of 21-hydroxylase activity. Although no mRNA transcripts from *CYP21P* have been identified in adrenal mRNA by Northern blot analysis, there is a suggestion that the promoter may have weak activity as a recombination event replacing the *CYP21* promoter with that of the pseudogene produced an enzyme with partial activity as determined *in vivo* (Killeen *et al.*, 1991). Deficiency of CYP21 is the most common cause of congenital adrenal hyperplasia and recombinations and gene conversion between *CYP21* and the pseudogene account for more than 90% of disease pathology at this locus (for a review see Miller, 1994). Unequal cross-over occurring during recombination can cause deletions and duplications and it is not unusual to find individuals with multiple copies of C4 and *CYP21P* alleles; up to five have been described although the number of copies of *CYP21* stays relatively constant at two per individual analysed. Deletions of *CYP21* are characterized by the 5′ end of the pseudogene fused to the 3′ end of *CYP21* creating a hybrid gene (Rumsby *et al.*, 1986, White *et al.*, 1988). Other deletions lead to loss of the pseudogene and the C4A gene (Garlepp *et al.*, 1986). Analysis of paired leukocyte and sperm samples from normal individuals indicated that recombination events leading to deletions/duplications occurred only during meiosis but gene conversion, in which part of the *CYP21* sequence is replaced by that of the pseudogene, could also take place during mitotic division in somatic cells and gametes (Tusie-Luna and White, 1995). This exchange of genetic information seems to be a two-way process with evidence for *CYP21* sequences in the pseudogene (Higashi *et al.*, 1988).

To date a deletion has never been described in which the 3′ end of the *CYP21* gene is lost. This finding may reflect the fact that another pair of duplicated genes are found in the same region of chromosome 6 but encoded by the opposite strand. These genes, XA and XB (Miller *et al.*, 1992) overlap the 3′ ends of *CYP21P* and *CYP21*, respectively. Deletions of XA occur in 14% of alleles (along with *CYP21P*) but deletions affecting *CYP21* do not extend into the XB domain suggesting that this gene may be essential for life. The XB gene produces two transcripts of 3.5 and 1.8 kb in adrenal and Leydig cell tumour tissue (Morel *et al.*, 1989), the protein product of these transcripts resembling the extracellular matrix protein, tenascin.

Binding motifs for a number of nuclear binding proteins have been identified in the promoter region of *CYP21* which will be discussed later (see Section 8.3.2).

8.2.5 11β-Hydroxylase (CYP11B1)

The final steps in cortisol and aldosterone biosynthesis are catalysed by two enzymes, 11β-hydroxylase (CYP11B1) and aldosterone synthase (CYP11B2), respectively; these are described in Chapter 9 and will therefore not be discussed further here.

8.2.6 17β-Hydroxysteroid dehydrogenase (17HSD)

The human 17HSD gene family has at least four members, numbered 1 to 4 in order of their isolation, which differ in their tissue distribution and the type of substrate preferred. The proteins are all members of the short-chain alcohol dehydrogenase (SCAD) superfamily but have limited homology to each other, for example the 17HSD2 enzyme has a greater similarity to D-β-hydroxybutyrate dehydrogenase than to 17HSD1 (Krozowski, 1994). The enzymes catalyse the reduction of steroids to active metabolites in the gonads, but an oxidative process to inactive steroid metabolites is favoured in peripheral tissues. The characteristics of the enzyme types are shown in *Table 8.1*.

The type 1 enzyme is expressed in the cytosol of a variety of steroidogenic and peripheral tissues and has a 100-fold greater affinity for oestrogens than for the C19 androgens. The protein was originally purified from placenta and the gene encoding it mapped to chromosome 17q11–21 (Luu-The *et al.*, 1989b), the region associated with familial breast cancer, which made it an early candidate for this disease before the identification of the BRCAI gene (Miki *et al.*, 1994). The gene encoding the placental enzyme, HSD17BP1 (originally EDH17B2), is 3.2 kb long, contains six exons and is located in tandem with a second gene, HSD17BP1, with which it has 89% overall similarity (Peltoketo *et al.*, 1992), although the function of the latter is currently unknown. There are two transcription start points for HSD17BP1 and mRNA of 1.3 and 2.3 kb are produced (Luu-The *et al.*, 1989b). The smaller species, found in steroidogenic cells from the placenta and ovary, is related to the enzyme content of these tissues (Lewintre *et al.*, 1994; Poutanen *et al.*, 1992) whereas the 2.3 kb mRNA has been found in all tissues but is not related

Table 8.1. Characteristics of the human isoenzymes of 17HSD

Type	Gene	Chromosome	Exons	Protein	Tissue distribution	Preferred substrate
1	HSD17B1 HSD17BP1(pseudogene)	17q21	6	327 aa	Wide	Oestradiol
2	HSD17B2	16q24.1–24.2	7	387 aa	Placental	E1 ↔E2 Δ^4↔testosterone Δ^4↔DHT
3	HSD17B3	9q22	11	310 aa	Testis	Δ^4↔testosterone

aa, amino acids.

to the presence of 17HSD enzyme activity. At least five sequence polymorphisms have been described in HSD17B1, four of which have no apparent effect on enzyme activity. Three of the five changes, two in the non-coding exon 1 and one in exon 6, are in linkage disequilibrium (Peltoketo *et al.*, 1994, Simard *et al.*, 1993b). A polymorphism involving a transversion from adenine to cytosine at position -27 in the putative TATA box of HSD17B1 reduced promoter activity by approximately 45% *in vitro* (Peltoketo *et al.*, 1994).

The 17HSD1 protein exists as a dimer of two identical 34 kDa subunits (Lin *et al.*, 1992) and contains the five domains common to members of the SCAD family in which the A domain is thought to interact with the co-factor and the D domain is part of the active site (Krozowski, 1994). Three highly conserved amino acids, serine 142, tyrosine 155 and lysine 159, are located in the active site of the enzyme (Krozowski, 1992). The enzyme is expressed in placenta and granulosa cells of developing follicles and granulosa–lutein cells of the corpus luteum (Sawetawan *et al.*, 1994). The reaction catalysed by 17HSD1 is reversible but favours the reduction of oestrone to oestradiol. Expression is regulated by cAMP and protein kinase C pathways in addition to modulation by progestins during the menstrual cycle. Experimental studies however have shown that cAMP can both stimulate and downregulate the enzyme suggesting that the effect is dependent on the cell type.

The type 2 enzyme is predominantly localized in the placental microsomes but is also found in the endometrium during the late phase of the cycle. The enzyme has equal affinity for androgens and oestrogens and catalyses the interconversion of androstenedione–testosterone and oestrone–oestradiol but generally favours oxidation reactions. It has also been shown to have progesterone 20α-hydroxylase activity converting 20α-progesterone to progesterone in the placenta and endometrium (Wu *et al.*, 1993). The gene, which has been mapped to chromosome 16q24 (Casey *et al.*, 1994; Durocher *et al.*, 1995), contains seven exons and spans more than 40 kb. Two different transcripts have been identified: the major 1.45 kb transcript present in placenta encodes a protein of 387 amino acids (17HSD2A) including a 60-amino-acid hydrophobic terminus characteristic of a transmembrane signal anchor (Wu *et al.*, 1993). A second transcript was also identified from a placental library which was identical to the first except for a 109 bp insertion between nucleotide 747 and 748 (Labrie *et al.*, 1995). Analysis of genomic DNA revealed that the 1.45 kb transcript was derived from exons 1 to 3 and 6 to 7 whereas the alternative form, designated 17HSD2B, included the sequence of exons 4 and 5 (Labrie *et al.*, 1995). The sequence of this second transcript predicts a protein of 291amino acids as a result of a shift in the translational reading frame and has the N-terminal signal–anchor motif associated with membrane anchoring but lacks the C-terminal motif, Lys-Lys-Lys, associated with retention of the 17HSD2A protein in the lumen of the endoplasmic reticulum (Labrie *et al.*, 1995). More importantly, the protein lacks the consensus sequence common to all members of the SCAD superfamily. 17HSD2B sequences were not detected by Northern blot analysis of endometrial mRNA but were detected by RT–PCR of the same tissue. The functional significance of this second transcript, if any, remains to be revealed. A number of potential regulatory elements have been identified in the promoter region of the type 2 gene but await functional analysis.

The type 3 enzyme is restricted to the testis and deficiencies of this enzyme are associated with male pseudohermaphroditism. This microsomal enzyme favours the reduction of androstenedione but can also utilize oestrone and dehydroepiandrosterone as substrates. The gene has 11 exons spanning more than 60 kb on chromosome 9q22 and encodes a protein of 310 amino acids (Geissler *et al.*, 1994). A number of mutations have been identified in the 17HSD3 gene in DNA from patients with 17β-hydroxysteroid dehydrogenase deficiency (Geissler *et al.*, 1994) confirming the role of this enzyme in testosterone biosynthesis. However, the high degree of genital ambiguity exhibited by patients with complete deficiency of the type 3 enzyme suggests that the high circulating levels of androstenedione produced by such individuals *in utero* might be converted to testosterone by peripheral enzymes such as the type 1 form thus allowing a variable degree of Wolffian duct differentiation. The alternative explanation for *in utero* virilization in these patients is to propose that the fetal androgen receptor has different affinities for androgenic ligands but there is currently no evidence to support this hypothesis. At puberty, patients with 17HSD3 deficiency undergo virilization as a result of the peripheral conversion of androstenedione to testosterone.

8.2.7 5α-Reductase

The conversion of testosterone to the potent androgen 5α-dihydrotestosterone is catalysed by the enzyme 5α-reductase with NADPH as cofactor. The importance of this enzyme in male development is demonstrated by the impaired development of the external genitalia in genetic males with 5α-reductase deficiency (for a review see Randall, 1994). The enzyme activity has been located in a wide variety of tissues including the liver and prostate (Andersson *et al.*, 1989), sweat and sebaceous glands (Luu-The *et al.*, 1994) and genital skin fibroblasts (Moore *et al.*, 1975) but, because the protein is membrane-bound and of fairly low activity, it has proved difficult to purify. There are however at least two isoenzymes, the one from genital skin fibroblasts and prostate showing a pH optimum around 5.5, and that from scalp skin with a broader activity profile in the alkaline pH. A cDNA from rat liver was first identified (Andersson *et al.*, 1989) which was used to screen a human prostate cDNA library. The gene identified, 5α-reductase 1 (SRD5A1), when transiently expressed in Cos cells had greatest activity at physiological pH and not at the acidic pH optimum expected. A second 5α-reductase gene was subsequently isolated from the prostate library which had the expected acidic pH optimum. This gene, 5α-reductase 2 (SRD5A2), encodes a hydrophobic protein of 254 amino acids with 50% homology to the type 1 form (Andersson *et al.*, 1991). The two enzymes differ in pH optima, substrate affinity and response to finasteride, which inhibits the type 2 enzyme but not 5α-reductase 1 (Jenkins *et al.*, 1992).

Comparison of the SRD5A1 and SRD5A2 sequences show that both genes have five exons and identical splice sites, but the length of intronic sequence differs and the exons share from 43.8% to 64.1% homology with each other (Labrie *et al.*, 1992). SRD5A1 has been mapped to chromosome 5p and an intronless pseudogene (SRD5AP1) to Xq24-qter (Jenkins *et al.*, 1991) while a combination of fluorescence *in situ* hybridization and somatic cell hybrids mapped SRD5A2 to chromosome 2p23 (Thigpen *et al.*, 1992). Deletions and mutations in the 5α-reductase 2 gene have been found in patients with 5α-reductase deficiency (Andersson *et al.*, 1991; Thigpen *et al.*, 1992) confirming the role of this gene in male development.

The study of naturally occurring mutations in the SRD5A2 gene identified two mutations, G196S and G34R, which affected the affinity of the enzyme for the two substrates, testosterone and NADPH. The protein containing G196S had a decreased affinity for the co-factor while G34R had a K_m for testosterone 15-fold greater than the normal enzyme (Thigpen *et al.*, 1992). Nearly all naturally occurring mutations were found to alter the pH optimum of the enzyme to a more basic pH similar to that of the type I enzyme (Thigpen *et al.*, 1992; Wigley *et al.*, 1994).

The role of the SRD5A2 gene in prostate development and growth has made it a candidate gene for prostate cancer and the search is on for polymorphic variants which may explain differences in the incidence of prostate cancer in certain ethnic groups. The distribution of a polymorphism in the 3′-untranslated region of the gene, which consists of variable numbers of [TA] repeats (Davis and Russell, 1993), has been determined in low-risk Asian–Americans and high-risk African–Americans with the finding of a particular group of alleles restricted to the African–Americans (Reichardt *et al.*, 1995). Whether these allelic differences are significant in terms of prostate cancer remains to be tested.

5α-Reductase is also implicated in the development of other androgen-dependent disorders such as acne and idiopathic hirsutism. The 5α-reductase cDNA was cloned from keratinocytes and fibroblasts and found to be identical to the type 1 gene (Luu-The *et al.*, 1994) with a small amount of the type 2 gene product detected by RT–PCR. The expression of the type 1 enzyme in the epidermis, sweat and sebaceous glands (Luu-The *et al.*, 1994) along with 3βHSD and 17HSD (Dumont *et al.*, 1992) suggests that androgen production in the skin from precursors such as dehydroepiandrosterone could exert a major paracrine effect on neighbouring cells via androgen receptors.

8.2.8 Aromatase (CYP19)

Conversion of the C19 steroids to oestrogens is catalysed by the aromatase enzyme which can use a variety of substrates including 16α-hydroxylated androgens, androstenedione and testosterone to produce the oestrogens oestriol, oestrone and oestradiol, the type of oestrogen produced being dependent on the substrate available in a particular tissue. CYP19 is encoded by a single gene which has been localized to chromosome 15p21.1 (Chen *et al.*, 1988) and is expressed in a variety of tissues including ovarian granulosa cells (McNatty *et al.*, 1976), placenta (Ryan, 1959), testicular Sertoli (Fritz *et al.*, 1976) and Leydig cells (Valladares and Payne, 1979), and adipose tissue (Grodin *et al.*, 1973). The nine coding exons, denoted II to X, span 36 kb of genomic DNA with the ATG initiation codon in exon II (Harada *et al.*, 1990; Means *et al.*, 1989; Toda *et al.*, 1990). CYP19 transcripts from various tissues have different 5′-termini as a result of transcription from different promoters and the presence of untranslated exons (*Figure 8.3*). However, as a result of splicing of the untranslated exons at a common 3′ splice junction upstream of exon II, the mature mRNA encodes the same protein in all tissues.

The six untranslated exons are denoted I.1 to I.5 and 2a (*Figure 8.3*), I.1 being approximately 35 kb upstream of exon II. The majority of placental transcripts contain exon I.1 (Means *et al.*, 1989) with a less abundant population containing I.2. In contrast, CYP19 from ovarian corpus luteum (Jenkins *et al.* 1993) was found to be transcribed using a promoter at the 5′ end of exon II while the type of promoter used in adipose tissue seems to be dependent on the regional location with exon I.4-specific sequences observed in breast and I.3-specific sequences in upper and lower body fat (Mahendroo *et al.*, 1993). What mechanism regulates the promoter selected is unknown although it is possible to change the promoter *in vitro* by altering the culture conditions. For example, adipose stromal cells cultured in the presence of dibutyryl cAMP but without fetal calf serum utilize promoter II and transcripts with exon I.4 are absent. However, when the same cells are maintained in dexamethasone and fetal calf serum the converse applies (Mahendroo *et al.*, 1993).

8.3 Regulation of steroidogenesis

8.3.1 Regulation of cholesterol uptake

Cholesterol can be synthesized from acetate in all steroidogenic tissues but the favoured route is to obtain the steroid from dietary sources via low density

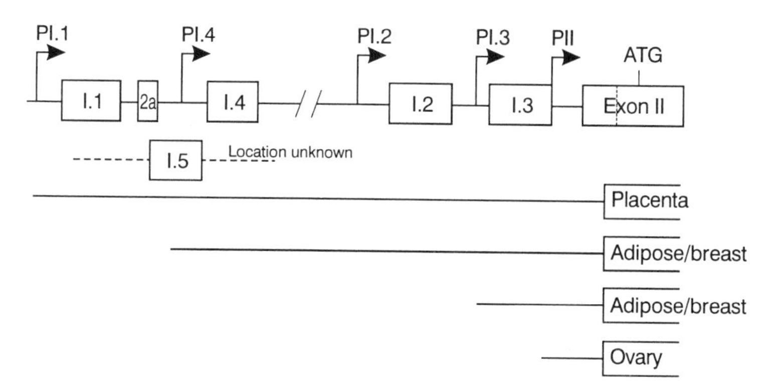

Figure 8.3. The aromatase promoter region and tissue-specific transcripts. The position of exon I.5 is unknown. P, promoter.

lipoprotein (LDL) (reviewed in Brown *et al.*, 1982). The cholesterol esters in LDL are hydrolysed by cholesterol ester hydrolase and the free cholesterol in the cytosol is then available for uptake into the mitochondria by a process which requires association with a carrier protein. A number of candidates have been proposed for this role including the steroidogenesis activator peptide, a 30-amino-acid peptide isolated from a rat Leydig cell tumour (Pedersen and Brownie, 1987) whose synthesis had the necessary prerequisites of induction by cAMP and inhibition by cycloheximide. However, it is not clear how a relatively hydrophilic protein would interact with an enzyme complex in the mitochondrial membrane.

Sterol carrier protein 2 (SCP2) is a 13.2 kDa protein which enhances the movement of cholesterol from lipid droplets to mitochondria (Chanderbhan *et al.*, 1982). Co-expression of SCP2 with CYP11A and ferredoxin in Cos cells stimulated pregnenolone synthesis over that seen with expression of the steroidogenic enzymes alone (Yahamoto *et al.*, 1991). The protein is abundant in steroidogenic tissues and antibodies to SCP2 have been shown to reduce steroid synthesis (Chanderbhan *et al.*, 1986). It is, however, slow to respond to cAMP stimulation (Trzeciak *et al.*, 1987) suggesting that it is not the primary factor responding to acute trophic hormone stimulation.

A third candidate, the mitochondrial benzodiazepine receptor (PBR), is found in a number of cells including steroidogenic tissues (Lin *et al.*, 1993b). The receptor has three components, an 18 kDa subunit which is located on the outer mitochondrial membrane and which binds benzodiazepine, a voltage-dependent anion channel in the outer mitochondrial membrane and an adenine nucleotide carrier which is found on the inner mitochondrial membrane. Binding of ligand to the receptor facilitates cholesterol transport into the mitochondria (Papadopoulos, 1993) perhaps by the formation of contact sites between the two membranes. However, PBR responds rather slowly to ACTH stimulation (Cavallaro *et al.*, 1993) suggesting that it is unlikely to be involved in the acute regulation of steroidogenesis which takes place within minutes.

The fourth candidate for cholesterol transporter is a 30 kDa hormonally induced protein, the steroidogenic acute regulatory protein (StAR) (*Figure 8.4*), which was isolated from mouse MA10 Leydig cells (Clark *et al.*, 1994) and exhibited the required characteristics of rapid cAMP inducibility and cycloheximide sensitivity). StAR is synthesized as a 37 kDa precursor containing a mitochondrial targeting sequence which is subsequently cleaved to produce the mature 30 kDa form found in the mitochondrial matrix (Gradi *et al.*, 1995). The mouse (Clark *et al.*, 1994) and human (Sugawara *et al.*, 1995a) cDNAs have been isolated and their predicted protein sequences share 87% homology.

The human StAR gene has been mapped to chromosome 8p11.2 (Sugawara *et al.*, 1995a) and contains seven exons spanning approximately 8 kb (Sugawara *et al.*, 1995b) with a pseudogene on chromosome 13 (Sugawara *et al.*, 1995a). Northern blot analysis showed that the gene is expressed in ovary, testis and kidney but not in placenta or brain, and encodes a major 1.6 kb transcript with less abundant transcripts of 4.4 and 7.5 kb in ovary and testis, respectively (Sugawara *et al.*, 1995a). StAR expression increased 3-fold following 1–2 h treatment of human granulosa cells with dibutyryl cAMP and induced pregnenolone secretion by Cos cells in the presence of CYP11A (Sugawara *et al.*, 1995a). Final confirmation of the

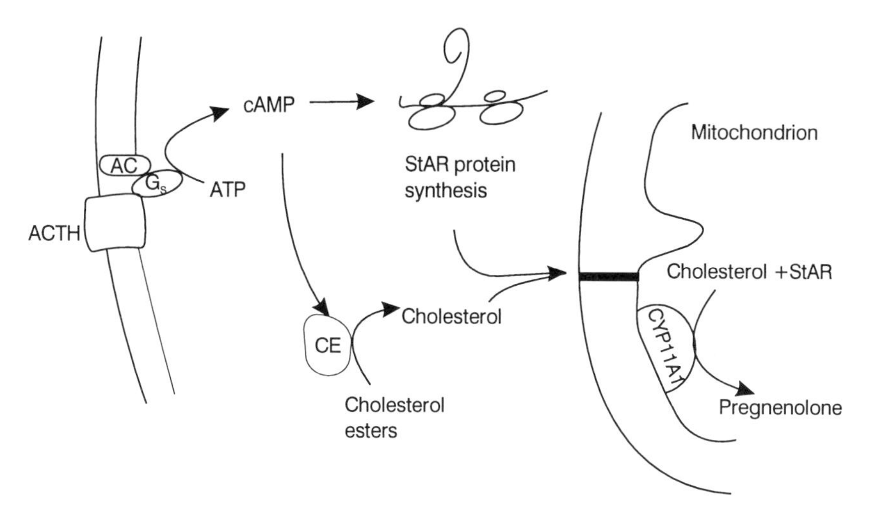

Figure 8.4. Stimulation of cholesterol uptake into mitochondria in response to acute ACTH stimulation. The response to ACTH is mediated by cAMP which induces translation of StAR and hydrolysis of cholesterol esters by cholesterol esterase (CE). Cholesterol is taken into the mitochondrion in a StAR-dependent process which is as yet undefined. AC, adenylate cyclase; G_s, G-protein.

role of this protein in adrenal and gonadal steroidogenesis came from the identification of mutations in the StAR gene from patients with congenital lipoid adrenal hyperplasia, a condition characterized by failure to produce any adrenal steroids in response to ACTH stimulation. Analysis of testicular mRNA from two patients identified premature stop codons which would create proteins truncated at 93 and 28 amino acids, respectively, from the C-terminal end of the protein (Lin *et al.*, 1995). Both mutant genes were completely inactive in a Cos-1 cell expression system although pregnenolone could be synthesized if 20α-hydroxycholesterol were used as substrate. This steroid can diffuse into the mitochondria, bypassing the cholesterol transport system entirely, and thereby confirms the role of the StAR protein in the transport of cholesterol. The initial rapid response of steroidogenesis to cAMP did not require transcription implying that the StAR protein may be translated from pre-existing mRNA (Stocco and Clark, 1996).

8.3.2 Regulation of transcription

Transcription is the result of the interaction of *trans*-acting factors with elements in the promoter region of genes (for description see Chapter 1). Basal transcription elements, for example TATA box sequences, have been identified in almost all of the 5′ promoter regions of steroidogenic enzymes studied to date, one exception being the CYP19 I.4 promoter (see Section 8.2.8). Transcription of these genes is induced by chronic stimulation with trophic hormones in a process mediated by increased intracellular cAMP. With such a co-ordinated response one might expect the genes to have a common system of activation but this is not the case. The classical cAMP response element (CRE) which has the sequence

TGACGTCA, binds a protein known as the CRE-binding protein (CREB) whose synthesis is induced by cAMP. Visual inspection of the DNA sequence can be used to identify putative regulatory sequences but the functional relevance of these sequences must be tested by expression studies. Thus, the CRE has been found in *CYP11B1* (Hashimoto *et al.*, 1992), *CYP11A1* (Watanabe *et al.*, 1994) and *CYP17* promoters but is non-functional in the latter. Other *cis* elements which imply cAMP responsiveness were identified by deletion analysis of promoter regions followed by isolation and identification of the *trans*-acting binding proteins which interact with these sequences. Interestingly, the DNA sequence conferring cAMP responsiveness is not necessarily the same in different tissues as shown in *Figure 8.5* for the *CYP11A1* gene.

One of the more exciting developments in this field was the identification of a nuclear receptor which binds to promoter elements on each steroid hydroxylase to mediate their co-ordinate expression. This protein, steroidogenic factor 1 (SF-1, also known as adrenal 4 binding protein, Ad4BP) is the product of the FTZ-F1 gene mapped to human chromosome 9q33 (Taketo *et al.*, 1995) and is a homologue of the *Drosophila* fushi-tarazu factor 1 transcription factor. A second transcript (designated embryonal long terminal repeat binding protein), the result of alternative promoter usage and 3′ splicing of FTZ-F1, has been isolated from embryonal carcinoma cells. The significance and function of this second transcript is unclear. The SF-1 protein is 53 kDa (Lala *et al.*, 1992; Morohashi *et al.*, 1992) with a zinc finger motif at the N-terminal end which mediates binding to the DNA sequence, and it is expressed in all steroidogenic tissues in a constitutive manner.

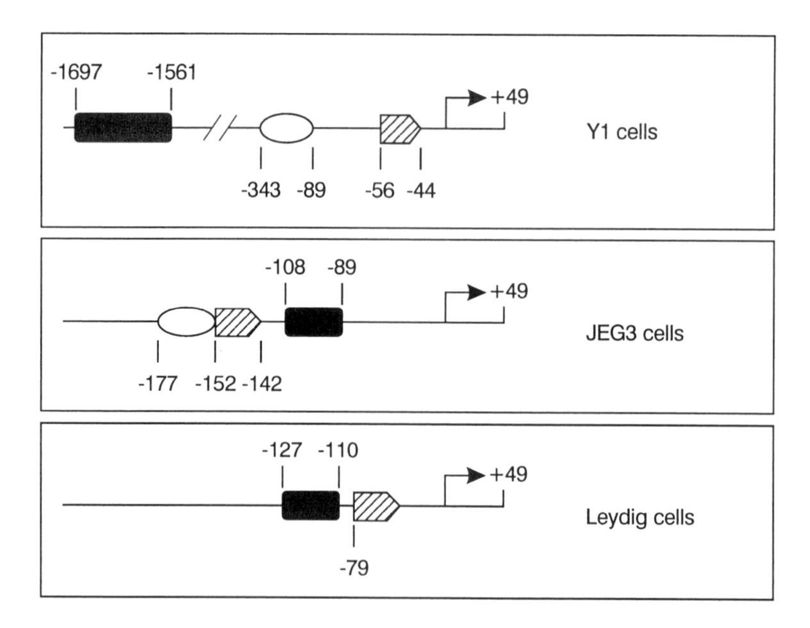

Figure 8.5. *Cis* regulatory elements identified in human CYP11A1 in adrenal Y1, placental JEG3 and Leydig cells. Sequences required for basal expression shown as hatched symbols, solid rectangles denote cAMP-responsive sequences and open circles denote negative regulatory elements. Numbering is relative to the transcription start site.

SF-1 and a homologue nerve growth factor 1B (NGF1-B), are members of the steroid/thyroid hormone nuclear receptor superfamily and recognize the same DNA core sequence, AGGTCA, the recognition sequence for the oestrogen receptor half site (Chapter 4), but differ in their requirements at the 5′ end, NGF1-B requiring two adenines, that is AA AGGTCA, and SF-1 binding most strongly to (C/T)CAAGG(T/C)CA (Morohashi *et al.*, 1992). The proteins are regarded as orphan receptors (because no ligand has been identified) and they bind to DNA as monomers, rather than the more typical dimer seen with the other nuclear receptors. Interaction between SF-1 and the inducible NGF1-B has also been proposed (Wilson *et al.*, 1993), the two proteins binding to the same -65 nucleotide element in the CYP21 promoter to increase transcription in response to ACTH and stress (Lala *et al.*, 1992; Wilson *et al.*, 1993).

The same core binding site is shared by the chick ovalbumin upstream promoter transcription factor (COUP-TF) protein and has been found in one of the two cAMP-responsive sequences, CRS2, of the *CYP17* promoter. Overexpression of SF-1 or COUP-TF separately in Cos cells showed that SF-1 could increase transcription of a CRS2–reporter gene construct while COUP-TF alone had no effect. Simultaneous expression of the two proteins reduced SF-1-induced gene expression suggesting that COUP-TF represses SF-1-stimulated transcription perhaps as a result of competition for an overlapping but not identical DNA-binding site (Bakke and Lund, 1995).

The roles of SF-1 and NGF1-B *in vivo* were determined by the generation of knockout mice. The offspring of heterozygous *Ftz*-F1-disrupted mice had no adrenal glands or gonads and had female internal genitalia (Sadovsky *et al.*, 1995). These mice died within 8 days of birth and were found to have reduced corticosterone and elevated ACTH levels. Intra-uterine steroid levels were however normal suggesting that placental CYP11A was not under the regulation of SF-1 in spite of SF-1 expression being detectable in the placenta. SF-1 binding elements have also been found in the DAX-1 gene (Burris *et al.*, 1995) (see Section 10.2.2), a nuclear hormone receptor whose function is necessary for adrenal development (Zanaria *et al.*, 1994), and in the promoter region of the mouse StAR gene, indicating that SF-1 has a number of levels of influence including a major role in sexual differentiation. Unfortunately, the absence of adrenal and gonadal tissues prevents *in vivo* testing of other effects of SF-1 on steroid synthesis.

The effect of deleting NGF1-B *in vivo* was far less dramatic. Unlike SF-1-deficient mice, NGF1-B knockouts continued to thrive and reproduce normally and appeared to suffer no impairment of corticosterone response to ACTH. These results suggest that, in spite of the obvious changes in NGF1-B transcription in response to stress and ACTH stimulation, sufficient other activators of CYP21 transcription are present to compensate for the loss of NGF1-B (Crawford *et al.*, 1995).

8.3.3 Tissue-specific expression

The mechanism by which steroidogenesis is limited to a small number of tissues is not understood. The first line of tissue specificity must be the presence or absence of receptors for the various trophic hormones. For example, FSH receptors are present on ovarian granulosa but not thecal cells, LH receptors are on theca and Leydig cells, and ACTH receptors are limited to the adrenal gland. The absence of transcription factors such as SF-1 from non-steroidogenic cells provides a second level of control. DNA footprint analysis (Section 2.3.4) can identify nuclear proteins which interact with promoter sequences and activate the gene. However, it is not clear whether such proteins are constitutively active in all tissues in which they are found or require activation, for example by phosphorylation, which may or may not occur in a particular cell type *in vivo*. A 55 kDa protein, complex VII, has been isolated from placenta which appears to act as a *trans*-acting factor necessary for placental expression of *CYP11A1*(Hum *et al.*, 1995). The protein, which interacts with a region of the CYP11A1 promoter known to be important for basal transcription of the gene in the placenta (*Figure 8.5*), was not found in adrenal tissue. Transcription modulators may also be found some way from the transcription start site. An adrenal-specific transcript designated Z, which arises from a cryptic promoter in the intron between exons 35 and 36 of the C4 gene, has recently been identified which may be relevant to the adrenal specific locus control region for *CYP21* (Tee *et al.*, 1995).

8.4 Developmental regulation of steroidogenesis

8.4.1 Adrenal gland

The fetal adrenal is composed of three zones, the outer definitive zone or neocortex, the transitional zone and the inner fetal zone, each with a characteristic steroid output reflecting differences in expression of the steroidogenic enzymes. CYP11A1 and CYP17 mRNA and protein were detected only in the fetal and transitional zones and not in the definitive zone in mid-gestation fetuses (Mesiano *et al.*, 1993). 3βHSD activity was deficient in all zones before 24 weeks gestation and remained undetectable in the fetal zone, but in later gestation was found in the definitive and transitional zones (Pelletier *et al.*, 1992). In late gestation, with the onset of CYP11A1 expression, the mineralocorticoid pathway in the definitive zone was fully functional suggesting that this zone may be the precursor of the zona glomerulosa. The resulting pattern of expression of steroidogenic enzymes in the transitional and fetal zones in the second half of pregnancy suggests that the transitional zone may be analogous to the adult zona fasciculata while the fetal zone, with no 3βHSD, produces dehydroepiandrosterone (sulphate) and after regression in the neonate may form the zona reticularis. What regulates 3βHSD expression in these zones is unknown but could reflect the presence of an inhibitor of the enzyme in the fetal zone or an activator of the gene in the other two zones.

Following birth and the regression of the fetal zone, adrenal steroidogenesis is primarily concerned with glucocorticoid and mineralocorticoid synthesis, adrenal androgens remaining low until adrenarche when there is an increase in 17,20-lyase activity and the output of dehydroepiandrosterone and androstenedione increases. The reason for this change in enzyme activity is still unclear.

8.4.2 Testicular development

Studies on mouse gonads show that all the mRNAs necessary for the conversion of cholesterol to androgen are expressed early in gestation prior to sexual differentiation (Greco and Payne, 1994) with some expression of CYP19 suggesting that the fetal testis has the capacity to synthesize oestrogens as well as androgens. CYP11A1 and CYP17 mRNAs were most abundant in the human fetal testis from 14 to 16 weeks gestation and then declined to approximately 30% of peak values at 25 weeks (Voutilainen and Miller, 1986). 3βHSD enzyme was detected by immunocytochemistry in the cytoplasm of the interstitial cells of the testis at 22 weeks gestation, the amount of enzyme increasing with gestational age (Pelletier *et al.*, 1992). In the postnatal testis, no immunostaining for this enzyme was observed until puberty when androgen production by the testis is stimulated by LH. The production of testosterone which continues to rise through gestation, falling just after birth and staying low until puberty, mirrors the expression of steroidogenic enzymes.

CYP19 is expressed in Leydig cells but not in the Sertoli cells of the adult testis. Neoplastic Sertoli cells have, however, been shown to switch on CYP19 using the same PII promoter as normal testicular tissue (Bulun *et al.*, 1993). Whether this is the result of upregulation of positive *trans*-acting regulatory elements or downregulation of inhibitory factors awaits clarification.

8.4.3 Ovarian steroidogenesis

Fetal ovaries exhibit little steroidogenic activity with no expression of *CYP17*, *CYP11A1* and *CYP19* mRNAs and only a small amount of 3βHSD detected using a sensitive RT–PCR procedure (Greco and Payne, 1994) confirming the earlier histochemical analysis of ovarian tissue (Pelletier *et al.*, 1992). After birth the ovaries are quiescent until puberty when oestrogen and progesterone are synthesized in a cyclical fashion.

The regulation of ovarian steroidogenesis during the menstrual cycle is multifactorial and requires the cooperation of two cell types, the theca and granulosa cells, which differ in terms of their hormone responsiveness and steroidogenic potential. The thecal cells produce progesterone and the androgens dehydroepiandrosterone and androstenedione in response to stimulation by LH. The granulosa cells are predominantly oestrogen-producing cells but express no CYP17 and therefore utilize thecal cell androgen as a substrate for aromatization by CYP19 with subsequent conversion of oestrone to oestradiol by 17HSD type 1. They also synthesize substantial amounts of progesterone on stimulation by FSH. The progression of the follicles from pre-antral, through antral to pre-ovulatory follicles is accompanied by an increase in the expression of CYP11A1. The ovulatory surge of LH, which occurs mid-cycle, triggers a change in ovarian function including termination of follicular development, ovulation and initiation of luteinization of the granulosa and thecal cells. It is also accompanied by an increase in CYP11A1 activity and progesterone production (Doody *et al.*, 1990). At this point, the regulation of CYP11A1 in luteinized granulosa cells changes from an inducible, cAMP-dependent process to a constitutive, cAMP-independent expression (Oonk *et al.*, 1989). CYP19 is undetectable until late in the follicular phase consistent with the rise of oestrogen seen at this time.

8.4.4 Placental steroidogenesis

The human placenta is an active steroidogenic tissue converting cholesterol to pregnenolone, progestins and oestrogens in response to cAMP stimulation. The tissue is deficient in CYP17 activity and utilizes circulating C19 precursors such as dehydroepiandrosterone sulphate as substrates for 17HSD1, 17HSD2 and aromatase. The regulation of steroidogenesis has been studied in JEG-3 trophoblast cells in culture. Both 17HSD type 2 and 3βHSD-I respond to cAMP stimulation in a dose- and time-dependent manner (Tremblay and Beaudoin, 1993) but the effect is not mediated by SF-1. This protein is present in the placenta but normal CYP11A1 levels were found in placenta from SF-1 knockout mice (Sadovsky *et al.*, 1995) suggesting that some other mechanism of transcriptional activation is used in this tissue.

8.4.5 Neurosteroidogenesis

The expression of CYP11A1 in the brain results in *de novo* biosynthesis of neurosteroids, a class of steroids which interact with γ-aminobutyric acid_A and *N*-methyl-D-aspartate receptors and can have profound effects on behaviour and memory. Examples of these steroids are dehydroepiandrosterone sulphate and allopregnenolone. Neurosteroid production, like steroidogenesis in other tissues, can be induced by cAMP but the regulation of steroidogenesis appears to be different in the brain. Studies on the regulation of CYP11A1 in rat glial C6 cells showed that these cell lines lacked specific nuclear factors, such as SF-1, found in steroidogenic tissues (Zhang *et al.*, 1995) and identified different *cis*-acting elements involved in cAMP-dependent regulation. SF-1 mRNA and protein has however been found in normal pituitary tissue and may regulate gonadotroph differentiation (Asa *et al.*, 1996).

8.5 Conclusion

The molecular analysis of the steroid biosynthetic enzymes has uncovered a plethora of gene families and diverse regulatory systems. While a number of clinical paradoxes can now be explained as a result of these discoveries, much still remains to be clarified with respect to the regulation and tissue-specific expression of these genes.

References

Adamovich TB, Pikuleva IA, Chashchin VL, Usanov SA. (1989) Selective chemical modification of cytochrome P-450scc lysine residues. Identification of lysines involved in the interaction with adrenodoxin. *Biochim. Biophys. Acta* **996:** 247–253.

Andersson S, Bishop RW, Russell DW. (1989) Expression cloning and regulation of steroid 5α-reductase, an enzyme essential for male sexual differentiation. *J. Biol. Chem.* **264:** 16249–16255.

Andersson S, Berman DM, Jenkins EP, Russell DW. (1991) Deletion of steroid 5α-reductase 2 gene on male pseudohermaphroditism. *Nature* **354:** 159–161.

Asa SL, Bamberger A-M, Cao B, Wong M, Parker KL, Ezzat S. (1996) The transcription activator steroidogenic factor-1 is preferentially expressed in the human pituitary gonadotroph. *J. Clin. Endocrinol. Metab.* **81:** 2165–2170.

Bakke M and Lund J. (1995) Mutually exclusive interactions of two nuclear orphan receptors determine activity of a cyclic adenosine 3′,5′-monophosphate-responsive sequence in the bovine CYP17 gene. *Mol. Endocrinol.* **9:** 327–339.

Berube D, Luu-The V, Lachance Y, Gagne R, Labrie F. (1989) Assignment of the human 3β-hydroxysteroid dehydrogenase gene to the p13 band of chromosome 1. *Cytogenet. Cell Genet.* **52:** 199–200.

Black AM, Szklarz GD, Harikrishna JA, Lin D, Wolf CR, Miller WL. (1993) Regulation of proteins in the cholesterol side-chain cleavage system in JEG-3 and Y-1 cells. *Endocrinology* **132:** 539–545.

Brown MS, Kovanen PT, Goldstein JL. (1982) Receptor mediated uptake of lipoprotein-cholesterol and its utilisation for steroidogenesis and cholesterol metabolism in steroidogenic glands. *Endocr. Rev.* **3:** 299–329.

Bulun SE, Rosenthal IM, Brodie AMH, Inkster SE, Zeller WP, Di Genge AM, Frasier SD, Kilgore MW, Simpson ER. (1993) Use of tissue-specific promoters in the regulation of aromatase cytochrome P450 gene expression in human testicular and ovarian sex cord tumours, as well as in normal fetal and adult gonads. *J. Clin. Endocrinol. Metab.* **77:** 1616–1621.

Burris TP, Guo W, Le T, McCabe ER. (1995) Identification of a putative steroidogenic factor-1 response element in the DAX-1 promoter. *Biochem. Biophys. Res. Commun.* **214:** 576–581.

Carey AH, Waterworth D, Patel K, White D, Little J, Novelli P, Franks S, Williamson R. (1994) Polycystic ovaries and premature male pattern baldness are associated with one allele of the steroid metabolism gene CYP17. *Hum. Mol. Genet.* **3:** 1873–1876.

Carroll MC, Campbell RD, Porter RR. (1985) The mapping of 21-hydroxylase genes adjacent to complement component C4 genes in HLA, the major histocompatability complex. *Proc. Natl Acad. Sci. USA* **82:** 521–525.

Casey ML, MacDonald PC, Andersson S. (1994)17 beta-hydroxysteroid dehydrogenase type 2: chromosomal assignment and progestin regulation of gene expression in human endometrium. *J. Clin. Invest.* **94:** 2135–2141.

Cavallaro S, Pani L, Guidotti A, Costa E. (1993) ACTH-induced mitochondrial DBI receptor (MDR) and diazepam binding inhibitor (DBI) expression in adrenals of hypophysectomized rats is not cause–effect related to its immediate steroidogenic action. *Life Sci.* **53:** 1137–1147.

Chanderbhan R, Noland BJ, Scallen TJ, Vahouny GV. (1982) Sterol carrier protein 2: Delivery of cholesterol from adrenal lipid droplets to mitochondria for pregnenolone synthesis. *J. Biol. Chem.* **257:** 8928–8934.

Chanderbhan RF, Kharroubi AT, Noland BJ, Scallen TJ, Vahouny GV. (1986) SCP2: Further evidence for its role in adrenal steroidogenesis. *Endocr. Res.* **12:** 351–360.

Chen S, Besman MJ, Sparkes RS, Zollman S; Klisak I; Mohandas T; Hall PF, Shively JE. (1988) Human aromatase: cDNA cloning, Southern blot analysis, and assignment of the gene to chromosome 15. *DNA* **7:** 27–38.

Cherradi N, Chambaz EM, Defaye G. (1995) Organisation of 3 beta-hydroxysteroid dehydrogenase/isomerase and cytochrome P450scc into a catalytically active molecular complex in bovine adrenocortical mitochondria. *J. Steroid Biochem. Mol. Biol.* **55:** 507–514.

Chung B-C, Matteson KJ, Voutilainen R, Hohandas TK, Miller WL. (1986) Human cholesterol side-chain cleavage enzyme, P450scc: cDNA cloning, assignment of the gene to chromosome 15, and expression in the placenta. *Proc. Natl Acad. Sci. USA* **83:** 8962–8966.

Chung B, Picado-Leonard J, Haniu M, Bienkowski M, Hall PF, Shivley JE, Miller WL. (1987) Cytochrome P450c17 (steroid 17α-hydroxylase/17,20-lyase). Cloning of human adrenal and testis cDNAs indicate the same gene is expressed in both tissues. *Proc. Natl Acad. Sci. USA* **84:** 407–411.

Clark BJ, Wells J, King SR, Stocco DM. (1994) The purification, cloning and expression of a novel luteinizing hormone-induced mitochondrial protein in MA-10 mouse Leydig tumour cells. *J. Biol. Chem.* **269:** 28314–28322.

Coghlan VM, Vickery LE. (1991) Site-specific mutations in human ferredoxin that affect binding to ferredoxin reductase and cytochrome P450scc. *J. Biol. Chem.* **266:** 18606–18612.

Crawford PA, Sadovsky Y, Woodson K, Lee SL, Millbrandt J. (1995) Adrenocortical function and regulation of the steroid 21-hydroxylase gene in NGF1-B deficient mice. *Mol. Cell. Biol.* **15:** 4331–4336.

Davis DL, Russell DW. (1993) Unusual length polymorphism in human steroid 5α-reductase type 2 gene (SRD5A2). *Hum. Mol. Genet.* **2:** 820.

Doody KJ, Lorence MC, Mason JI, Simpson ER. (1990) Expression of messenger ribonucleic acid species encoding steroidogenic enzymes in human follicles and corporea lutea throughout the menstrual cycle. *J. Clin. Endocrinol. Metab.* **70:** 1041–1045.

Dumont M, Luu-The V, Dupond E, Pelletier G, Labrie F. (1992) Characterization, expression, and immunohistochemical localization of 3β-hydroxysteroid dehydrogenase/Δ5-Δ4 isomerase in human skin. *J. Invest. Dermatol.* **99:** 415–421.

Durocher F, Morissette J, Labrie Y, Labrie F, Simard J. (1995) Mapping of the HSD17B2 gene encoding type II 17 beta-hydroxysteroid dehydrogenase close to D16S422 on chromosome 16q24/1-q24.2. *Genomics* **25:** 724–726.

Fisher CW, Shet MS, Caudle DL, Martin-Wixtrom CA, Estabrook RW. (1992) High-level expression in Escherichia coli of enzymatically active fusion proteins containing the domains of mammalian cytochromes P450 and NADPH-P450 reductase flavoprotein. *Proc. Natl Acad. Sci. USA* **89:** 10817–10821.

Fritz IB, Griswold MD, Lousi BF, Dorrington JH. (1976) Similarity of responses of cultured Sertoli cells to cholera toxin and FSH. *Mol. Cell. Endocrinol.* 5: 289–294.

Garlepp MJ, Wilton AN, Dawkins RL, White PC. (1986) Rearrangement of 21-hydroxylase genes in disease-associated MHC supratypes. *Immunogenetics* **23:** 100–105.

Geissler WM, Davis DL, Wu L *et al.* (1994) Male pseudohermaphroditism caused by mutations of testicular 17β-hydroxysteroid dehydrogenase 3. *Nature Genet.* **7:** 24–39.

Gradi A, Tang-Wai R, McBride HM, Chu LL, Shore GC, Pelletier J. (1995) The human steroidogenic acute regulatory (StAR) gene is expressed in the urogenital system and encodes a mitochondrial polypeptide. *Biochim. Biophys. Acta* **1258:** 228–233.

Greco TL, Payne AH. (1994) Ontogeny of expression of the genes for steroidogenic enzymes P450 side-chain cleavage, 3β-hydroxysteroid dehydrogenase, P450 17α-hydroxylase/C17,10 lyase and P450 aromatase in fetal mouse gonads. *Endocrinology* **135:** 262–268.

Grodin JM, Siiteri PK, MacDonald PC. (1973) Source of estrogen production in the postmenopausal woman. *J. Clin. Endocrinol. Metab.* **36:** 207–214.

Harada N, Yamada K, Saito K, Kibe N, Dohmae S, Takagi Y. (1990) Structural characterization of the human estrogen synthetase (aromatase) gene. *Biochem. Biophys. Res. Commun.* **166:** 365–372.

Hashimoto T, Morohashi K, Takayama K, Honda S, Wada T, Handa H, Omura T. (1992) Cooperative transcription activation between Ad1, a CRE-like element and other elements in the CYP11B gene promoter. *J. Biochem.* **112:** 573–575.

Higashi Y, Yoshioka H, Yamane M, Gotoh O, Fujii-Kuriyama Y. (1986) Complete nucleotide sequence of two steroid 21-hydroxylase genes tandemly arranged in human chromosome: a pseudogene and a genuine gene. *Proc. Natl Acad. Sci. USA* **83:** 2841–2845.

Higashi Y, Tanae A, Inoue H, Fujii-Kuriyama Y. (1988) Evidence for frequent gene conversion in the steroid 21-hydroxylase P-450(C21) gene: implications for steroid 21-hydroxylase deficiency. *Am. J. Hum. Genet.* **42:** 17–25.

Hornsby PJ, Aldern KA. (1984) Steroidogenic enzyme activities in cultured human definitive zone adrenocortical cells: comparison with bovine adrenocortical cells and resultant differences in adrenal androgen synthesis. *J. Clin. Endocrinol. Metab.* **58:** 121–127.

Hum DW, Aza-Blanc P, Miller WL. (1995) Characterization of placental transcriptional activation of the human gene for P450scc. *DNA Cell. Biol.* **14:** 451–463.

Jenkins C, Michael D, Mahendroo M, Simpson ER. (1993) Exon-specific Northern analysis and rapid amplification of cDNA ends (RACE) reveal that the proximal promoter II (PII) is responsible for aromatase cytochrome P450 (CYP19) expression in human ovary. *Mol. Cell. Endocrinol.* **97:** R1-R6.

Jenkins EP, Hsieh CL, Milatovich A, Normington K, Berman DM, Francke U, Russell DW. (1991) Characterization and chromosomal mapping of a human steroid 5 alpha-reductase gene and pseudogene and mapping of the mouse homologue. *Genomics* **11:** 1102–1112.

Jenkins EP, Andersson S, Imperato-McGinley J, Wilson JD, Russell DW. (1992) Genetic and pharmacological evidence for more than one human steroid 5α-reductase *J. Clin. Invest.* **89:** 293–300.

Killeen AA, Sane KS, Orr HT. (1991) Molecular and endocrine characterization of a mutation involving a recombination between the steroid 21-hydroxylase functional gene and pseudogene. *J. Steroid. Biochem. Mol. Biol.* **38:** 677–686.

Kitamura M, Buczko E, Dufau ML. (1991) Dissociation of hydroxylase and lyase activities by site-directed mutagenesis of the rate P450–17α. *Mol. Endocrinol.* **5:** 1373–1380.

Koh Y, Bucko E, Dufau ML. (1993) Requirement of phenylalanine 343 for the preferential Δ4 -lyase versus Δ5-lyase activity of rat CYP17. *J. Biol. Chem.* **268:** 18267–18271.

Krozowski Z. (1992) At the cutting edge: 11β-hydroxysteroid dehydrogenase and the short chain alcohol dehydrogenase (SCAD) superfamily. *Mol. Cell. Endocrinol.* **84:** C25-C31.

Krozowski Z. (1994) The short-chain alcohol dehydrogenase superfamily: variations on a common theme. *J. Steroid. Biochem. Mol. Biol.* **51:** 125–130.

Labrie F, Sugimoto Y, Luu-The V, Simard J, Lachance Y, Bachvarov D, Leblanc G, Durocher F, Paquet N. (1992) Structure of human type II 5 alpha-reductase gene. *Endocrinology* **131:** 1572–1573.

Labrie Y, Durocher F, Lachance Y, Turgeon C, Simard J, Labrie C, Labrie F. (1995) The human type II 17β-hydroxysteroid dehydrogenase gene encodes two alternatively spliced mRNA species. *DNA Cell. Biol.* **14:** 849–861.

Lachance Y, Luu-The V, Verrault H, Dumont M, Rheaume E, Leblanc G, Labrie F. (1991) Structure of human type II 3β-hydroxysteroid dehydrogenase/Δ^5-Δ^4 isomerase (3β-HSD) gene: adrenal and gonadal specificity. *DNA Cell. Biol.* **10:** 701–711.

Lala DS, Rice DA, Parker KL. (1992) Steroidogenic factor 1, a key regulator of steroidogenic enzyme expression, is the mouse homolog of fushi tarazu-factor 1. *Mol. Endocrinol.* **6:** 1249–1258.

Lambeth JD, Pember SO. (1983) Reduction properties of the substrate-associated cytochrome and relation of the reduction states of heme and iron-sulphur centers to association of the proteins. *J. Biol. Chem.* **258:** 5596–5602.

Lewintre JE, Orava M, Peltoketo H, Vihko R. (1994) Characterization of 17β-hydroxysteroid dehydrogenase type 1 in choriocarcinoma cells: regulation by basic fibroblast growth factor. *Mol. Cell. Endocrinol.* **104:** 1–9.

Lin D, Black SM, Nagahama Y, Miller WL. (1993a) Steroid 17α-hydroxylase and 17,20-lyase activities of P450c17: Contributions of serine 106 and P450 reductase. *Endocrinology* **132:** 2498–2506.

Lin D, Chang YJ, Strauss JF III, Miller WL. (1993b) The human peripheral benzodiazepine receptor gene: cloning and characterization of alternative splicing in normal tissues and in a patient with congenital adrenal hyperplasia. *Genomics* **18:** 643–650.

Lin D, Sugawara T, Strauss JF, Clark BJ, Stocco DM, Saenger P, Rogol A, Miller WL. (1995) Role of steroidogenic acute regulatory protein in adrenal and gonadal steroidogenesis. *Science* **267:** 1828–1831.

Lin S-X, Yang F, Jin J-Z, Breton R, Zhu D-W, Luu-The V, Labrie F. (1992) Subunit identity of the dimeric 17β-hydroxysteroid dehydrogenase from human placenta. *J. Biol. Chem.* **267:** 16182–16187.

Lorence MC, Corbin CJ, Kamimura N, Mahendroo MS, Mason JI. (1990a) Structural analysis of the gene encoding human 3β-hydroxysteroid dehydrogenase/$\Delta^{5\rightarrow4}$ isomerase. *Mol. Endocrinol.* **4:** 1850–1855.

Lorence MC, Murry BA, Trant JM, Mason JI. (1990b) Human 3β-hydroxysteroid dehydrogenase/$\Delta^5\rightarrow\Delta^4$ isomerase from placenta: expression in nonsteroidogenic cells of a protein that catalyses the dehydrogenation/isomerization of C21 and C19 steroids. *Endocrinology* **126:** 2493–2498.

Luu-The V, Lachance U, Labrie C, Leblanc G, Thomas JL, Strickler RC, Labrie F. (1989a) Full length cDNA structure and deduced amino acid sequence of human 3β-hydroxy-5-ene steroid dehydrogenase. *Mol. Endocrinol.* **3:** 1310–1312.

Luu-The V, Labrie C, Zhao HF *et al.* (1989b) Characterization of cDNAs for human estradiol 17β-hydroxysteroid dehydrogenase and assignment of the gene to chromosome 17: evidence of two mRNA species with distinct 5′-termini in human placenta. *Mol. Endocrinol.* **3:** 1301–1309.

Luu-The V, Takahashi M, de-Launoit Y, Dumont M, Lachance Y, Labrie F. (1991) Evidence for distinct dehydrogenase and isomerase sites within a single 3 beta-hydroxysteroid dehydrogenase/5-ene-4-ene isomerase protein. *Biochemistry* **30:** 8861–8865.

Luu-The V, Sugimoto Y, Libertad P, Labrie Y, Solache IL, Singh M, Labrie F. (1994) Characterization, expression and immunohistochemical localization of 5α-reductase in human skin. *J. Invest. Dermatol.* **102:** 221–226.

Mahendroo MS, Mendelson CR, Simpson ER. (1993) Tissue-specific and hormonally controlled alternative promoters regulate aromatase cytochrome P450 gene expression in human adipose tissue. *J. Biol. Chem.* **268:** 19463–19470.

Mason JI. (1993) The 3β-hydroxysteroid dehydrogenase gene family of enzymes. *Trends Endocrinol. Metab.* **4:** 199–203.

Matteson KJ, Picado-Leonard J, Chung B-C, Mohandas TK, Miller WL. (1986) Assignment of the gene for adrenal P450c17(steroid 17α-hydroxylase/17,20 lyase) to human chromosome 10. *J. Clin. Endocrinol. Metab.* **63:** 789–791.

McBride MW, Russell AJ, Vass K, Forster V, Burridge SM, Morrison N, Boyd E, Ponder BA, Sutcliffe RG. (1995) New members of the 3 beta-hydroxysteroid dehydrogenase gene family. *Mol. Cell. Probes* **9:** 121–128.

McNatty KP, Baird DT, Bolton A, Chambers P, Corker CS, MacLean H. (1976) Concentrations of oestrogens and androgens in human ovarian venous plasma and follicular fluid throughout the menstrual cycle. *J. Endocrinol.* **71:** 77–85.

Means GD, Mahendroo MS, Corbin CJ, Mathis MJ, Powell FE, Mendelson CR, Simpson ER. (1989) Structural analysis of the gene encoding human aromatase cytochrome P-450, the enzyme responsible for estrogen biosynthesis. *J. Biol. Chem.* **264:** 19385–19391.

Mellon SH, Deschepper CF. (1993) Neurosteroid biosynthesis: genes for adrenal steroidogenic enzymes are expressed in the brain. *Brain Res.* **629:** 283–292.

Mesiano S, Coulter CL, Jaffe RB. (1993) Localization of cytochrome P450 cholesterol side-chain cleavage, cytochrome P450 17α-hydroxylase/17,20 lyase, and 3β-hydroxysteroid dehydrogenase isomerase steroidogenic enzymes in human and rhesus monkey fetal adrenal glands: reappraisal of functional zonation. *J. Clin. Endocrinol. Metab.* **77:** 1184–1189.

Miki Y, Swenson J, Shattuck-Eidens D *et al.* (1994) A strong candidate for the breast and ovarian cancer susceptibility gene BRCA1. *Science* **266:** 66–71.

Miller WL. (1994) Genetics, diagnosis and management of 21-hydroxylase deficiency. *J. Clin. Endocrinol. Metab.* **78:** 241–246.

Miller WL, Gitelman SE, Bristow J, Morel Y. (1992) Analysis of the duplicated human C4/P450c21/X gene cluster. *J. Steroid. Biochem. Mol. Biol.* **43:** 961–971.

Morel Y, Bristow J, Gitelman SE, Miller WL. (1989) Transcript encoded on the opposite strand of the human steroid 21-hydroxylase/complement C4 gene locus. *Proc. Natl Acad. Sci. USA* **86:** 6582–6586.

Moore CCD, Brentano ST, Miller WL. (1990) Human P450scc gene transcription is induced by cyclic AMP and repressed by 12-O-tetradecanoylphorbol-13-acetate and A23187 by independent cis-elements. *Mol. Cell. Biol.* **10:** 6013–6023.

Moore RJ, Griffin JE, Wilson JD. (1975) Diminished 5α-reductase activity in extracts of fibroblasts cultured from patients with familial incomplete male pseudohermaphroditism, type 2. *J. Biol. Chem.* **250:** 7168–7172.

Morissette J, Rheaume E, Leblanc JF, Luu-The V, Labrie F, Simard J. (1995) Genetic linkage mapping of HSD3B1 and HSD3B2 encoding human types I and II 3 beta-hydroxysteroid dehydrogenase/delta 5-delta 4-isomerase close to D1S514 and the centromeric D1Z5 locus. *Cytogenet. Cell. Genet.* **69:** 59–62.

Morohashi K, Honda S, Inomata Y, Handa H, Omura T. (1992) A common trans-acting factor, Ad4-binding protein, to the promoters of steroidogenic P-450s. *J. Biol. Chem.* **267:** 17913–17919.

Nelson DR, Kamataki T, Waxman D *et al.* (1993) The P450 superfamily: update on new sequences, gene mapping accession numbers, early trivial names of enzymes and nomenclature. *DNA Cell. Biol.* **12:** 1–51.

Oonk RB, Krasnow JS, Beattie WG, Richards JS. (1989) Cyclic AMP-dependent and independent regulation of cholesterol side chain cleavage cytochrome P-450 (P450scc) in rat ovarian granulosa cells and corpus luteum. *J. Biol. Chem.* **264:** 21934–21942.

Papadopoulos V. (1993) Peripheral benzodiazepine/diazepam binding inhibitor receptor: biological role in steroidogenic cell function. *Endocr. Rev.* **14:** 222–240.

Pedersen RC, Brownie AC. (1987) Steroidogenesis activator polypeptide isolated from a rat Leydig cell tumor. *Science* **236:** 188–190.

Pelletier G, Dupont E, Simard J, Luu-The V, Belanger A, Labrie F. (1992) Ontogeny and subcellular localisation of 3β-hydroxysteroid dehydrogenase (3β-HSD) in the human and rat adrenal, ovary and testis. *J. Steroid. Biochem. Mol. Biol.* **43:** 441–467.

Peltoketo H, Isomaa V, Maentausta O, Vihko R. (1992) Genomic organisation and DNA sequences of human 17β-hydroxysteroid dehydrogenase genes and flanking region: localisation of multiple alu sequences and putative cis-acting elements. *Eur. J. Biochem.* **209:** 459–466.

Peltoketo H, Piao Y, Mannermaa A, Ponder BAJ, Isomaa V, Poutanen M, Winqvist R, Vihko R. (1994) A point mutation in the putative TATA box, detected in nondiseased individuals and patients with hereditary breast cancer, decreases promoter activity of the 17β-hydroxysteroid dehydrogenase type 1 gene 2 (EDH17B2) *in vitro*. *Genomics* **23:** 250–252.

Poutanen M, Moncharmont B, Vihko R. (1992) 17β-hydroxysteroid dehydrogenase gene expression in human breast cancer cells: regulation of expression by a progestin. *Cancer Res.* **52:** 290–294.

Randall VA. (1994) Role of 5α-reductase in health and disease. In: *Baillière's Clinical Endocrinology and Metabolism* (eds MC Sheppard, PM Stewart). Baillière Tindall, London, pp. 405–431.

Reichardt JKV, Makridakis N, Henderson BE, Yu MC, Pike MC, Ross RK. (1995) Genetic variability of the human SRD5A2 gene: implications for prostate cancer risk. *Cancer Res.* **55:** 3973–3975.

Reyland ME. (1993) Protein kinase C is a tonic negative regulator of steroidogenesis and steroid hydroxylase gene expression in Y1 adrenal cells and functions independently of protein kinase A. *Mol. Endocrinol.* **7:** 1021–1030.

Rheaume E, Lachance Y, Zhao HF, Breton N, Dumont M, Laurloit Y, Trudel C, Luu-The V, Simard J, Labrie F. (1991) Structure and expression of a new complementary DNA encoding the almost exclusive 3 beta-hydroxysteroid dehydrogenase/delta 5-delta 4-isomerase in human adrenals and gonads. *Mol. Endocrinol.* **5:** 1147–1157.

Rheaume E, Simard J, Morel Y, Mebarki F, Zachmann M, Forest MG, New MI, Labrie F. (1992) Congenital adrenal hyperplasia due to point mutations in the type II 3β-hydroxysteroid dehydrogenase gene. *Nature Genet.* **1:** 239–245.

Rodrigues NR, Dunham I, Yu CY, Carroll MC, Porter RR, Campbell RD. (1987) Molecular characterization of the HLA-linked steroid 21-hydroxylase B gene from an individual with congenital adrenal hyperplasia. *EMBO J.* **6:** 1653–1661.

Rumsby G, Carroll MC, Porter RR, Grant DB, Hjelm M. (1986) Deletion of the steroid 21-hydroxylase and complement C4 genes in congenital adrenal hyperplasia. *J. Med. Genet.* **23:** 204–209.

Ryan KJ. (1959) Biological aromatization of steroids. *J. Biol. Chem.* **234:** 268–272.

Ryan T, Chamberlin M. (1983) Transcription analyses with heteroduplex trp attenuator templates indicate that the transcript stem and loop structure serves as termination signal. *J. Biol. Chem.* **258:** 4690–4693.

Sadovsky Y, Crawford PA, Woodson KG, Polish JA, Clements MA, Tourtellotte LM, Simburgerik, Milbrandt J. (1995) Mice deficient in the orphan receptor steroidogenic factor 1 lack adrenal glands and gonads but express P450 side-chain cleavage enzyme in the placenta and have normal embryonic serum levels of corticosteroids. *Proc. Natl Acad. Sci. USA* **92:** 10939–10943.

Sanchez R, Rheaume E, Laflamme N, Rosenfield RL, Labrie F, Simard J. (1994) Detection and functional characterization of the novel missense mutation Y254D in type II 3β-hydroxysteroid dehydrogenase (3βHSD) gene of a female patient with nonsalt-losing 3βHSD deficiency. *J. Clin. Endocrinol. Metab.* **78:** 561–567.

Sawetawan C, Milewich L, Word RA, Carr BR, Rainey WE. (1994) Compartmentalization of type 1 17beta-hydroxysteroid oxidoreductase in the human ovary. *Mol. Cell. Endocrinol.* **99:** 161–168.

Simard J, Rheaume E, Sanchez R *et al.* (1993a) Molecular basis of congenital adrenal hyperplasia due to 3β-hydroxysteroid dehydrogenase deficiency. *Mol. Endocrinol.* **7:** 716–728.

Simard J, Feunteun J, Lenoir G *et al.* (1993b) Genetic mapping of the breast–ovarian cancer syndrome to a small interval on chromosome 17q12–21: exclusion of candidate genes EDH17B2 and RARA. *Hum. Mol. Genet.* **2:** 1193–1199.

Sparkes RS, Klisak I, Miller WL. (1991) Regional mapping of genes encoding human steroidogenic enzymes: P450scc to 15q23–24, adrenodoxin to 11q22; adrenodoxin reductase to 17q24-q25; and P450c17 to 10q24-q25. *DNA Cell. Biol.* **10:** 359–365.

Stocco DM, Clark BJ. (1996) Role of the steroidogenic acute regulatory protein (StAR) in steroidogenesis. *Biochem. Pharmacol.* **51:** 197–205.

Sugawara T, Holt JA, Driscoll D *et al.* (1995a) Human steroidogenic acute regulatory protein: Functional activity in Cos-1 cells, tissue-specific expression, and mapping of the structural gene to 8p11.2 and a pseudogene to chromosome 13. *Proc. Natl Acad. Sci. USA* **92:** 4778–4782.

Sugawara T, Lin D, Holt JA, Martin KO, Javitt NB, Miller WL, Strauss JF. (1995b) Structure of the human steroidogenic acute regulatory protein (StAR) gene: StAR stimulates mitochondrial cholesterol 27-hydroxylase activity. *Biochemistry* **34:** 12506–12512.

Taketo M, Parker KL, Howard TA, Tsukiyama T, Wong M, Niwa O, Morton CC, Miron PM, Seldin M F. (1995) Homologs of Drosophila Fushi-Tarazu factor 1 map to mouse chromosome 2 and human chromosome 9q33. *Genomics* **25:** 565–567.

Techatraisak K, Conway GC, Rumsby G. (1996) Frequency of a polymorphism in the regulatory region of the 17α-hydroxylase-17,20 lyase (CYP17) gene in hyperandrogenic states. *Clin. Endocrinol.* **45:** (in press).

Tee MK, Babalola GO, Aza PB, Speek M, Gitelman SE, Miller WL. (1995) A promoter within intron 35 of the human C4A gene initiates abundant adrenal-specific transcription of a 1 kb RNA: location of a cryptic CYP21 promoter element. *Hum. Mol. Genet.* **4:** 2109–2116.

Thigpen AE, Davis DL, Milatovich A, Mendonda BB, Imperato-McGinley J, Griffin JE, Francke U, Wilson JD, Russell DW. (1992) Molecular genetics of steroid 5α-reductase 2 deficiency. *J. Clin. Invest.* **90:** 799–809.

Thomas JL, Nash WE, Myers RP, Crankshaw MW, Strickler RC. (1993) Affinity radiolabeling identifies peptides and amino acids associated with substrate binding in human placental 3β-hydroxysteroid Δ^5-dehydrogenase. *J. Biol. Chem.* **268:** 18507–18512.

Thomas JL, Frieden C, Nash WE, Strickler RC. (1995) An NADH-induced conformational change that mediates the sequential 3 beta-hydroxysteroid dehydrogenase/isomerase activities is supported by affinity labeling and the time-dependent activation of isomerase. *J. Biol. Chem.* **270:** 21003–21008.

Toda K, Terashima M, Kawamoto T *et al.* (1990) Structural and functional characterization of human aromatase P450 gene. *Eur. J. Biochem.* **193:** 559–565.

Tremblay Y, Beaudoin C. (1993) Regulation of 3β-hydroxysteroid dehydrogenase messenger ribonucleic acid levels by cyclic adenosine 3′5′-monophosphate and phorbol myristate acetate in human choriocarcinoma cells. *Mol. Endocrinol.* **7:** 355–364.

Trzeciak WH, Simpson ER, Scallen TJ, Vahouny GV, Waterman MR. (1987) Studies on the synthesis of SCP-2 in rat adrenal cortical cells in monolayer culture. *J. Biol. Chem.* **262:** 3713–3720.

Tusie-Luna M-T, White PC. (1995) Gene conversions and unequal crossovers between CYP21 (steroid 21-hydroxylase gene) and CYP21P involve different mechanisms. *Proc. Natl Acad. Sci. USA* **92:** 10796–10800.

Valladares LE, Payne AH. (1979) Induction of testicular aromatization by luteinizing hormone in mature rats. *Endocrinology* **105:** 431–436.

Voutilainen R, Miller WL. (1986) Developmental expression of genes for the steroidogenic enzymes P450scc (20,22-desmolase), P450c17 (17α-hydroxylase/17,20 lyase), and P450c21 (21-hydroxylase) in the human fetus. *J. Clin. Endocrinol. Metab.* **63:** 1145–1150.

Voutilainen R, Tapanainen J, Chung B, Matteson KJ, Miller WL. (1986) Hormonal regulation of P450scc (20,22 desmolase) and P450c17 (17α-hydroxylase/17,20 lyase) in cultured human granulosa cells. *J. Clin. Endocrinol. Metab.* **63:** 202–207.

Watanabe N, Inoue H, Fujii-Kuriyama Y. (1994) Regulatory mechanisms of cAMP-dependent and cell-specific expression of human steroidogenic cytochrome P450scc (CYP11A1) gene. *Int. J. Biochem.* **222:** 825–834.

White PC, New MI, Dupont B. (1986) Structure of human steroid 21-hydroxylase genes. *Proc. Natl Acad. Sci. USA* **83:** 5111–5115.

White PC, Vitek A, Dupont B, New MI. (1988) Characterization of frequent deletions causing steroid 21-hydroxylase deficiency. *Proc. Natl Acad. Sci. USA* **85:** 4436–4440.

Wigley WC, Prihoda JS, Mowszowicz I, Mendonca BB, New MI, Wilson JD, Russell DW. (1994) Natural mutagenesis study of the human steroid 5alpha-reductase 2 isozyme. *Biochemistry* **33:** 1265–1270.

Wilson TE, Fahrner TJ, Milbrandt J. (1993) The orphan receptors NGF1-B and steroidogenic factor 1 establish monomer binding as a third paradigm of nuclear receptor–DNA interaction. *Mol. Cell. Biol.* **13:** 5794–5804.

Wu L, Einstein M, Geissler WM, Chan HK, Elliston KO, Andersson S. (1993) Expression cloning and characterization of human 17 beta-hydroxysteroid dehydrogenase type 2, a microsomal enzyme possessing 20alpha-hydroxysteroid dehydrogenase activity. *J. Biol. Chem.* **268:** 12964–12969.

Yahamoto R, Kallen CB, Babalola GO, Rennert H, Billheimer JT, Strauss JF. (1991) Cloning and expression of a cDNA encoding human sterol carrier protein 2. *Proc. Natl Acad. Sci. USA* **88:** 463–467.

Yanagibashi K, Hall PF. (1986) Role of electron transport in the regulation of the lyase activity of C21 side-chain cleavage P-450 from porcine adrenal and testicular microsomes. *J. Biol. Chem.* **261:** 8429–8433.

Zanaria E, Muscatelli F, Bardoni B *et al.* (1994) An unusual member of the nuclear hormone receptor superfamily responsible for X-linked adrenal hypoplasia congenital. *Nature* **372:** 635–641.

Zhang P, Rodriguez H, Mellon SH. (1995) Transcriptional regulation of P450scc gene expression in neural and steroidogenic cells: implications for regulation of neurosteroidogenesis. *Mol. Endocrinol.* **9:** 1571–1582.

9

Genetics of hypertension and the renin–angiotensin–aldosterone system

Perrin C. White

9.1 Introduction

Hypertension is a significant underlying cause of cardio- and cerebrovascular morbidity and mortality and much effort has been devoted to identifying genetic risk factors for its development. Thus far, all known genetic causes of human hypertension involve dysregulation of sodium resorption in the renal distal tubule. The bulk of sodium resorption from the glomerular filtrate occurs in the proximal tubule and the thin ascending limb of the loop of Henle. Regulated sodium resorption takes place in the distal convoluted tubule and cortical collecting duct (Berry *et al.*, 1996). The first step in resorption is passive diffusion through sodium-permeable channels in the apical membranes of epithelial cells lining the tubules. This process is dependent on an electrochemical gradient generated by a sodium/potassium ATPase located in the basolateral cell membrane.

The most important regulator of sodium resorption is the renin–angiotensin–aldosterone system. The classic regulatory pathway of this system begins in the renal juxtaglomerular apparatus, where stretch receptors in the afferent arteriole respond to a fall in intravascular volume by stimulating secretion of renin, a proteolytic enzyme that cleaves angiotensinogen to produce a decapeptide, angiotensin I (*Figure 9.1*). This is converted by a widely distributed angiotensin-converting enzyme to the octapeptide, angiotensin II. Angiotensin II binds to a membrane receptor of the G-protein-coupled type (Curnow *et al.*, 1992; Murphy *et al.*, 1991; Sasaki *et al.*, 1991) on the surface of zona glomerulosa cells, activating phospholipase C which hydrolyses phosphatidylinositol biphosphate, producing inositol triphosphate and diacylglycerol. The latter substances raise intracellular calcium concentrations (Berridge, 1993) and rapidly activate aldosterone biosynthesis through effects on steroidogenic enzymes.

Molecular Endocrinology, edited by G. Rumsby and S.M. Farrow.

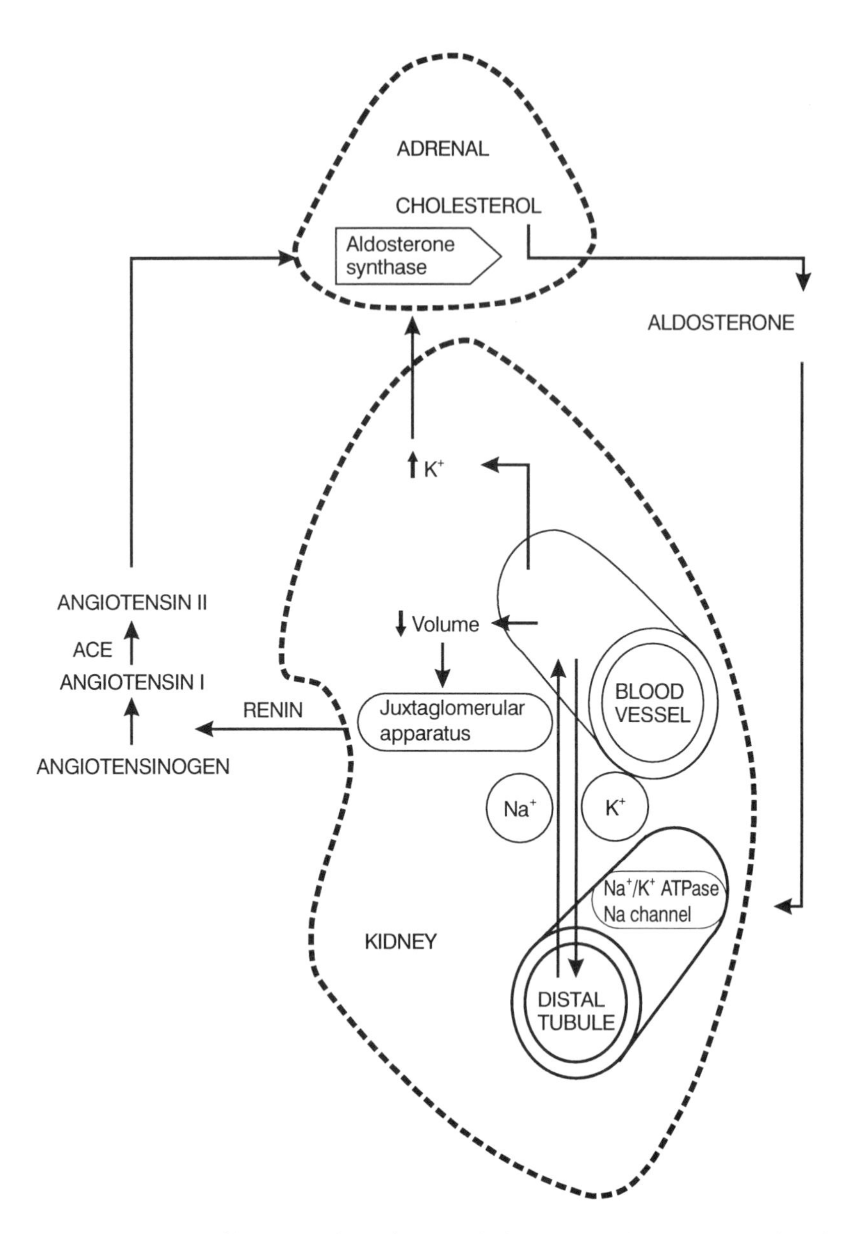

Figure 9.1. Regulation of intravascular volume and plasma potassium concentrations by the renin–angiotensin–aldosterone system. Decreased intravascular volume increases renin secretion by the juxtaglomerular apparatus leading to increased conversion of angiotensinogen to angiotensin II. Angiotensin II acts on the adrenal zona glomerulosa to increase the activity of aldosterone synthase and therefore aldosterone secretion. Hyperkalaemia also increases activity of aldosterone synthase. Aldosterone acts on the renal distal tubule to increase sodium (Na^+) resorption and potassium (K^+) excretion. ACE, angiotensin-converting enzyme; Na^+/K^+ ATPase, sodium–potassium-dependent ATPase.

Aldosterone, in turn, increases the apparent number of sodium channels in the epithelium of the distal tubule. This may reflect an increase in the percentage of time that each channel stays open (Duchatelle *et al.*, 1992), and/or an increase in the actual number of channels (Palmer and Frindt, 1992), although the signalling pathways by which aldosterone acts are not well understood. Aldosterone also increases synthesis of the sodium/potassium ATPase (Horisberger and Rossier, 1992).

All of the genes of the renin–angiotensin–aldosterone system have been cloned. Thus far, the genes encoding renin (Jeunemaitre *et al.*, 1992b), angiotensin-converting enzyme (Jeunemaitre *et al.*, 1992a) and the angiotensin II receptor (Bonnardeaux *et al.*, 1994) have not been demonstrated to be involved in the development of any form of hypertension and will not be discussed further in this chapter. The gene for angiotensinogen is linked to the development of essential hypertension, whereas mutations in regulatory subunits of the sodium channel itself, or in any of several steroid-metabolizing enzymes, cause syndromes of hypertension inherited as monogenic traits. These genes are reviewed in the following sections.

9.2 Angiotensinogen

9.2.1 Structure and function of angiotensinogen

Angiotensinogen is a glycoprotein synthesized mainly in the liver but also in adipose tissue, the brain, kidney and the circulatory system (Campbell and Habener, 1986). It is secreted directly without storage in secretory granules. Angiotensinogen normally circulates at levels near the K_m of renin for this substrate, and thus under some circumstances its concentration may be rate limiting for its conversion to angiotensin I. Therefore, genetic polymorphisms influencing angiotensinogen levels might be expected to have significant effects on activation of the renin–angiotensin system and thereby affect blood pressure.

Human pre-angiotensinogen contains 485 amino acids including a signal peptide that is cleaved to yield the secreted mature angiotensinogen protein of 452 amino acids (Kageyama *et al.*, 1984). The first 10 amino acids of the mature protein are cleaved by renin to angiotensin I (Printz *et al.*, 1977). The remaining des(Ang I)angiotensinogen moiety does not have any apparent biological function. Angiotensinogen is 21% identical in amino acid sequence to the serine protease inhibitor α_1-antitrypsin and is also related in its structure to corticosteroid-binding globulin and ovalbumin (Doolittle, 1983). However, it is not known to have any ability to inhibit serine proteases.

The angiotensinogen (*AGT*) gene is located on chromosome 1q42–43 (Isa *et al.*, 1990). It consists of five exons spaced over 13 kb. Exon 2 contains the majority of coding sequences including the region encoding angiotensin I (Gaillard *et al.*, 1989; Tanaka *et al.*, 1984).

9.2.2 Regulation of the angiotensinogen gene (AGT)

Most circulating angiotensinogen is synthesized in the liver. Factors that increase *AGT* expression include glucocorticoids, oestrogen, thyroid hormone, inflammatory cytokines, β_2-adrenergic agonists and angiotensin II. DNA elements involved in both basal and inducible expression have been identified by transfection of *AGT*–reporter constructs in cultured hepatocellular carcinoma cells. For example, a *cis*-acting basal element is located 115–145 bp upstream from the transcriptional start site. This element contains binding sites for NF-1 and C/EBP transcription factors, both of which transactivate expression of reporter constructs (Zhao *et al.*, 1993). A proximal element at –96 to –52 binds a liver-specific transcription factor which acts synergistically with ubiquitous nuclear factors binding at –6 to +22 to increase expression (Tamura *et al.*, 1994a). There are cell-specific enhancers, including an 80 bp element in exon 5 and a 24 bp element in the 3′ untranslated region (Nibu *et al.*, 1994a). The 80 bp element interacts with both liver-specific and ubiquitous nuclear proteins (Nibu *et al.*, 1994b). Inflammatory cytokines such as interleukin-1 and tumour necrosis factor increase transcription of the *AGT* gene through interaction of the nuclear factors C/EBP and NF-κB with elements located 615–440 bp upstream of the start of transcription. This region also contains glucocorticoid response elements (Brasier *et al.*, 1994). Post-transcriptional regulation is also important; for example, angiotensin II increases angiotensinogen mRNA stability (Klett *et al.*, 1988).

Angiotensinogen is also synthesized in a number of other tissues where it may function in a paracrine manner. Synthesis is increased during adipocyte differentiation (Tamura *et al.*, 1993, 1994b), a process that is controlled by a 14 bp element located 1000 bp upstream of the start of transcription (McGehee *et al.*, 1993) which binds a specific nuclear factor, differentiation-specific element-binding protein (McGehee and Habener, 1995). The expression of *AGT* is increased by cyclical mechanical stretch in cultured rat cardiomyocytes, apparently mediated in a positive feedback loop by angiotensin II (Shyu *et al.*, 1995).

9.2.3 Linkage of **AGT** *to essential hypertension*

Essential hypertension has been linked to *AGT* in hypertensive siblings. The evidence for linkage is strongest for young, severely hypertensive individuals requiring several drugs for adequate control, and it has been observed in American and northern European kindreds (Caulfield *et al.*, 1994; Jeunemaitre *et al.*, 1992c) and also in African Caribbeans (Caulfield *et al.*, 1995) using highly polymorphic loci tightly linked to *AGT*. In addition, two intragenic polymorphisms T235 (in place of M) and M174 (in place of T) are associated with hypertension in some populations (Jeunemaitre *et al.*, 1992c; Caulfield *et al.*, 1994). Moreover, T235 has a higher frequency in those African and Japanese populations that are prone to develop hypertension. T235 is also a risk factor for the development of hypertension of pregnancy, atherosclerosis and coronary heart disease (Ishigami *et al.*, 1995; Katsuya *et al.*, 1995).

Although T235 is associated with higher blood levels of angiotensinogen in both hypertensive and normotensive individuals, there is no evidence at this time of any functional differences between proteins carrying M235 and T235. This amino acid is not within the portion of the polypeptide that constitutes angiotensin I or II. As M174 and T235 are in linkage disequilibrium, it seems likely that these polymorphisms are markers for one or more other associated polymorphisms that actually affect angiotensinogen levels.

One sequence variant, L10F, that actually does lead to altered kinetics of cleavage of angiotensinogen to angiotensin I is located at the site at which renin cleaves angiotensinogen and was identified in a patient with pre-eclampsia (Inoue *et al.*, 1995).

9.3 Steroid 11β-hydroxylase isozymes

9.3.1 Zonal location of steroid biosynthesis

Cortisol is synthesized from cholesterol in the zona fasciculata of the adrenal cortex. This process requires five enzymatic conversions (see Section 8.2): cleavage of the cholesterol side chain to yield pregnenolone, 17α-hydroxylation and 3β-dehydrogenation to 17-hydroxyprogesterone, and successive hydroxylations at the 21 and 11β positions. A '17-deoxy' pathway is also active in the zona fasciculata, in which 17α-hydroxylation does not occur and the final product is normally corticosterone. The same 17-deoxy pathway is active in the adrenal zona glomerulosa, which contains no 17α-hydroxylase activity. However, corticosterone is not the final product in the zona glomerulosa; instead, corticosterone is successively hydroxylated and oxidized at the 18 position to yield aldosterone.

9.3.2 Steroid 11β-hydroxylase isozymes

Humans have two isozymes that are responsible for cortisol and aldosterone biosynthesis, respectively, 11β-hydroxylase (CYP11B1) and aldosterone synthase (CYP11B2). These isozymes are mitochondrial cytochromes P450 located in the inner membrane on the matrix side. Each is synthesized with 503 amino acid residues, but a signal peptide is cleaved in mitochondria to yield the mature protein of 479 residues (Yanagibashi *et al.*, 1986). The sequences of the proteins are 93% identical (Mornet *et al.*, 1989).

The CYP11B1 isozyme 11β-hydroxylates 11-deoxycorticosterone to corticosterone and 11-deoxycortisol to cortisol, as determined by expressing the corresponding cDNAs in cultured cells (Curnow *et al.*, 1991; Kawamoto *et al.*, 1990a, 1992) and after actual purification from aldosterone-secreting tumours (Ogishima *et al.*, 1991). It can also convert 11-deoxycorticosterone to 18-hydroxy, 11-deoxycorticosterone. However, CYP11B1 18-hydroxylates corticosterone poorly and cannot convert corticosterone into aldosterone. In contrast, CYP11B2 has strong 11β-hydroxylase activity, but also 18-hydroxylates and then 18-oxidizes corticosterone and cortisol to aldosterone and 18-oxocortisol, respectively. When deoxycorticosterone is converted to aldosterone, the same steroid molecule probably remains bound to the enzyme for all three conversions without release of the intermediate products.

The structural features responsible for the differences in activities of these isozymes have not been elucidated but must reside in the C-terminal halves of the enzymes based on studies of patients with glucocorticoid-suppressible hyperaldosteronism (see Section 9.3.5).

In humans, CYP11B1 and CYP11B2 are encoded by two genes (Mornet *et al.*, 1989) on chromosome 8q21–22 (Chua *et al.*, 1987; Wagner *et al.*, 1991). Each contains nine exons spread over approximately 7 kb of DNA. The nucleotide sequences of these genes are 95% identical in coding sequences and about 90% identical in introns. The two genes are approximately 40 kb apart (Lifton *et al.*, 1992b; Pascoe *et al.*, 1992) and *CYP11B2* is 5′ to *CYP11B1*.

9.3.3 Regulation of the CYP11B *genes*

Cortisol synthesis is primarily controlled by adrenocorticotrophin (ACTH) (Waterman and Simpson, 1989), which acts through a specific G-protein-coupled receptor (Mountjoy *et al.*, 1992) to increase levels of cAMP. cAMP has short-term (minutes to hours) effects on cholesterol desmolase activity but longer term (hours to days) effects on transcription of genes encoding the enzymes required to synthesize cortisol (John *et al.*, 1986). The transcriptional effects occur at least in part through increased activity of protein kinase A (Wong *et al.*, 1989) which phosphorylates transcriptional regulatory factors, not all of which have been identified. Conversely, the rate of aldosterone synthesis, which is normally 100–1000-fold less than that of cortisol synthesis, is regulated mainly by angiotensin II (via phospholipase C) and potassium levels with ACTH having only a short-term effect (Quinn and Williams, 1988).

Thus, if CYP11B1 and CYP11B2 are respectively required for cortisol and aldosterone synthesis, their genes should be regulated quite differently. The human *CYP11B1* gene is expressed at high levels in normal adrenal glands (Mornet *et al.*, 1989), and transcription of this gene is appropriately regulated by cAMP (Kawamoto *et al.*, 1990b). *CYP11B2* transcripts cannot be detected by hybridization to Northern blots of normal adrenal RNA (Mornet *et al.*, 1989), although such transcripts have been detected in normal adrenal RNA using the more sensitive technique of reverse transcriptase–polymerase chain reaction (RT–PCR) (Curnow *et al.*, 1991). *CYP11B2* transcripts are present at increased levels in aldosterone-secreting tumours (Curnow *et al.*, 1991; Kawamoto *et al.*, 1990a). In primary cultures of human zona glomerulosa cells, angiotensin II markedly increases levels of both *CYP11B1* and *CYP11B2* transcripts whereas ACTH increases *CYP11B1* mRNA levels only and more effectively than angiotensin II (Curnow *et al.*, 1991). However, *in situ* hybridization suggests that *CYP11B1* expression is limited to the zona fasciculata (Pascoe *et al.*, 1995).

The difference in regulation of *CYP11B1* and *CYP11B2* is presumably due to the extensive divergence between the 5′ regions flanking these genes. Both human genes include a TATA box variant, a palindromic cAMP response element (CRE), and several recognition sites for steroidogenic factor 1 (SF-1) (Mornet *et al.*, 1989). SF-1 sites appear in the regulatory regions of all steroid hydroxylase genes expressed in the adrenal cortex and the gonads (see Section 8.3.2) (Morohashi *et al.*, 1992; Rice *et al.*, 1991). Although the human transcriptional regulatory regions have not yet been analysed in detail, all of these sequences are required for normal transcription of murine and bovine *CYP11B* genes (Hashimoto *et al.*, 1992; Honda *et al.*, 1990; Mouw *et al.*, 1989; Rice *et al.*, 1989). Factors binding TATA boxes (Kao *et al.*, 1990) and CREs (Meyer and Habener, 1993) were previously identified. SF-1 (also called Ad4BP) is an orphan nuclear receptor; that is, it is a member of the steroid and thyroid hormone receptor superfamily but its ligand is not known. Intriguingly, it is closely related in its predicted amino acid sequence to FTZ-F1, a

Drosophila protein that regulates transcription of a homeobox gene, *fushi-tarazu* (Honda *et al.*, 1993; Lala *et al.*, 1992).

As yet, the genetic elements responsible for the differential regulation of *CYP11B1* and *CYP11B2* in the zonae fasciculata and glomerulosa have not been identified.

9.3.4 Steroid 11β-hydroxylase deficiency

Steroid 11β-hydroxylase deficiency is the second most common cause of congenital adrenal hyperplasia, the inherited inability to synthesize cortisol (see Chapter 8). In most populations, 11β-hydroxylase deficiency comprises approximately 5–8% of cases of congenital adrenal hyperplasia (Zachmann *et al.*, 1983) and thus it occurs in approximately 1 in 200 000 births. A large number of cases of 11β-hydroxylase deficiency have been reported in Israel among Jewish immigrants from Morocco; the incidence in this group is currently estimated to be 1/5000–1/7000 births (Rosler *et al.*, 1992).

In 11β-hydroxylase deficiency, 11-deoxycortisol and deoxycorticosterone are not efficiently converted to cortisol and corticosterone, respectively. Decreased production of glucocorticoids reduces their feedback inhibition on the hypothalamus and anterior pituitary, increasing secretion of ACTH. This stimulates the zona fasciculata of the adrenal cortex to overproduce steroid precursors proximal to the blocked 11β-hydroxylase step. Thus, 11β-hydroxylase deficiency can be diagnosed by detecting high basal or ACTH-stimulated levels of deoxycorticosterone and/or 11-deoxycortisol in the serum, or increased excretion of the tetrahydro metabolites of these compounds in a 24 h urine collection. Obligate heterozygous carriers of 11β-hydroxylase deficiency alleles (e.g. parents) have no consistent biochemical abnormalities detectable even after stimulation of the adrenal cortex with intravenous ACTH (Pang *et al.*, 1980), consistent with an autosomal recessive mode of inheritance.

Approximately two-thirds of patients with the severe, 'classic' form of 11β-hydroxylase deficiency have high blood pressure (Rosler *et al.*, 1982, 1992), often beginning in the first few years of life (Mimouni *et al.*, 1985). Although the hypertension is usually of mild to moderate severity, left ventricular hypertrophy and/or retinopathy have been observed in up to one-third of patients, and deaths from cerebrovascular accidents have been reported (Hague and Honour, 1983; Rosler *et al.*, 1992). Other signs of mineralocorticoid excess such as hypokalaemia and muscle weakness or cramping occur in a minority of patients and are not well correlated with blood pressure. Plasma renin activity is usually suppressed and levels of aldosterone are consequently low even though the ability to synthesize aldosterone is actually unimpaired.

The cause of hypertension in 11β-hydroxylase deficiency is not well understood. It might be assumed that it is caused by elevated serum levels of deoxycorticosterone but blood pressure and deoxycorticosterone levels are poorly correlated in patients (Rosler *et al.*, 1982; Zachmann *et al.*, 1983). In addition, this steroid has only weak mineralocorticoid activity when administered to humans or other animals. Perhaps other metabolites of deoxycorticosterone are responsible for the development of hypertension. The 18-hydroxy and 19-nor metabolites of deoxycorticosterone are thought to be more potent mineralocorticoids (Griffing *et al.*, 1983), but consistent elevation of these steroids in 11β-hydroxylase deficiency has not been documented. Moreover, synthesis of these steroids requires hydroxylations within the adrenal that are probably mediated primarily by CYP11B1 (Ohta *et al.*, 1988). This is unlikely to take place efficiently in 11β-hydroxylase deficiency.

In addition to hypertension, patients with 11β-hydroxylase deficiency often exhibit signs of androgen excess. This occurs because accumulated cortisol precursors in the adrenal cortex are shunted (through the activity of 17α-hydroxylase/17,20-lyase) into the pathway of androgen biosynthesis, which is active in the human adrenal in both sexes. Signs of androgen excess distinguish 11β-hydroxylase deficiency from 17α-hydroxylase deficiency, in which patients also tend to become hypertensive but are unable to synthesize sex steroids. This is discussed in more detail in Chapter 8.

Deficiency of 11β-hydroxylase results from mutations in *CYP11B1* (*Figure 9.2*). At this time, more than 20 mutations have been identified in patients with classic 11β-hydroxylase deficiency (Curnow *et al.*, 1993; Geley *et al.*, 1996; Helmberg *et al.*, 1992; Naiki *et al.*, 1993; Skinner *et al.*, 1996; White *et al.*, 1991). In Moroccan Jews, a group that has a high prevalence of 11β-hydroxylase deficiency, almost all affected alleles carry the same mutation, R448H (White *et al.*, 1991). This probably represents a founder effect, but this mutation has occurred independently in other ethnic groups, and another mutation of the same residue (R448C) has also been reported (Geley *et al.*, 1996). This apparent mutational 'hotspot' contains a CpG dinucleotide. Such dinucleotides are prone to methylation of the cytosine followed by deamination to TpG; several other mutations in CYP11B1 (T318M, R374Q, R384Q) are of this type.

These and almost all other missense mutations identified thus far are in regions of known functional importance (Nelson and Strobel, 1988; Poulos, 1991; Ravichandran *et al.*, 1993) and abolish enzymatic activity (Curnow *et al.*, 1993). For example, R448 is adjacent to C450 which is a ligand of the haem iron atom of

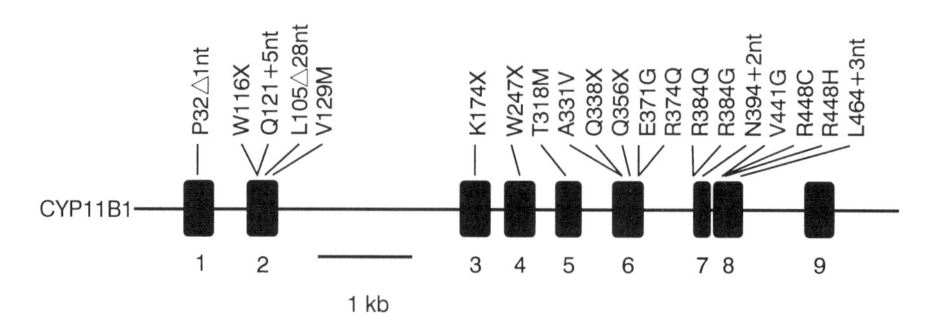

Figure 9.2. Diagram of the *CYP11B1* gene showing the location of mutations causing 11β-hydroxylase deficiency. The gene is drawn to scale as marked. Exons are represented by numbered boxes. Mutations are in single letter code. X, nonsense (stop) mutation; +, insertion of nucleotides (nt); Δ, deletion of nucleotides. For example, W116X represents a nonsense mutation of Trp-116.

this cytochrome P450 enzyme. T318M modifies an absolutely conserved residue that is thought to be critical for proton transfer to the bound oxygen molecule (Ravichandran *et al.*, 1993). E371G and R374Q also mutate highly conserved residues and may affect binding of adrenodoxin. R384Q is in a region that may form part of the substrate binding pocket (Ravichandran *et al.*, 1993). Almost all cytochrome P450 enzymes have a basic residue (H or R) at this or the immediately adjacent position (Nelson and Strobel, 1988). Finally, V441G is adjacent to the highly conserved haem binding region, and this mutation may change the secondary structure of the protein.

Other mutations found in patients with the classic form of the disease are nonsense or frameshift mutations that also abolish enzymatic activity. One, a nonsense mutation at codon 247 (W247X), has been identified in several unrelated kindreds in Austria and also probably represents a founder effect (Geley *et al.*, 1996).

Although patients with the classic form of the disease apparently completely lack 11β-hydroxylase activity, they differ significantly in the severity of the various signs and symptoms of their disease. There is not a strong correlation between severity of hypertension and biochemical parameters such as plasma levels of the 11β-hydroxylase substrates, deoxycortisol and deoxycorticosterone, and urinary excretion of tetrahydro-deoxycortisol (Rosler *et al.*, 1992; White *et al.*, 1991). Moreover, there is no consistent correlation between the severity of hypertension and degree of virilization. These phenotypic variations must be governed by factors outside the *CYP11B1* locus.

9.3.5 Glucocorticoid-suppressible hyperaldosteronism

Glucocorticoid-suppressible hyperaldosteronism (also called dexamethasone-suppressible hyperaldosteronism or glucocorticoid-remediable aldosteronism) is a form of hypertension inherited in an autosomal dominant manner with high penetrance (New and Peterson, 1967; Sutherland *et al.*, 1966). It is characterized by moderate hypersecretion of aldosterone, suppressed plasma renin activity, and rapid reversal of these abnormalities after administration of glucocorticoids. It is a rare disorder, with less than two dozen kindreds documented as of 1985, but until recently the absence of reliable biochemical or genetic markers made this disease difficult to ascertain.

Hypokalaemia is usually mild and may be absent. Absolute levels of aldosterone secretion are usually moderately elevated in the untreated state but may be within normal limits. Plasma renin activity is strongly suppressed, so that the ratio of aldosterone secretion to renin activity is always abnormally high. Levels of 18-hydroxycortisol and 18-oxocortisol are elevated to 20–30 times normal (Connell *et al.*, 1986; Gomez-Sanchez *et al.*, 1988; Stockigt and Scoggins, 1987). The ratio of urinary excretion of tetrahydro metabolites of 18-oxocortisol to those of aldosterone exceeds 2.0 whereas this ratio averages 0.2 in normal individuals. Elevation of 18-oxocortisol is the most consistent and reliable biochemical marker of the disease, although it may also be elevated in cases of primary aldosteronism (Hamlet *et al.*, 1988). This steroid may be of pathophysiologic significance; it is an agonist for the mineralocorticoid receptor and has been shown to raise blood pressure in animal studies (Hall and Gomez-Sanchez, 1986).

Once an affected individual has been identified in a kindred, additional cases may be ascertained within that kindred using biochemical (18-oxocortisol levels) or genetic (see below) markers (Rich *et al.*, 1992). It is apparent from these studies that affected individuals have blood pressures that are markedly elevated as compared to unaffected individuals in the same kindred, although some patients may in fact have normal blood pressures. Even young children typically have blood pressures greater than the 95th percentile for age, and most are frankly hypertensive before the age of 20. The hypertension is often of only moderate severity and blood pressures exceeding 180/120 are unusual. Associated signs of hypertension are frequent including left ventricular hypertrophy on the electrocardiogram and retinopathy. Some affected kindreds have remarkable histories of early (before age 45) death from strokes in many family members (O'Mahony *et al.*, 1989; Rich *et al.*, 1992). Steroid biosynthesis is otherwise normal so that affected individuals have normal growth and sexual development.

Most laboratory and clinical abnormalities are suppressed by treatment with glucocorticoids, whereas infusion of ACTH exacerbates these problems (Ganguly *et al.*, 1984; Oberfield *et al.*, 1981). This suggests that aldosterone is being inappropriately synthesized in the zona fasciculata and is being regulated by ACTH. Moreover, 18-hydroxycortisol and 18-oxocortisol, the steroids that are characteristically elevated in this disorder, are 17α-hydroxylated analogues of 18-hydroxycorticosterone and aldosterone, respectively. Because 17α-hydroxylase is not expressed in the zona glomerulosa, the presence of large amounts of a 17α-hydroxy, 18-oxo-steroid suggests that an enzyme with 18-oxidase activity (i.e. aldosterone synthase, CYP11B2) is abnormally expressed in the zona fasciculata (White, 1991).

All patients with glucocorticoid-suppressible hyperaldosteronism have the same type of mutation, a chromosome that carries three *CYP11B* genes instead of the normal two (*Figure 9.3*) (Lifton *et al.*, 1992a,b; Pascoe *et al.*, 1992). The middle gene on this chromosome is a chimaera with 5′ and 3′ ends corresponding to *CYP11B1* and *CYP11B2*, respectively. The chimaeric gene is flanked by presumably normal *CYP11B2* and *CYP11B1* genes. In all kindreds analysed thus far, the breakpoints (the points of transition between *CYP11B1* and *CYP11B2* sequences) are located between intron 2 and exon 4. As the breakpoints are not identical in different kindreds, these must represent independent mutations.

The chromosomes carrying chimaeric genes are presumably generated by unequal crossing over. The high homology and proximity of the *CYP11B1* and *CYP11B2* genes makes it possible for them to become misaligned during meiosis. If this occurs, crossing over between the misaligned genes creates two chromosomes, one of which carries one *CYP11B* gene (i.e. a deletion) whereas the other carries three *CYP11B* genes.

The invariable presence of a chimaeric gene in patients with this disorder suggests that this gene is regulated like *CYP11B1* (expressed at high levels in the zona fasciculata and regulated primarily by ACTH) because it has transcriptional regulatory sequences identical to those of *CYP11B1*. If the chimaeric gene has enzymatic activity similar to that of *CYP11B2*, a single copy of such an abnormally regulated

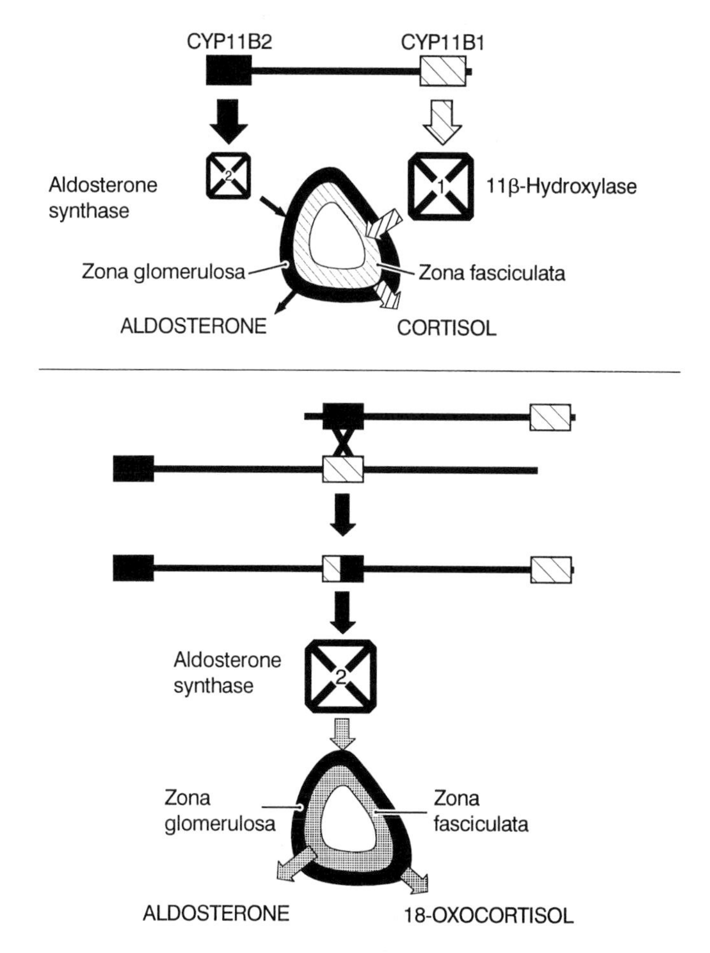

Figure 9.3. Schematic representation of *CYP11B1* and *CYP11B2* genes. Unequal crossing over generates a chimaeric *CYP11B1/2* gene that has aldosterone synthase activity but is expressed in the zona fasciculata. This causes glucocorticoid-suppressible hyperaldosteronism. Squares marked '1' and '2' denote the enzymes CYP11B1 and CYP11B2, respectively.

gene should be sufficient to cause the disorder, consistent with the known autosomal dominant mode of inheritance of this syndrome. Recently, abnormal expression of the chimaeric gene in the zona fasciculata was directly demonstrated by *in situ* hybridization studies of an adrenal gland from a patient with this disorder (Pascoe *et al.*, 1995).

The limited region in which cross-over breakpoints have been observed in glucocorticoid-suppressible hyperaldosteronism alleles suggests that there are functional constraints on the structures of chimaeric genes able to cause this disorder. One obvious constraint is that sufficient *CYP11B2*-coding sequences must be present in the chimaeric gene so that the encoded enzyme actually has aldosterone synthase (i.e. 18-hydroxylase and 18-oxidase) activity. As determined by expressing chimaeric cDNAs in cultured cells, chimaeric enzymes with amino-termini from CYP11B1 and carboxyl-termini from CYP11B2 have 18-oxidase activity only if at least the region encoded by exons 5–9 corresponds to CYP11B2. If the sequence of exon 5 corresponds instead to CYP11B1, the enzyme has 11β-hydroxylase but no 18-oxidase activity (Pascoe *et al.*, 1992). This is entirely consistent with the observation that no breakpoints in glucocorticoid-suppressible hyperaldosteronism alleles occur after exon 4. The chimaeric enzymes either have strong 18-oxidase activity or none detectable and there does not appear to be any location of cross-over that yields an enzyme with an intermediate level of 18-oxidase activity. Thus, there is no evidence for allelic variation in this disorder (i.e. variations in clinical severity are unlikely to be the result of different cross-over locations).

Presumably the transcriptional regulatory region of the chimaeric gene must correspond completely to that of *CYP11B1* or the chimaeric gene will not be expressed at sufficiently high levels in the zona fasciculata to cause the disorder. Although transcriptional regulatory elements in the *CYP11B* genes have not been completely defined, the fact that no breakpoints have been detected before intron 2 in glucocorticoid-suppressible hyperaldosteronism alleles suggests that there is a transcriptional enhancer in exon 1–intron 2 of *CYP11B1* or, conversely, a silencer in this region of *CYP11B2*.

Although it originally seemed possible that a 'mild' form of glucocorticoid-suppressible hyperaldosteronism might be a common aetiology of essential hypertension, the lack of allelic variation in this disorder makes this unlikely. However, other polymorphisms in the 5′-flanking region of *CYP11B2* have been documented (Lifton *et al.*, 1992b; White and Slutsker, 1995), although none has been shown to affect expression of the gene. It has also been suggested that polymorphisms in the coding sequence of *CYP11B2* might increase the aldosterone synthase activity of the enzyme and thus might be a risk factor for hypertension (Fardella *et al.*, 1995). Although such polymorphisms have been documented in rats, none have been found to occur naturally in humans.

Other factors such as kallikrein levels may affect the development of hypertension in this disorder (Dluhy and Lifton, 1995). One study found that blood pressure in persons with glucocorticoid-suppressible hyperaldosteronism is higher when the disease is inherited from the mother than when it is paternally inherited (Jamieson *et al.*, 1995). It is theoretically possible that the gene is imprinted (i.e. the maternal and paternal copies are expressed differently), but it seems more likely that exposure of the fetus to elevated levels of maternal aldosterone subsequently exacerbates the hypertension.

It is notable that many kindreds with glucocorticoid-suppressible hyperaldosteronism are of Anglo-Irish extraction (Lifton *et al.*, 1992a,b; Pascoe *et al.*, 1992). Moreover, the chromosomes carrying chimaeric genes tend to occur in association with specific polymorphisms in the *CYP11B* genes (Lifton *et al.*, 1992b), even though the duplications generating the chimaeric genes are apparently independent events. This suggests that one of these polymorphisms is, or is in linkage disequilibrium with, a structural polymorphism that predisposes to unequal crossing over during meiosis. Such features might include sequences similar to *chi* sites in bacteriophage lambda; this type of sequence has been postulated to increase the frequency of recombination in the *CYP21* genes (Amor *et al.*, 1988).

9.4 11β-Hydroxysteroid dehydrogenase

9.4.1 Structure and function of the mineralocorticoid receptor

Although membrane receptors for aldosterone may exist (Wehling, 1995), most effects of aldosterone are mediated by a specific nuclear receptor referred to as the mineralocorticoid or 'type 1 steroid' receptor. This receptor is expressed at high levels in renal distal convoluted tubules and cortical collecting ducts but also in other mineralocorticoid target tissues, including salivary glands and the colon. It is also found at multiple sites in the brain and at low levels in the myocardium and in the peripheral vasculature.

The mineralocorticoid receptor has a high degree of sequence identity with the glucocorticoid or 'type 2' receptor (Arriza *et al.*, 1987) but they share less than 15% similarity in their amino-terminal domains. This region in the glucocorticoid receptor, is known to interact with other nuclear transcription factors, and presumably has a similar function in the mineralocorticoid receptor (Pearce and Yamamoto, 1993). The centre of the molecule contains a DNA-binding domain consisting of two 'zinc fingers' and is also involved in dimerization of liganded receptors. The amino acid sequences of the mineralocorticoid and glucocorticoid receptors are 94% identical in this region. The carboxyl-terminus is a ligand-binding domain that is 57–60% identical in amino acid sequence in the two receptors.

As is the case with the glucocorticoid receptor, the unliganded mineralocorticoid receptor is located mainly in the cytoplasm (Robertson *et al.*, 1993). On binding of ligand, the receptor is translocated to the nucleus where it dimerizes and binds to hormone response elements (see Chapter 4) in the 5′-flanking regions of specific genes, thus increasing their transcription.

When cDNA encoding the mineralocorticoid receptor was cloned and expressed (Arriza *et al.*, 1987), it became apparent that this receptor had very similar binding affinities for aldosterone and for glucocorticoids such as corticosterone and cortisol. These results were consistent with observations that the rat hippocampal mineralocorticoid receptor had identical affinities for aldosterone and corticosterone (Krozowski and Funder, 1983), but it was difficult to reconcile with the fact that corticosterone and cortisol are relatively weak mineralocorticoids *in vivo*.

A mechanism conferring ligand specificity on the mineralocorticoid receptor was deduced from studies of two conditions in which the normal specificity of the receptor is lost; namely, the syndrome of apparent mineralocorticoid excess (AME) and liquorice intoxication. These conditions have three features in common: cortisol acts as a much stronger mineralocorticoid agonist than is normally the case, the plasma half-life of cortisol is prolonged, and urinary excretion of cortisone metabolites is decreased relative to excretion of cortisol metabolites, implying that 11β-hydroxysteroid dehydrogenase (11HSD), the enzyme catalysing the conversion of cortisol to cortisone, is decreased in activity.

It was proposed (Stewart *et al.*, 1987) that whereas AME represented a congenital deficiency of this enzyme, liquorice intoxication resulted from pharmacological inhibition of 11HSD. In either case, intra-renal concentrations of cortisol would be abnormally high as a consequence of deficient metabolism by 11HSD and would thus saturate mineralocorticoid receptors. This model was refined by studies of tissue distribution of 11HSD activity (Edwards *et al.*, 1988; Funder *et al.*, 1988). Activity was high in tissues such as the kidney and parotid gland in which mineralocorticoid receptors were specific for aldosterone, and low in the hippocampus and heart, tissues in which glucocorticoids are able to bind this receptor. Moreover, inhibition of 11HSD in various tissues by active components of liquorice resulted in loss of ligand specificity as evidenced by increased ability of tissues or cytosolic extracts to bind glucocorticoids.

Thus, it was hypothesized (Edwards *et al.*, 1988; Funder *et al.*, 1988) that oxidation by 11HSD of cortisol or corticosterone to cortisone or 11-dehydrocorticosterone, respectively, represented the physiological mechanism conferring specificity for aldosterone upon the mineralocorticoid receptor (*Figure 9.4*). Although cortisol and corticosterone bind the receptor avidly *in vitro*, cortisone and 11-dehydrocorticosterone are poor agonists for this receptor. Aldosterone is a poor substrate for 11HSD because, in solution, its 11-hydroxyl group is normally in a hemiacetal conformation with the 18-aldehyde group and is therefore not available for dehydrogenation.

9.4.2 Isozymes of 11HSD

There are two distinct isozymes of 11HSD. Both are members of the 'short-chain dehydrogenase' family including 17β-hydroxysteroid dehydrogenase (see Section 8.2.6). These enzymes all have a highly conserved nucleotide co-factor-binding domain near the amino-terminus; the co-factor functions as an electron acceptor for dehydrogenation (NAD^+ or $NADP^+$) and as an electron donor for reduction (NADH or NADPH). Completely conserved tyrosine and lysine residues toward the carboxyl-terminus function in catalysis. Conservative substitutions of either of these residues destroy enzymatic activity in a number of related enzymes (Chen *et al.*, 1993; Ensor and Tai, 1993; Ghosh *et al.*, 1991; Obeid and White, 1992).

X-ray crystallographic studies of a related enzyme, 3α,20β-hydroxysteroid dehydrogenase from *Streptomyces hydrogenans*, demonstrated that the conserved tyrosine and lysine residues are located near the pyridine ring of the co-factor in a cleft presumed to be the substrate-binding site (Ghosh *et al.*, 1991). In 11HSD, these two residues may facilitate transfer of a hydride ion (a proton plus two electrons) from the 11α position to $NADP^+$ or NAD^+. It is hypothesized that the ε-amino group of

---

Figure 9.4. Schematic of mineralocorticoid action. (Top) A normal mineralocorticoid target cell in a renal cortical collecting duct. Aldosterone occupies nuclear receptors (MR) that bind to hormone response elements (HRE), increasing transcription of genes and directly or indirectly increasing activities of apical sodium (Na) channels and the basolateral sodium–potassium (Na^+/K^+) ATPase. This increases resorption of sodium from and excretion of potassium into the tubular lumen. Cortisol, which circulates at higher levels than aldosterone, cannot occupy the receptor because it is oxidized to cortisone by 11β-hydroxysteroid dehydrogenase (11HSD). ER, endoplasmic reticulum. (Bottom) A cell from a patient with the syndrome of apparent mineralocorticoid excess. Because 11HSD is absent, cortisol inappropriately occupies mineralocorticoid receptors, leading to increased gene transcription, increased activity of sodium channels and the Na^+/K^+ ATPase, increased resorption of sodium and excretion of potassium, and hypertension.

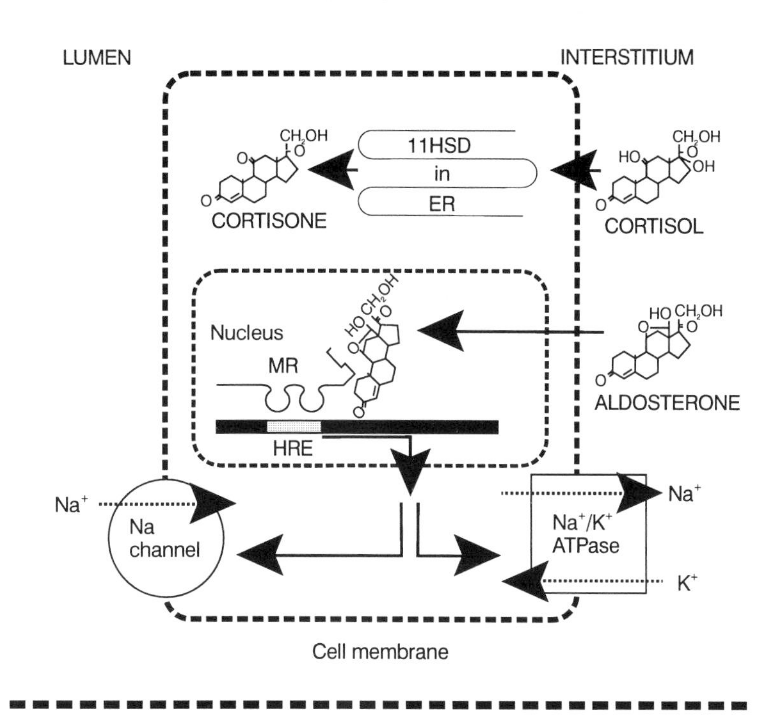
NORMAL
LUMEN
INTERSTITIUM
11HSD
in
ER
CORTISONE
CORTISOL
Nucleus
MR
HRE
ALDOSTERONE
Na+
Na
channel
Na+/K+
ATPase
K+
Cell membrane

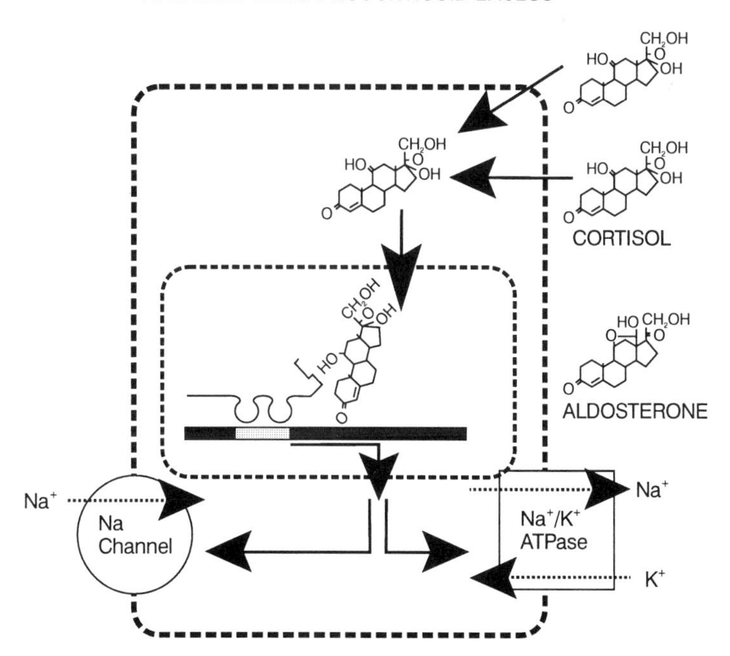
APPARENT MINERALOCORTICOID EXCESS
CORTISOL
ALDOSTERONE
Na+
Na
Channel
Na+/K+
ATPase
K+

the lysine deprotonates the phenolic group of the tyrosine. Deprotonation of a phenolic group in aqueous solution normally has a pK_a of about 10, but the local alkaline milieu provided by lysine lowers the apparent pK_a of the phenolic group of tyrosine into the physiologic range. The deprotonated phenolic group then removes a proton from the 11β-hydroxyl group of the steroid, leaving a negative charge on the 11 position of the steroid nucleus. This allows transfer of the 11α hydrogen (as a hydride) to the pyridine group of the co-factor.

The first isozyme of 11HSD that was characterized, termed the liver (L) or type I isozyme, was originally isolated from rat liver microsomes (Lakshmi and Monder, 1988) and the corresponding cDNA was cloned (Agarwal *et al.*, 1989). It requires $NADP^+$ as a co-factor and has an affinity for steroids in the micromolar range. Although the enzyme purified from rat liver functions only as a dehydrogenase, the recombinant enzyme expressed from cloned cDNA exhibits both 11β-dehydrogenase and the reverse oxidoreductase activity (conversion of 11-dehydrocorticosterone to corticosterone) when expressed in mammalian cells (Agarwal *et al.*, 1989) suggesting that the reductase activity is destroyed during purification from the liver.

Several lines of evidence suggest that this isozyme does not play a significant role in conferring ligand specificity on the mineralocorticoid receptor. It is expressed at highest levels in the liver, which does not respond to mineralocorticoids, and although it is expressed at high levels in the rat kidney (Agarwal *et al.*, 1989), it is expressed at much lower levels in human kidneys (Tannin *et al.*, 1991). Even in rat kidney, immunoreactivity to the protein is observed primarily in proximal tubules and not in distal tubules and collecting ducts, the sites of mineralocorticoid action (Rundle *et al.*, 1989). Finally, when the *HSD11L* (*HSD11B1*) gene encoding this isozyme was cloned (Tannin *et al.*, 1991) and examined for mutations in patients with AME, none were found (Nikkila *et al.*, 1993).

Accordingly, a second isozyme was sought in mineralocorticoid target tissues. Evidence for an NAD^+-dependent isozyme was obtained from histochemical studies of rat kidney (Mercer and Krozowski, 1992). In isolated rabbit kidney cortical collecting duct cells, 11HSD was detected in the microsomal fraction (Rusvai and Naray-Fejes-Toth, 1993). This activity was almost exclusively NAD^+ dependent and had a very high affinity for steroids (K_m for corticosterone of 26 nM). There was almost no reduction of 11-dehydrocorticosterone to corticosterone, suggesting that, unlike the L isozyme, the kidney (K) or type II isozyme only catalysed dehydrogenation. The enzyme in the human placenta had similar characteristics (Brown *et al.*, 1993); it was NAD^+ dependent and had K_m values for steroids in the nanomolar range. Similar activities were noted in sheep kidney (Yang and Yu, 1994) and in many human fetal tissues (Stewart *et al.*, 1994).

Thus far, the K isozyme has not been purified to homogeneity in active form from any source. This rendered the cloning of the corresponding cDNA more difficult. It was eventually accomplished by expression screening strategies in which pools of clones were assayed for their ability to confer NAD^+-dependent 11HSD activity on *Xenopus* oocytes or cultured mammalian cells. Positive pools were divided into smaller pools and rescreened until a single positive clone was identified. Both sheep (Agarwal *et al.*, 1994) and human (Albiston *et al.*, 1994) cDNA encoding this isoform were isolated in this manner. The recombinant K isozyme has properties that are virtually identical to the activity found in mineralocorticoid target tissues. The recombinant enzyme functions exclusively as a dehydrogenase; no reductase activity is detectable with either NADH or NADPH as a co-factor (Agarwal *et al.*, 1994; Albiston *et al.*, 1994). It has an almost exclusive preference for NAD^+ as a co-factor and a very high affinity for glucocorticoids. The K isozyme is expressed in mineralocorticoid target tissues, particularly the kidney, and in human placenta, whereas it is not detected in the liver.

The predicted amino acid sequence of the K isozyme is only 21% identical to that of the L isozyme of 11HSD. The enzyme consists of 404 amino acid residues. The corresponding gene, termed *HSD11K* or *HSD11B2*, is located on chromosome 16q22 (Agarwal *et al.*, 1995). It consists of five exons spaced over approximately 6 kb (*Figure 9.5*). This organization differs from *HSD11L*, suggesting that the two isozymes are only distantly related.

9.4.3 Clinical features of the syndrome of AME

AME is an inherited syndrome in which children develop hypertension, hypokalaemia and low plasma renin activity. Other clinical features include moderate intra-uterine growth retardation and postnatal failure to thrive. Consequences of the often severe hypokalaemia include nephrocalcinosis, nephrogenic diabetes insipidus and rhabdomyolysis. Complications of hypertension have included cerebrovascular accidents, and several patients have died during infancy or adolescence. Several affected sibling pairs have been reported but parents have usually been asymptomatic, suggesting that AME is a genetic disorder with an autosomal recessive mode of inheritance.

A low salt diet or blockade of mineralocorticoid receptors with spironolactone ameliorates the hypertension whereas ACTH and hydrocortisone exacerbate it. Levels of all known mineralocorticoids are low (Oberfield *et al.*, 1983; Ulick *et al.*, 1979). These findings suggest that cortisol (i.e. hydrocortisone) acts as a stronger mineralocorticoid than is normally the case. Indeed, patients with AME have abnormal cortisol metabolism. Cortisol half-life in plasma is prolonged from

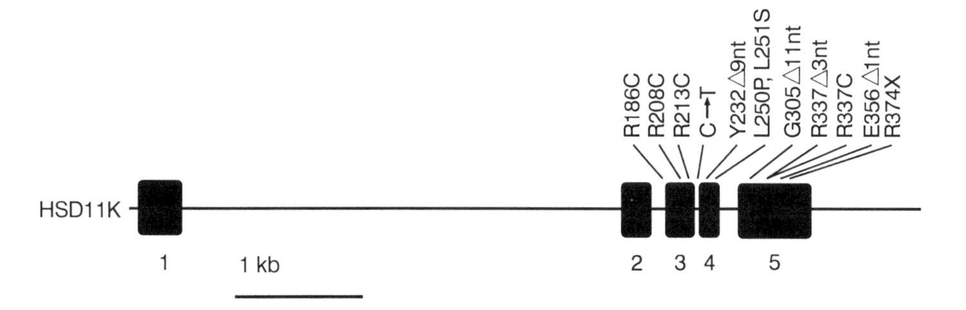

Figure 9.5. Diagram of the *HSD11K* gene showing locations of mutations causing apparent mineralocorticoid excess. Exons are represented by numbered boxes. See nomenclature in *Figure 9.2* legend.

approximately 80 to 120–190 min (Ulick *et al.*, 1979). Very low levels of cortisone metabolites are excreted in the urine as compared with cortisol metabolites, indicating a marked deficiency in 11HSD, the enzyme catalysing the conversion of cortisol to cortisone. This has been assayed directly by administering 11α[^{3}H]-cortisol to subjects and measuring the production of tritiated water. Most often the deficiency is measured as an increase in the sum of the urinary concentrations of tetrahydrocortisol and allo-tetrahydrocortisol, divided by the concentration of tetrahydrocortisone, abbreviated (THF+aTHF)/THE. However, 11-reduction is unimpaired; labeled cortisone administered to patients is excreted entirely as cortisol and other 11β-reduced metabolites (Shackleton *et al.*, 1985).

Similar but milder abnormalities occur with liquorice intoxication (Stewart *et al.*, 1987). The active component of liquorice, glycyrrhetinic acid, inhibits 11HSD in isolated rat kidney microsomes (Monder *et al.*, 1989). Thus, it appears that liquorice intoxication is a reversible pharmacological counterpart to the inherited syndrome of apparent mineralocorticoid excess.

Juvenile hypertension, marked hypokalaemia and suppressed plasma renin activity are also found in Liddle's syndrome caused by activating mutations in the regulatory subunits of the sodium channel (see Section 9.5.4). However, Liddle's syndrome has an autosomal dominant mode of inheritance, and whereas it can be treated by blockade of the epithelial sodium channel with amiloride or triamterine, blockade of the mineralocorticoid receptor with spironolactone is not effective.

9.4.4 Detection of mutations in HSD11K *in patients with AME*

Thus far, 11 different mutations have been identified in the *HSD11K* gene in 15 kindreds with AME (*Figure 9.5*). These mutations all affect enzymatic activity or pre-mRNA splicing, thus confirming the hypothesis that 11HSD protects the mineralocorticoid receptor from high concentrations of cortisol (Mune and White, 1996; Mune *et al.*, 1995; Stewart *et al.*, 1996; Wilson *et al.*, 1995a,b). Only one patient has been a compound heterozygote for two different mutations, whereas all other patients have carried homozygous mutations. This suggests that the prevalence of AME mutations in the general population is low, so that the disease is found mostly in limited populations in which inbreeding is relatively high. Six kindreds are of Native American origin. Three from Minnesota or Canada carry the same mutation (L250S,L251P), consistent with a founder effect, but the others are each homozygous for a different mutation. The reason for the relatively high prevalence of this very rare disease among Native Americans is not immediately apparent.

Of the mutations identified thus far, two shift the reading frame of translation, a third deletes three amino acids including the catalytic tyrosine residue (Y232) and one is a nonsense mutation. One mutation in the third intron leads to skipping of the fourth exon during processing of pre-mRNA (Mune *et al.*, 1995). As the fourth exon encodes the catalytic site, the resulting enzyme is again presumably inactive. The other six mutations have been introduced into cDNA and expressed in cultured cells to determine their effects. One is completely inactive and one has only a trace of activity. The others are all partially active in cultured cells with one, R337C, having greater than 50% of normal activity (Mune and

White, 1996). Only R337C is partially active in lysed cells [although Obeyesekere *et al.* (1995) reported this mutation to be inactive in cell lysates, they did not utilize conditions which would promote enzyme stability]. Thus, most mutations of this enzyme adversely affect protein stability once cells are lysed; this has been confirmed by Western blots (Mune and White, 1996). Both the wild-type enzyme and most mutants are concentrated in the nucleus as determined by Western blots of cell fractions. This may reflect the enzyme's function in protecting the nuclear mineralocorticoid receptor from excessive concentrations of cortisol.

Although the number of patients with AME is small, sufficient data now exist to demonstrate a statistically significant correlation between degree of enzymatic impairment and biochemical severity as measured by the precursor:product ratio, (THF+aTHF)/THE (Mune and White, 1996). This correlation is most obvious for the partially active mutants. We assume that R337C is the only significant mutation in the patients who carry it, even though only one exon of the gene was sequenced (Wilson *et al.*, 1995b). If so, a 50% impairment of enzymatic activity is apparently sufficient to compromise metabolism of cortisol in the kidney, suggesting that there is very little excess capacity to metabolize cortisol in this organ. This seems to raise a paradox, because AME is a recessive disorder and heterozygous carriers, who would be expected to have 50% of normal activity, are asymptomatic. Altered stability or kinetic properties of the R337C mutant may be important, including alterations in enzyme inhibition by end product (i.e. cortisone or corticosterone) or by other circulating steroids.

Because of the small numbers of patients, and the possible confounding effects of prior antihypertensive therapy, it is difficult to correlate biochemical severity with measures of clinical severity, although anecdotal reports suggest that mutations that do not destroy activity may be associated with milder disease (Mune *et al.*, 1995; Wilson *et al.*, 1995b). With the elucidation of the molecular genetic basis of this disorder, ascertainment of additional cases may permit these questions to be answered.

9.4.5 **HSD11K** *as a candidate locus for essential hypertension*

Whereas apparent 11HSD deficiency causes severe hypertension, it is reasonable to hypothesize that milder decreases in enzymatic activity might be associated with common 'essential' hypertension. Patients with AME are often born with a mild to moderate degree of intra-uterine growth retardation. Although the reason for this is not known, it seems likely that deficiency of 11HSD in the placenta permits excessive quantities of maternal glucocorticoids to cross the placenta and thus inhibit fetal growth (Reinisch *et al.*, 1978). Thus, a hypothetical mild form of 11HSD deficiency might also present with low birthweight and subsequent hypertension (Edwards *et al.*, 1993). In rats, placental 11HSD activity is inversely correlated with placental weight and directly correlated with term fetal weight (Benediktsson *et al.*, 1993). In human population studies, most of which are retrospective, low birthweight and increased placental weight are indeed risk factors for subsequent development of adult hypertension (Barker *et al.*, 1990). Although variations in 11HSD might in principle be responsible for this correlation, a recent study in humans (Stewart *et al.*, 1995) did not find such a correlation between placental 11HSD activity and placental weight. A weak but significant positive correlation was observed between 11HSD activity and fetal birthweight, but a subsequent larger study of the identical population was unable to confirm this (unpublished observations). Thus, the currently available data do not support the idea that low 11HSD activity is a risk factor for low birthweight in humans who do not suffer from AME. Of course, this does not rule out a possible effect of genetically determined mild variations in 11HSD activity upon blood pressure. Molecular studies of *HSD11K* should unambiguously determine if this gene is frequently involved in the development of hypertension. These might include linkage studies and a search for frequent polymorphisms in *HSD11K* that might be associated with the development of hypertension.

9.5 The amiloride-inhibitable epithelial sodium channel

9.5.1 Regulation of the epithelial sodium channel

The sodium channel is regulated by several parallel pathways (Bubien *et al.*, 1994). Protein kinase A phosphorylates the channel and increases single channel open probability. Cholera toxin can thus increase channel activity through cAMP signalling pathways. In addition, G-proteins (e.g. $G\alpha_{i3}$, see Chapter 3) interact directly with the channel to inhibit it. Pertussin toxin increases channel activity by ADP-ribosylating $G\alpha_i$, causing it to associate with $G\beta\gamma$ subunits and thus inhibiting its association with the channel. The protein kinase A- and G-protein-mediated pathways interact in a complex but usually mutually inhibitory manner.

9.5.2 Genes encoding subunits of the sodium channel

The first cDNA encoding a vertebrate sodium channel subunit was isolated from rat colon by expression screening using *Xenopus* oocytes, looking for an amiloride-sensitive current when sodium was added to the extracellular buffer. The 'α'-subunit encoded by this clone was predicted to contain 698 amino acids with two transmembrane segments (*Figure 9.6*). The structure of this protein was similar to that of several mechanoreceptor proteins in the nematode *Caenorhabditis elegans* (Canessa *et al.*, 1993). The large central domain between these transmembrane segments contains several potential sites for N-glycosylation. The C-terminal region is rich in proline residues. This region binds to the *src* homology 3 (SH3) domain of α-spectrin and mediates localization to the apical side of epithelial cells (Rotin *et al.*, 1994).

Additional regulatory subunits were also isolated by expression screening, looking for clones that acted synergistically with that encoding the α-subunit to increase the sodium current in *Xenopus* oocytes. Clones encoding two additional subunits, β and γ, were identified. These subunits had a similar organization to that of the α-subunit and were each 34–35% identical to it in amino acid sequence; the β- and γ-subunits contained 638 and 650 amino acids, respectively. When the corresponding RNA was injected into *Xenopus* oocytes, the β- and γ-subunits did not generate a sodium current either alone or together, but when co-injected with the α-subunit each increased the sodium current by 3–5-fold over

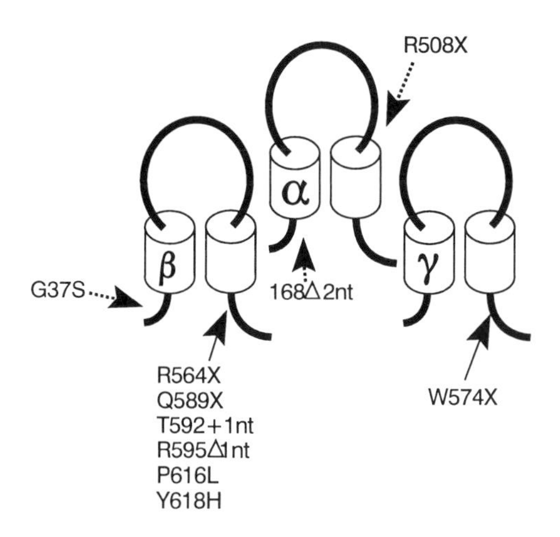

Figure 9.6. Schematic of the epithelial sodium channel. Three subunits are shown; the actual number of subunit molecules in a single functional channel is not known. The cylinders represent transmembrane domains, but the cell membrane is not shown. The cytoplasmic domains are toward the bottom of the figure. Mutations causing Liddle's syndrome are marked by solid arrows, and mutations causing pseudohypoaldosteronism are marked by broken arrows. See nomenclature in *Figure 9.2* legend.

that observed with injection of α-subunit RNA alone. When RNAs for all three subunits were injected together, the observed current was more than 100 times greater than that seen with the α-subunit alone. Thus, β and γ are regulatory subunits (Canessa *et al.*, 1994). All three subunits are apparently expressed in the same cells (Duc *et al.*, 1994). Although the optimum stoichiometry has not been determined, it seems likely that the functional sodium channel contains the α-, β- and γ-subunits in equimolar ratios. The channel could consist of one of each subunit and thus contain six transmembrane domains, but other regulated channels such as the cystic fibrosis transmembrane conductance regulator and the sodium–calcium exchanger contain 12 transmembrane domains. If the sodium channel has a similar organization, this would require two of each subunit.

Recently cDNA encoding a novel δ-subunit was isolated. This subunit is expressed in gonads, pancreas and brain but not in the kidney. The δ-subunit forms an active complex with the β- and γ-subunits and thus can substitute for the α-subunit, although its properties differ with regard to conductance, ionic specificity and sensitivity to amiloride (Waldmann *et al.*, 1995).

9.5.3 Clinical features of Liddle's syndrome

Patients with Liddle's syndrome, an inherited autosomal dominant disorder with high penetrance, have signs of mineralocorticoid excess including hypertension and usually, but not always, hypokalaemic alkalosis. However, levels of aldosterone and other known mineralocorticoids are subnormal. Blockade of the mineralocorticoid receptor with spironolactone does not ameliorate these problems, but sodium restriction combined with the potassium-sparing diuretic triamterine is effective in reducing blood pressure and increasing serum potassium levels (Botero-Velez *et al.*, 1994; Liddle *et al.*, 1963). Although affected individuals excrete relatively high amounts of potassium in the urine, the electrolyte composition of saliva is normal.

A number of affected subjects have died in their 30s of hypertensive vascular disease. The index case developed chronic renal failure necessitating a renal transplant. After transplantation, her hypokalaemia resolved and her blood pressure decreased to near normal levels.

These features suggest that there is a structural abnormality in the kidney in this disorder rather than abnormal levels of any circulating humoral factor. The improvement in signs of mineralocorticoid excess with triamterine, an agent that inhibits the sodium channel in the renal distal tubule and collecting duct, suggests that the abnormality is one of regulation of the sodium channel. This has been demonstrated by genetic studies.

9.5.4 Genetics of Liddle's syndrome

In the original kindred reported by Liddle *et al.* (1963), the disease is completely linked genetically to polymorphic markers in and near the *SCNN1B* gene encoding the β-subunit of the epithelial sodium channel on chromosome 16p12–13 (Shimkets *et al.*, 1994). In this kindred, the gene carries a nonsense mutation (R564X) that upon translation leaves both transmembrane domains intact but deletes almost the entire C-terminal cytoplasmic domain (*Figure 9.6*). The identical mutation has been detected in another kindred. Expression of the truncated β-subunit in *Xenopus* oocytes together with wild-type α- and γ-subunits increased channel activity threefold over that seen when the wild-type β-subunit was used (Schild *et al.*, 1995). Three other kindreds have nonsense or frameshift mutations in the same region (Q589X, T592+1nt, R595Δ1nt) (Shimkets *et al.*, 1994). Missense mutations have also been identified, P616L occurring *de novo* (Hansson *et al.*, 1995b) and Y618H in a larger kindred (Tamura *et al.*, 1996). Expression of β-subunits carrying each of the missense mutations in *Xenopus* oocytes again leads to markedly greater stimulation of sodium channel activity than is seen with the wild-type subunit. These findings suggest that an important regulatory site in the β-subunit exists in the cytoplasmic C-terminal domain including P616-Y618, and that mutation or deletion of this site leads to constitutive activation of the receptor. Deletion of the corresponding domain of the γ-subunit also constitutively activates the sodium channel when the truncated subunit is expressed in *Xenopus* oocytes. Such a mutation in the *SCNN1G* gene on chromosome 12p13.1-pter that encodes the γ-subunit has been detected in a kindred with Liddle's syndrome and leads to an identical clinical phenotype (Hansson *et al.*, 1995a). Thus, the β- and γ-subunits apparently act equivalently and synergistically to regulate the epithelial sodium channel.

Because activating mutations in the genes encoding the β- and γ-subunits cause severe hypertension, it is plausible that milder activating mutations might underly some cases of essential hypertension. At this time, however, no studies have demonstrated any linkage of the *SCNN1B* or *SCNN1G* genes to milder forms of hypertension.

9.5.5 Pseudohypoaldosteronism

If activating mutations in the regulatory subunits of the epithelial sodium channel lead to signs and symptoms of aldosterone excess with suppressed renin and aldosterone levels, one might expect that mutations that cause loss of function of one of the three subunits of the receptor would lead, conversely, to signs of mineralocorticoid deficiency such as hyponatraemia, hypovolaemia, hyperreninaemia and hyperkalaemia. Pseudohypoaldosteronism is such a condition.

Patients with pseudohypoaldosteronism present shortly after birth with signs of mineralocorticoid deficiency but with markedly elevated plasma concentrations of aldosterone and renin. Because these patients are resistant to mineralocorticoids, they must be treated with sodium chloride supplements. Signs and symptoms of aldosterone resistance decrease with age in many patients so that concentrations of aldosterone fall and salt supplements may be discontinued when a patient is a few years old. Other patients, however, are resistant to therapy and they may die in infancy from hyperkalaemia.

Sporadic, autosomal dominant and recessive forms of pseudohypoaldosteronism are recognized. In patients with the autosomal dominant form, aldosterone resistance is confined to the renal tubule and is less severe than in patients with the autosomal recessive form. In patients with the recessive form, the resistance is expressed in both the renal tubules and other aldosterone target tissues including salivary and sweat glands and the colon (Kuhnle *et al.*, 1995).

Because patients were resistant to mineralocorticoids, it was initially assumed that the defect lay in the mineralocorticoid receptor itself. This hypothesis was supported by studies demonstrating reduced numbers of mineralocorticoid receptors in peripheral blood mononuclear leukocytes (Armanini *et al.*, 1995). However, no mutation affecting function was detected in any of seven unrelated patients with pseudohypoaldosteronism (Arai *et al.*, 1995; Komesaroff *et al.*, 1994; Zennaro *et al.*, 1994). Presumably, patients with this disorder have decreased numbers of mineralocorticoid receptors because elevated blood levels of aldosterone downregulate the receptor.

The autosomal recessive form of the disorder segregates with the genes encoding the various subunits of the epithelial sodium channel (Strautnieks *et al.*, 1996), and nonsense, frameshift and missense mutations in the α- or β-subunits have been identified in five kindreds with this disorder (Chang *et al.*, 1996) (*Figure 9.6*). Thus, Liddle's syndrome and pseudohypoaldosteronism represent allelic variants.

Acknowledgement

Supported by grants DK37867 and DK42169 from the National Institutes of Health.

References

Agarwal AK, Monder C, Eckstein B, White PC. (1989) Cloning and expression of rat cDNA encoding corticosteroid 11 beta-dehydrogenase. *J. Biol. Chem.* **264:** 18939–18943.

Agarwal AK, Mune T, Monder C, White PC. (1994) NAD+-dependent isoform of 11 beta hydroxysteroid dehydrogenase: cloning and characterization of cDNA from sheep kidney. *J. Biol. Chem.* **269:** 25959–25962.

Agarwal AK, Rogerson FM, Mune T, White PC. (1995) Gene structure and chromosomal localization of the human HSD11K gene encoding the kidney (type 2) isozyme of 11β-hydroxysteroid dehydrogenase. *Genomics* **29:** 195–199.

Albiston AL, Obeyesekere VR, Smith RE, Krozowski ZS. (1994) Cloning and tissue distribution of the human 11HSD type 2 enzyme. *Mol. Cell. Endocrinol.* **105:** R11–R17.

Amor M, Parker KL, Globerman H, New MI, White PC. (1988) Mutation in the CYP21B gene (Ile-172-Asn) causes steroid 21-hydroxylase deficiency. *Proc. Natl Acad. Sci. USA* **85:** 1600–1604.

Arai K, Tsigos C, Suzuki Y, Listwak S, Zachman K, Zangeneh F, Rapaport R, Chanoine JP, Chrousos GP. (1995) No apparent mineralocorticoid receptor defect in a series of sporadic cases of pseudohypoaldosteronism. *J. Clin. Endocrinol. Metab.* **80:** 814–817.

Armanini D, Karbowiak I, Zennaro CM, Zovato S, Pratesi C, De Lazzari P, Krozowski Z, Kuhnle U. (1995) Pseudohypoaldosteronism: evaluation of type I receptors by radioreceptor assay and by antireceptor antibodies. *Steroids* **60:** 161–163.

Arriza JL, Weinberger C, Cerelli G, Glaser TM, Handelin BL, Housman DE, Evans RM. (1987) Cloning of human mineralocorticoid receptor complementary DNA: structural and functional kinship with the glucocorticoid receptor. *Science* **237:** 268–275.

Barker DJ, Bull AR, Osmond C, Simmonds SJ. (1990) Fetal and placental size and risk of hypertension in adult life. *Br. Med. J.* **301:** 259–262.

Benediktsson R, Lindsay RS, Noble J, Seckl JR, Edwards CR. (1993) Glucocorticoid exposure in utero: new model for adult hypertension. *Lancet* **341:** 339–341.

Berridge MJ. (1993) Inositol trisphosphate and calcium signalling. *Nature* **361:** 315–325.

Berry CA, Ives HF, Rector FC. (1996) Renal transport of glucose, amino acids, sodium, chloride and water. In: *The Kidney*, 5th Edn. (ed. BM Brenner). W.B.Saunders, Philadelphia, pp. 334–370.

Bonnardeaux A, Davies E, Jeunemaitre X *et al.* (1994) Angiotensin II type 1 receptor gene polymorphisms in human essential hypertension. *Hypertension* **24:** 63–69.

Botero-Velez M, Curtis JJ, Warnock DG. (1994) Brief report: Liddle's syndrome revisited – a disorder of sodium reabsorption in the distal tubule. *New Engl. J. Med.* **330:** 178–181.

Brasier AR, Li J, Copland A. (1994) Transcription factors modulating angiotensinogen gene expression in hepatocytes. *Kidney Int.* **46:** 1564–1566.

Brown RW, Chapman KE, Edwards CR, Seckl JR. (1993) Human placental 11 beta-hydroxysteroid dehydrogenase: evidence for and partial purification of a distinct NAD-dependent isoform. *Endocrinology* **132:** 2614–2621.

Bubien JK, Jope RS, Warnock DG. (1994) G-proteins modulate amiloride-sensitive sodium channels. *J. Biol. Chem.* **269:** 17780–17783.

Campbell DJ, Habener JF. (1986) Angiotensinogen gene is expressed and differentially regulated in multiple tissues of the rat. *J. Clin. Invest.* **78:** 31–39.

Canessa CM, Horisberger JD, Rossier BC. (1993) Epithelial sodium channel related to proteins involved in neurodegeneration. *Nature* **361:** 467–470.

Canessa CM, Schild L, Buell G, Thorens B, Gautschi I, Horisberger JD, Rossier BC. (1994) Amiloride-sensitive epithelial Na+ channel is made of three homologous subunits. *Nature* **367:** 463–467.

Caulfield M, Lavender P, Farrall M, Munroe P, Lawson M, Turner P, Clark AJ. (1994) Linkage of the angiotensinogen gene to essential hypertension. *New Engl. J. Med.* **330:** 1629–1633.

Caulfield M, Lavender P, Newell-Price J *et al.* (1995) Linkage of the angiotensinogen gene locus to human essential hypertension in African Caribbeans. *J. Clin. Invest.* **96:** 687–692.

Chang SS, Grunder S, Hanukoglu A *et al.* (1996) Mutations in subunits of the epithelial sodium channel cause salt wasting with hyperkalaemic acidosis, pseudohypoaldosteronism type 1. *Nature Genet.* **12:** 248–253.

Chen Z, Jiang JC, Lin ZG, Lee WR, Baker ME, Chang SH. (1993) Site-specific mutagenesis of Drosophila alcohol dehydrogenase: evidence for involvement of tyrosine-152 and lysine-156 in catalysis. *Biochemistry* **32:** 3342–3346.

Chua SC, Szabo P, Vitek A, Grzeschik KH, John M, White PC. (1987) Cloning of cDNA encoding steroid 11 beta-hydroxylase (P450c11). *Proc. Natl Acad. Sci. USA* **84:** 7193–7197.

Connell JM, Kenyon CJ, Corrie JE, Fraser R, Watt R, Lever AF. (1986) Dexamethasone-suppressible hyperaldosteronism. Adrenal transition cell hyperplasia? *Hypertension* **8:** 669–676.

Curnow KM, Tusie-Luna MT, Pascoe L, Natarajan R, Gu JL, Nadler JL, White PC. (1991) The product of the CYP11B2 gene is required for aldosterone biosynthesis in the human adrenal cortex. *Mol. Endocrinol.* **5:** 1513–1522.

Curnow KM, Pascoe L, White PC. (1992) Genetic analysis of the human type-1 angiotensin II receptor. *Mol. Endocrinol.* **6:** 1113–1118.

Curnow KM, Slutsker L, Vitek J, Cole T, Speiser PW, New MI, White PC, Pascoe L. (1993) Mutations in the CYP11B1 gene causing congenital adrenal hyperplasia and hypertension cluster in exons 6, 7, and 8. *Proc. Natl Acad. Sci. USA* **90:** 4552–4556.

Dluhy RG, Lifton RP. (1995) Glucocorticoid-remediable aldosteronism (GRA): diagnosis, variability of phenotype and regulation of potassium homeostasis. *Steroids* **60:** 48–51.

Doolittle RF. (1983) Angiotensinogen is related to the antitrypsin-antithrombin-ovalbumin family. *Science* 222: 417–419.

Duc C, Farman N, Canessa CM, Bonvalet JP, Rossier BC. (1994) Cell-specific expression of epithelial sodium channel alpha, beta, and gamma subunits in aldosterone-responsive epithelia from the rat: localization by in situ hybridization and immunocytochemistry. *J. Cell. Biol.* **127:** 1907–1921.

Duchatelle P, Ohara A, Ling BN, Kemendy AE, Kokko KE, Matsumoto PS, Eaton DC. (1992) Regulation of renal epithellial sodium channels. *Mol. Cell. Biochem.* **114:** 27–34.

Edwards CR, Stewart PM, Burt D, Brett L, McIntyre MA, Sutanto WS, de Kloet ER, Monder C. (1988) Localisation of 11 beta-hydroxysteroid dehydrogenase – tissue specific protector of the mineralocorticoid receptor. *Lancet* 2: 986–989.

Edwards CR, Benediktsson R, Lindsay RS, Seckl JR. (1993) Dysfunction of placental glucocorticoid barrier: link between fetal environment and adult hypertension? *Lancet* **341:** 355–357.

Ensor CM, Tai HH. (1993) Site-directed mutagenesis of the conserved tyrosine-151 of human placental NAD+-dependent 15-hydroxyprostaglandin dehydrogenase yields a catalytically inactive enzyme. *Biochem. Biophys. Res. Commun.* **176:** 840–845.

Fardella CE, Rodriguez H, Hum DW, Mellon SH, Miller WL. (1995) Artificial mutations in P450c11AS (aldosterone synthase) can increase enzymatic activity: a model for low-renin hypertension? *J. Clin. Endocrinol. Metab.* **80:** 1040–1043.

Funder JW, Pearce PT, Smith R, Smith AI. (1988) Mineralocorticoid action: target tissue specificity is enzyme, not receptor, mediated. *Science* **242:** 583–585.

Gaillard I, Clauser E, Corvol P. (1989) Structure of human angiotensinogen gene. *DNA* **8:** 87–99.

Ganguly A, Weinberger MH, Guthrie GP, Fineberg NS. (1984) Adrenal steroid responses to ACTH in glucocorticoid-suppressible aldosteronism. *Hypertension* **6:** 563–567.

Geley S, Kapelari K, Johrer K, Peter M, Glatzl J, Vierhapper H, Sippell WG, White PC, Kofler R. (1996) CYP11B1 mutations causing congenital adrenal hyperplasia due to 11β-hydroxylase deficiency. *J. Clin. Endocrinol. Metab.* **81:** 2896–2901.

Ghosh D, Weeks CM, Grochulski P, Duax WL, Erman M, Rimsay RL, Orr JC. (1991) Three-dimensional structure of holo 3 alpha,20 beta-hydroxysteroid dehydrogenase: a member of a short-chain dehydrogenase family. *Proc. Natl Acad. Sci. USA* **88:** 10064–10068.

Gomez-Sanchez CE, Gill JR, Jr., Ganguly A, Gordon RD. (1988) Glucocorticoid-suppressible aldosteronism: a disorder of the adrenal transitional zone. *J. Clin. Endocrinol. Metab.* **67:** 444–448.

Griffing GT, Dale SL, Holbrook MM, Melby JC. (1983) 19-nor-deoxycorticosterone excretion in primary aldosteronism and low renin hypertension. *J. Clin. Endocrinol. Metab.* **56:** 218–221.

Hague WM, Honour JW. (1983) Malignant hypertension in congenital adrenal hyperplasia due to 11 beta-hydroxylase deficiency. *Clin. Endocrinol. (Oxf)* **18:** 505–510.

Hall CE, Gomez-Sanchez CE. (1986) Hypertensive potency of 18-oxocortisol in the rat. *Hypertension* **8:** 317–322.

Hamlet SM, Gordon RD, Gomez-Sanchez CE, Tunny TJ, Klemm SA. (1988) Adrenal transitional zone Steroids 18-oxo and 18-hydroxycortisol, useful in the diagnosis of primary aldosteronism, are ACTH-dependent. *Clin. Exp. Pharmacol. Physiol.* **15:** 317–322.

Hansson JH, Nelson-Williams C, Suzuki H *et al.* (1995a) Hypertension caused by a truncated epithelial sodium channel gamma subunit: genetic heterogeneity of Liddle syndrome. *Nature Genet.* **11:** 76–82.

Hansson JH, Schild L, Lu Y, Wilson TA, Gautschi I, Shimkets R, Nelson-Williams C, Rossier BC, Lifton RP. (1995b) A de novo missense mutation of the beta subunit of the epithelial sodium channel causes hypertension and Liddle syndrome, identifying a proline-rich segment critical for regulation of channel activity. *Proc. Natl Acad. Sci. USA* **92:** 11495–11499.

Hashimoto T, Morohashi K, Takayama K, Honda S, Wada T, Handa H, Omura T. (1992) Cooperative transcription activation between Ad1, a CRE-like element, and other elements in the CYP11B gene promoter. *J. Biochem. (Tokyo)* **112:** 573–575.

Helmberg A, Ausserer B, Kofler R. (1992) Frame shift by insertion of 2 basepairs in codon 394 of CYP11B1 causes congenital adrenal hyperplasia due to steroid 11 beta-hydroxylase deficiency. *J. Clin. Endocrinol. Metab.* **75:** 1278–1281.

Honda S, Morohashi K, Omura T. (1990) Novel cAMP regulatory elements in the promoter region of bovine P-450(11 beta) gene. *J. Biochem. (Tokyo)* **108:** 1042–1049.

Honda S, Morohashi K, Nomura M, Takeya H, Kitajima M, Omura T. (1993) Ad4BP regulating steroidogenic P-450 gene is a member of steroid hormone receptor superfamily. *J. Biol. Chem.* **268:** 7494–7502.

Horisberger JD, Rossier BC. (1992) Aldosterone regulation of gene transcription leading to control of ion transport. *Hypertension* **19:** 221–227.

Inoue I, Rohrwasser A, Helin C *et al.* (1995) A mutation of angiotensinogen in a patient with preeclampsia leads to altered kinetics of the renin-angiotensin system. *J. Biol. Chem.* **270:** 11430–11436.

Isa M, Boyd E, Morrison N, Harrap E, Clauser E, Connor JM. (1990) Assignment of the human angiotensinogen gene to chromjsome 1q42-q43 by nonisotopic *in situ* hybridization. *Genomics* **8:** 598–600.

Ishigami T, Umemura S, Iwamoto T *et al.* (1995) Molecular variant of angiotensinogen gene is associated with coronary atherosclerosis. *Circulation* **91:** 951–954.

Jamieson A, Slutsker L, Inglis GC, Fraser R, White PC, Connell JM. (1995) Glucocorticoid-suppressible hyperaldosteronism: effects of crossover site and parental origin of chimaeric gene on phenotypic expression. *Clin. Sci.* **88:** 563–570.

Jeunemaitre X, Lifton RP, Hunt SC, Williams RR, Lalouel JM. (1992a) Absence of linkage between the angiotensin converting enzyme locus and human essential hypertension. *Nature Genet.* **1:** 72–75.

Jeunemaitre X, Rigat B, Charru A, Houot AM, Soubrier F, Corvol P. (1992b) Sib pair linkage analysis of renin gene haplotypes in human essential hypertension. *Hum. Genet.* **88:** 301–306.

Jeunemaitre X, Soubrier F, Kotelevtsev YV *et al.* (1992c) Molecular basis of human hypertension: role of angiotensinogen. *Cell* **71:** 169–180.

John ME, John MC, Boggaram V, Simpson ER, Waterman MR. (1986) Transcriptional regulation of steroid hydroxylase genes by corticotropin. *Proc. Natl Acad. Sci. USA* **83:** 4715–4719.

Kageyama R, Ohkubo H, Nakanishi S. (1984) Primary structure of human preangiotensinogen deduced from the cloned cDNA sequence. *Biochemistry* **23:** 3603–3609.

Kao CC, Lieberman PM, Schmidt MC, Zhou Q, Pei R, Berk AJ. (1990) Cloning of a transcriptionally active human TATA binding factor. *Science* **248:** 1646–1650.

Katsuya T, Koike G, Yee TW *et al.* (1995) Association of angiotensinogen gene T235 variant with increased risk of coronary heart disease. *Lancet* **345:** 1600–1603.

Kawamoto T, Mitsuuchi Y, Ohnishi T *et al.* (1990a) Cloning and expression of a cDNA for human cytochrome P-450aldo as related to primary aldosteronism. *Biochem. Biophys. Res. Commun.* **173:** 309–316.

Kawamoto T, Mitsuuchi Y, Toda K *et al.* (1990b) Cloning of cDNA and genomic DNA for human cytochrome P-45011 beta. *FEBS Lett.* **269:** 345–349.

Kawamoto T, Mitsuuchi Y, Toda K *et al.* (1992) Role of steroid 11 beta-hydroxylase and steroid 18-hydroxylase in the biosynthesis of glucocorticoids and mineralocorticoids in humans. *Proc. Natl Acad. Sci. USA* **89:** 1458–1462.

Klett C, Hellmann W, Muller F. (1988) Angiotensin II controls angiotensinogen secretion at a pre-translational level. *J. Hypertens.* **6:** S442–S445.

Komesaroff PA, Verity K, Fuller PJ. (1994) Pseudohypoaldosteronism: molecular characterization of the mineralocorticoid receptor. *J. Clin. Endocrinol. Metab.* **79:** 27–31.

Krozowski ZS, Funder JW. (1983) Renal mineralocorticoid receptors and hippocampal corticosterone binding species have identical intrinsic steroid specificity. *Proc. Natl Acad. Sci. USA* **80:** 6056–6060.

Kuhnle U, Hinkel GK, Akkurt HI, Krozowski Z. (1995) Familial pseudohypoaldosteronism: a review on the heterogeneity of the syndrome. *Steroids* **60:** 157–160.

Lakshmi V, Monder C. (1988) Purification and characterization of the corticosteroid 11 beta-dehydrogenase component of the rat liver 11 beta-hydroxysteroid dehydrogenase complex. *Endocrinology* **123:** 2390–2398.

Lala DS, Rice DA, Parker KL. (1992) Steroidogenic factor I, a key regulator of steroidogenic enzyme expression, is the mouse homolog of fushi tarazu-factor I. *Mol. Endocrinol.* **6:** 1249–1258.

Liddle GW, Bledsoe T, Coppage WS. (1963) A familial renal disorder simulating primary aldosteronism but with negligible aldosterone secretion. *Trans. Assoc. Am. Physicians* **76:** 199–213.

Lifton RP, Dluhy RG, Powers M *et al.* (1992a) A chimaeric 11 beta-hydroxylase/aldosterone synthase gene causes glucocorticoid-remediable aldosteronism and human hypertension. *Nature* **355:** 262–265.

Lifton RP, Dluhy RG, Powers M *et al.* (1992b) Hereditary hypertension caused by chimaeric gene duplications and ectopic expression of aldosterone synthase. *Nature Genet.* **2:** 66–74.

McGehee RE Jr, Habener JF. (1995) Differentiation-specific element binding protein (DSEB) binds to a defined element in the promoter of the angiotensinogen gene required for the irreversible induction of gene expression during differentiation of 3T3-L1 adipoblasts to adipocytes. *Mol. Endocrinol.* **9:** 487–501.

McGehee RE Jr, Ron D, Brasier AR, Habener JF. (1993) Differentiation-specific element: a cis-acting developmental switch required for the sustained transcriptional expression of the angiotensinogen gene during hormonal-induced differentiation of 3T3-L1 fibroblasts to adipocytes. *Mol. Endocrinol.* **7:** 551–560.

Mercer WR, Krozowski ZS. (1992) Localization of an 11 beta hydroxysteroid dehydrogenase activity to the distal nephron. Evidence for the existence of two species of dehydrogenase in the rat kidney. *Endocrinology* **130:** 540–543.

Meyer TE, Habener JF. (1993) Cyclic adenosine 3′,5′-monophosphate response element binding protein (CREB) and related transcription-activating deoxyribonucleic acid-binding proteins. *Endocr. Rev.* **14:** 269–290.

Mimouni M, Kaufman H, Roitman A, Morag C, Sadan N. (1985) Hypertension in a neonate with 11 beta-hydroxylase deficiency. *Eur. J. Pediatr.* **143:** 231–233.

Monder C, Stewart PM, Lakshmi V, Valentino R, Burt D, Edwards CR. (1989) Licorice inhibits corticosteroid 11 beta-dehydrogenase of rat kidney and liver: in vivo and in vitro studies. *Endocrinology* **125:** 1046–1053.

Mornet E, Dupont J, Vitek A, White PC. (1989) Characterization of two genes encoding human steroid 11 beta-hydroxylase (P-450(11) beta). *J. Biol. Chem.* **264:** 20961–20967.

Morohashi K, Honda S, Inomata Y, Handa H, Omura T. (1992) A common trans-acting factor, Ad4-binding protein, to the promoters of steroidogenic P-450s. *J. Biol. Chem.* **267:** 17913–17919.

Mountjoy KG, Robbins LS, Mortrud MT, Cone RD. (1992) The cloning of a family of genes that encode the melanocortin receptors. *Science* **257:** 1248–1251.

Mouw AR, Rice DA, Meade JC, Chua SC, White PC, Schimmer BP, Parker KL. (1989) Structural and functional analysis of the promoter region of the gene encoding mouse steroid 11 beta-hydroxylase. *J. Biol. Chem.* **264:** 1305–1309.

Mune T, White PC. (1996) Apparent mineralocorticoid excess: genotype is correlated with biochemical phenotype. *Hypertension* **27:** 1193–1199.

Mune T, Rogerson FM, Nikkila H, Agarwal AK, White PC. (1995) Human hypertension caused by mutations in the kidney isozyme of 11 beta-hydroxysteroid dehydrogenase. *Nature Genet.* **10:** 394–399.

Murphy TJ, Alexander RW, Griendling KK, Runge MS, Bernstein KE. (1991) Isolation of a cDNA encoding the vascular type-1 angiotensin II receptor. *Nature* **351:** 233–236.

Naiki Y, Kawamoto T, Mitsuuchi Y, Miyahara K, Toda K, Orii T, Imura H, Shizuta Y. (1993) A nonsense mutation (TGG [Trp116]-TAG [Stop] in CYP11B1 causes steroid 11beta-hydroxylase deficiency. *J. Clin. Endocrinol. Metab.* **77:** 1677–1682.

Nelson DR, Strobel HW. (1988) On the membrane topology of vertebrate cytochrome P-450 proteins. *J. Biol. Chem.* **263:** 6038–6050.

New MI, Peterson RE. (1967) A new form of congenital adrenal hyperplasia. *J. Clin. Endocrinol. Metab.* **27:** 300–305.

Nibu Y, Takahashi S, Tanimoto K, Murakami K, Fukamizu A. (1994a) Identification of cell type-dependent enhancer core element located in the 3′-downstream region of the human angiotensinogen gene. *J. Biol. Chem.* **269:** 28598–28605.

Nibu Y, Tanimoto K, Takahashi S, Ono H, Murakami K, Fukamizu A. (1994b) A cell type-dependent enhancer core element is located in exon 5 of the human angiotensinogen gene. *Biochem. Biophys. Res. Commun.* **205:** 1102–1108.

Nikkila H, Tannin GM, New MI, Taylor NF, Kalaitzoglou G, Monder C, White PC. (1993) Defects in the HSD11 gene encoding 11β-hydroxysteroid dehydrogenase are not found in patients with apparent mineralocorticoid excess or 11-oxoreductase deficiency. *J. Clin. Endocrinol. Metab.* **77:** 687–691.

Obeid J, White PC. (1992) Tyr-179 and Lys-183 are essential for enzymatic activity of 11 beta-hydroxysteroid dehydrogenase. *Biochem. Biophys. Res. Commun.* **188:** 222–227.

Oberfield SE, Levine LS, Stoner E *et al.* (1981) Adrenal glomerulosa function in patients with dexamethasone-suppressible hyperaldosteronism. *J. Clin. Endocrinol. Metab.* **53:** 158–164.

Oberfield SE, Levine LS, Carey RM, Greig F, Ulick S, New MI. (1983) Metabolic and blood pressure responses to hydrocortisone in the syndrome of apparent mineralocorticoid excess. *J. Clin. Endocrinol. Metab.* **56:** 332–339.

Obeyesekere VR, Ferrari P, Andrews RK, Wilson RC, New MI, Funder JW, Krozowski ZS. (1995) The R337C mutation generates a high Km 11β-hydroxysteroid dehydrogenase type II enzyme in a family with apparent mineralocorticoid excess. *J. Clin. Endocrinol. Metab.* **80:** 3381–3383.

Ogishima T, Shibata H, Shimada H, Mitani F, Suzuki H, Saruta T, Ishimura Y. (1991) Aldosterone synthase cytochrome P-450 expressed in the adrenals of patients with primary aldosteronism. *J. Biol. Chem.* **266:** 10731–10734.

Ohta M, Fujii S, Ohnishi T, Okamoto M. (1988) Production of 19-oic-11-deoxycorticosterone from 19-oxo-11-deoxycorticosterone by cytochrome P-450(11)beta and nonenzymatic production of 19-nor-11-deoxycorticosterone from 19-oic-11-deoxycorticosterone. *J. Steroid Biochem.* **29:** 699–707.

O'Mahony S, Burns A, Murnaghan DJ. (1989) Dexamethasone-suppressible hyperaldosteronism: a large new kindred. *J. Hum. Hypertens.* **3:** 255–258.

Palmer LG, Frindt G. (1992) Regulation of apical membrane Na and K channels in rat renal collecting tubules by aldosterone. *Semin. Nephrol.* **12:** 37–43.

Pang S, Levine LS, Lorenzen F, Chow D, Pollack M, Dupont B, Genel M, New MI. (1980) Hormonal studies in obligate heterozygotes and siblings of patients with 11 beta-hydroxylase deficiency congenital adrenal hyperplasia. *J. Clin. Endocrinol. Metab.* **50:** 586–589.

Pascoe L, Curnow KM, Slutsker L, Connell JM, Speiser PW, New MI, White PC. (1992) Glucocorticoid-suppressible hyperaldosteronism results from hybrid genes created by unequal crossovers between CYP11B1 and CYP11B2. *Proc. Natl Acad. Sci. USA* **89:** 8327–8331.

Pascoe L, Jeunemaitre X, Lebrethon MC, Curnow KM, Gomez-Sanchez CE, Gasc JM, Saez JM, Corvol P. (1995) Glucocorticoid-suppressible hyperaldosteronism and adrenal tumours occurring in a single French pedigree. *J. Clin. Invest.* **96:** 2236–2246.

Pearce D, Yamamoto KR. (1993) Mineralocorticoid and glucocorticoid receptor activities distinguished by nonreceptor factors at a composite response element. *Science* **259:** 1161–1165.

Poulos TL. (1991) Modeling of mammalian P450s on basis of P450cam X-ray structure. *Meth. Enzymol.* **206:** 11–30.

Printz MP, Printz JM, Dworschack RT. (1977) Human angiotensinogen. Purification, partial characterization and a comparison with animal prohormones. *J. Biol. Chem.* **252:** 1654–1662.

Quinn SJ, Williams GH. (1988) Regulation of aldosterone secretion. *Ann. Rev. Physiol.* **50:** 409–426.

Ravichandran KG, Boddupalli SS, Hasemann CA, Peterson JA, Deisenhofer J. (1993) Crystal structure of hemoprotein domain of P450BM-3, a prototype for microsomal P450's. *Science* **261:** 731–736.

Reinisch J, Simon NG, Karwo WG. (1978) Prenatal exposure to prednisone in humans and animals retards intrauterine growth. *Science* **202:** 436–438.

Rice DA, Aitken LD, Vandenbark GR, Mouw AR, Franklin A, Schimmer BP, Parker KL. (1989) A cAMP-responsive element regulates expression of the mouse steroid 11 beta-hydroxylase gene. *J. Biol. Chem.* **264:** 14011–14015.

Rice DA, Mouw AR, Bogerd AM, Parker KL. (1991) A shared promoter element regulates the expression of three steroidogenic enzymes. *Mol. Endocrinol.* **5:** 1552–1561.

Rich GM, Ulick S, Cook S, Wang JZ, Lifton RP, Dluhy RG. (1992) Glucocorticoid-remediable aldosteronism in a large kindred: clinical spectrum and diagnosis using a characteristic biochemical phenotype. *Ann. Intern. Med.* **116:** 813–820.

Robertson NM, Schulman G, Karnik S, Alnemri E, Litwack G. (1993) Demonstration of nuclear translocation of the mineralocorticoid receptor (MR) using an anti-MR antibody and confocal laser scanning microscopy. *Mol. Endocrinol.* **7:** 1226–1239.

Rosler A, Leiberman E, Sack J, Landau H, Benderly A, Moses SW, Cohen T. (1982) Clinical variability of congenital adrenal hyperplasia due to 11 beta-hydroxylase deficiency. *Hormone Res.* **16:** 133–141.

Rosler A, Leiberman E, Cohen T. (1992) High frequency of congenital adrenal hyperplasia (classic 11 beta-hydroxylase deficiency) among Jews from Morocco. *Am. J. Med. Genet.* **42:** 827–834.

Rotin D, Bar-Sagi D, O'Brodovich H, Merilainen J, Lehto VP, Canessa CM, Rossier BC, Downey GP. (1994) An SH3 binding region in the epithelial Na+ channel (alpha rENaC) mediates its localization at the apical membrane. *EMBO J.* **13:** 4440–4450.

Rundle SE, Funder JW, Lakshmi V, Monder C. (1989) The intrarenal localization of mineralocorticoid receptors and 11 beta-dehydrogenase: immunocytochemical studies. *Endocrinology* **125:** 1700–1704.

Rusvai E, Naray-Fejes-Toth A. (1993) A new isoform of 11 beta-hydroxysteroid dehydrogenase in aldosterone target cells. *J. Biol. Chem.* **268:** 10717–10720.

Sasaki K, Yamamo Y, Bardhan S, Iwai N, Murray JJ, Hasegawa M, Matsuda Y, Inagami T. (1991) Cloning and expression of a complementary cDNA encoding a bovine adrenal angiotensin II type-1 receptor. *Nature* **351:** 230–233.

Schild L, Canessa CM, Shimkets RA, Gautschi I, Lifton RP, Rossier BC. (1995) A mutation in the epithelial sodium channel causing Liddle disease increases channel activity in the Xenopus laevis oocyte expression system. *Proc. Natl Acad. Sci. USA* **92:** 5699–5703.

Shackleton CH, Rodriguez J, Arteaga E, Lopez JM, Winter JS. (1985) Congenital 11 beta-hydroxysteroid dehydrogenase deficiency associated with juvenile hypertension: corticosteroid metabolite profiles of four patients and their families. *Clin. Endocrinol. (Oxf)* **22:** 701–712.

Shimkets RA, Warnock DG, Bositis CM *et al.* (1994) Liddle's syndrome: heritable human hypertension caused by mutations in the β subunit of the epithelial sodium channel. *Cell* **79:** 407–414.

Shyu KG, Chen JJ, Shih NL, Chang H, Wang DL, Lien WP, Liew CC. (1995) Angiotensinogen gene expression is induced by cyclical mechanical stretch in cultured rat cardiomyocytes. *Biochem. Biophys. Res. Commun.* **211:** 241–248.

Skinner CA, Rumsby G, Honour JW. (1996) Single strand conformation polymorphism analysis for detection of mutations in the CYP11B1 gene. *J. Clin. Endocrinol. Metab.* **81:** 2389–2393.

Stewart PM, Wallace AM, Valentino R, Burt D, Shackleton CH, Edwards CR. (1987) Mineralocorticoid activity of liquorice: 11-beta-hydroxysteroid dehydrogenase deficiency comes of age. *Lancet* **2:** 821–824.

Stewart PM, Murry BA, Mason JI. (1994) Type 2 11beta-hydroxysteroid dehydrogenase in human fetal tissues. *J. Clin. Endocrinol. Metab.* **78:** 1529–1532.

Stewart PM, Rogerson FM, Mason JI. (1995) Type 2 11 beta-hydroxysteroid dehydrogenase messenger ribonucleic acid and activity in human placenta and fetal membranes: its relationship to birth weight and putative role in fetal adrenal steroidogenesis. *J. Clin. Endocrinol. Metab.* **80:** 885–890.

Stewart PM, Krozowski ZS, Gupta A, Milford DV, Howie AJ, Sheppard MC, Whorwood CB. (1996) Hypertension in the syndrome of apparent mineralocorticoid excess due to mutation of the 11β-hydroxysteroid dehydrogenase type 2 gene. *Lancet* **347:** 88–91.

Stockigt JR, Scoggins BA. (1987) Long term evolution of glucocorticoid-suppressible hyperaldosteronism. *J. Clin. Endocrinol. Metab.* **64:** 22–26.

Strautnieks SS, Thompson RJ, Hanukoglu A, Dillon MJ, Hanukoglu I, Kuhnle U, Seckl JR, Gardiner SM, Chung E. (1996) Localization of pseudohypoaldosteronism genes to chromosome 16p12.2–13.11 and 12p13.1-pter by homozygosity mapping. *Hum. Mol. Genet.* **5:** 293–299.

Sutherland DJ, Ruse JL, Laidlaw JC. (1966) Hypertension, increased aldosterone secretion and low plasma renin activity relieved by dexamethasone. *Can. Med. Assoc. J.* **95:** 1109–1119.

Tamura H, Schild L, Enomoto N, Matsui N, Marumo F, Rossier BC. (1996) Liddle disease caused by a missense mutation of beta subunit of the epithelial sodium channel gene. *J. Clin. Invest.* **97:** 1780–1784.

Tamura K, Tanimoto K, Ishii M, Murakami K, Fukamizu A. (1993) Proximal and core DNA elements are required for efficient angiotensinogen promoter activation during adipogenic differentiation. *J. Biol. Chem.* **268:** 15024–15032.

Tamura K, Umemura S, Ishii M, Tanimoto K, Murakami K, Fukamizu A. (1994a) Molecular mechanism of transcriptional activation of angiotensinogen gene by proximal promoter. *J. Clin. Invest.* **93:** 1370–1379.

Tamura K, Umemura S, Iwamoto T *et al.* (1994b) Molecular mechanism of adipogenic activation of the angiotensinogen gene. *Hypertension* **23:** 364–368.

Tanaka T, Ohkubo H, Nakanishi S. (1984) Common structural organization of the angiotensinogen and α1-antitrypsin genes. *J. Biol. Chem.* **259:** 8063–8065.

Tannin GM, Agarwal AK, Monder C, New MI, White PC. (1991) The human gene for 11 beta-hydroxysteroid dehydrogenase. Structure, tissue distribution, and chromosomal localization. *J. Biol. Chem.* **266:** 16653–16658.

Ulick S, Levine LS, Gunczler P, Zanconato G, Ramirez LC, Rauh W, Rosler A, Bradlow HL, New MI. (1979) A syndrome of apparent mineralocorticoid excess associated with defects in the peripheral metabolism of cortisol. *J. Clin. Endocrinol. Metab.* **49:** 757–764.

Wagner MJ, Ge Y, Siciliano M, Wells DE. (1991) A hybrid cell mapping panel for regional localization of probes to human chromosome 8. *Genomics* **10:** 114–125.

Waldmann R, Champigny G, Bassilana F, Voilley N, Lazdunski M. (1995) Molecular cloning and functional expression of a novel amiloride-sensitive Na(+) channel. *J. Biol. Chem.* **270:** 27411–27414.

Waterman MR, Simpson ER. (1989) Regulation of steroid hydroxylase gene expression is multifactorial in nature. *Recent Prog. Hormone Res.* **45:** 533–563.

Wehling M. (1995) Nongenomic aldosterone effects: the cell membrane as a specific target of mineralocorticoid action. *Steroids* **60:** 153–156.

White PC. (1991) Defects in cortisol metabolism causing low-renin hypertension. *Endocr. Res.* **17:** 85–107.

White PC, Slutsker L. (1995) Haplotype analysis of CYP11B2. *Endocr. Res.* **21:** 437–442.

White PC, Dupont J, New MI, Leiberman E, Hochberg Z, Rosler A. (1991) A mutation in CYP11B1 (Arg-448-His) associated with steroid 11 beta-hydroxylase deficiency in Jews of Moroccan origin. *J. Clin. Invest.* **87:** 1664–1667.

Wilson RC, Harbison MD, Krozowski ZS *et al.* (1995a) Several homozygous mutations in the gene for 11β-hydroxysteroid dehydrogenase type 2 in patients with apparent mineralocorticoid excess. *J. Clin. Endocrinol. Metab.* **80:** 3145–3150.

Wilson RC, Krozowski ZS, Li K *et al.* (1995b) A mutation in the HSD11B2 gene in a family with apparent mineralocorticoid excess. *J. Clin. Endocrinol. Metab.* **80:** 2263–2266.

Wong M, Rice DA, Parker KL, Schimmer BP. (1989) The roles of cAMP and cAMP-dependent protein kinase in the expression of cholesterol side chain cleavage and steroid 11 beta-hydroxylase genes in mouse adrenocortical tumour cells. *J. Biol. Chem.* **264:** 12867–12871.

Yanagibashi K, Haniu M, Shively JE, Shen WH, Hall P. (1986) The synthesis of aldosterone by the adrenal cortex. Two zones (fasciculata and glomerulosa) possess one enzyme for 11 beta-, 18-hydroxylation, and aldehyde synthesis. *J. Biol. Chem.* **261:** 3556–3562.

Yang K, Yu M. (1994) Evidence for distinct isoforms of 11-beta-hydroxysteroid dehydrogenase in the ovine liver and kidney. *J. Steroid Biochem. Mol. Biol.* **49:** 245–250.

Zachmann M, Tassinari D, Prader A. (1983) Clinical and biochemical variability of congenital adrenal hyperplasia due to 11 beta-hydroxylase deficiency. A study of 25 patients. *J. Clin. Endocrinol. Metab.* **56:** 222–229.

Zennaro MC, Borensztein P, Jeunemaitre X, Armanini D, Soubrier F. (1994) No alteration in the primary structure of the mineralocorticoid receptor in a family with pseudohypoaldosteronism. *J. Clin. Endocrinol. Metab.* **79:** 32–38.

Zhao YY, Qasba P, Siddiqui MA, Kumar A. (1993) Multiple CCAAT binding proteins regulate the expression of the angiotensinogen gene. *Cell. Mol. Biol. Res.* **39:** 727–737.

10

Sex differentiation, gonadal development and reproductive function

Gerard S. Conway

10.1 Introduction

In this chapter, fertility is considered as the end product of two major complex phenomena. The first is a cascade of events that ensures sexual dimorphism and gonadal development. The second is a hormonal feedback system that controls gamete production and presentation. The synthesis and action of sex steroids is considered elsewhere (Chapters 4 and 8).

10.2 Sex differentiation and gonadal development

Sex differentiation, as explored by Jost (1953), was defined in terms of hormonal control. Jost discovered that the 'default' phenotype, in the absence of gonads, was female and he proposed that testicular hormones determined the diversion towards male differentiation. He demonstrated the major role of testosterone in determining the male phenotype and also predicted the existence of a second substance which would promote the regression of Mullerian structures. Later, cytogeneticists reported the association of the Y chromosome with maleness (Welshons and Russell, 1959) and the scene was set to identify the sex-determining factor on the Y chromosome responsible for the switch from female to male phenotype (*Figure 10.1*).

Much of the progress in the genetic approach to sex determination was made through the investigation of sex-reversed individuals – 46XY females and 46XX males. In particular, most 46XX males have some remaining Y chromosome material inherited from their fathers (Guellaen *et al.*, 1984). Analysis of these Y-unique sequences in 46XX males defined a critical 35 kb region of the Y chromosome

Molecular Endocrinology, edited by G. Rumsby and S.M. Farrow.

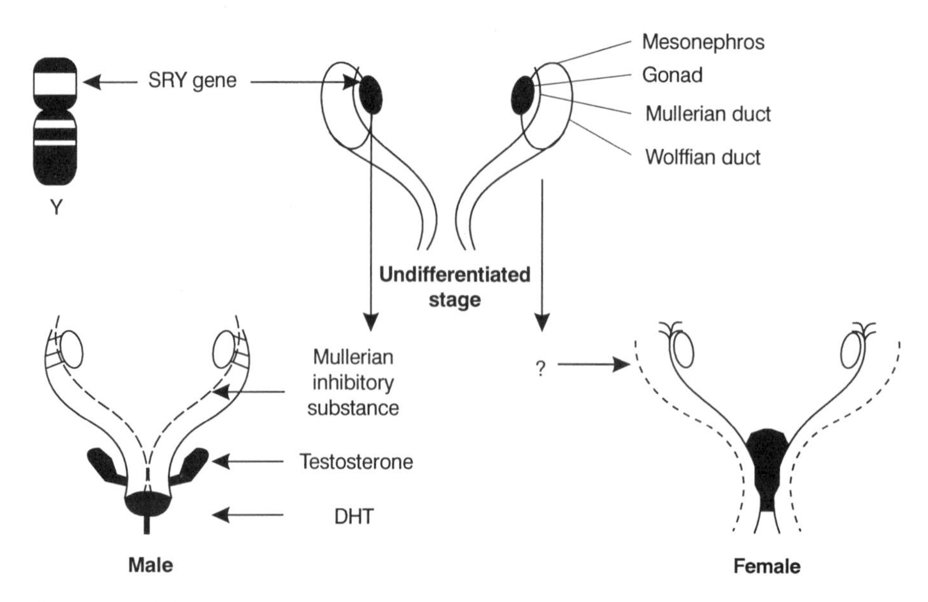

Figure 10.1. Scheme of the factors controlling sexual differentiation. DHT, dihydrotestosterone.

which correlated with maleness (Sinclair *et al.*, 1990). This region was cloned and 0.5 kb fragments were used to probe Southern blots of male and female DNA from various mammals in order to detect conserved sex-specific sequences. Only one gene emerged as male-specific – the SRY gene (see Section 10.2.2) (Gubbay *et al.*, 1990). Following the identification of the SRY gene, the field of sexual differentiation has again taken on an endocrine emphasis as the final downstream events of SRY action, which result in the male phenotype, are the two testicular hormones predicted by Jost – testosterone and Mullerian inhibiting substance (MIS) (*Figure 10.1*).

Testosterone production from Leydig cells promotes the development of Wolffian ducts (vas deferens, seminal vesicles and epididymis) and is converted into dihydrotestosterone (DHT) by the enzyme 5α-reductase (Section 8.2.7). DHT promotes the development of the prostate gland and external genitalia while MIS controls the regression of Mullerian ducts (uterus and Fallopian tubes).

A major focus of current research is defining the intermediary steps between SRY and the testicular hormones. A list of such intermediaries is growing quickly and includes factors which control androgen synthesis, such as steroidogenic factor 1 (SF-1) and the luteinizing hormone (LH) receptor. Other candidate intermediaries have been identified through the further study of 46XY females in whom no SRY mutations have been identified and through the study of individuals with gonadal dysgenesis. A scheme outlining the steps of sexual differentiation is presented in *Figure 10.2* and each of these components will be considered individually.

10.2.1 Genes contributing to early gonadal development

In this section genes that have been identified to have a role in both ovarian and testicular development are considered, these are SF-1 and Wilms' tumour gene (WT1).

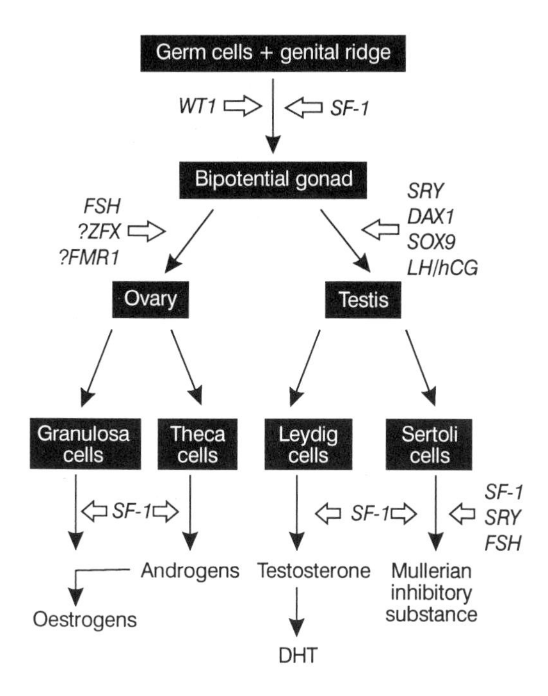

Figure 10.2. Scheme of the genes controlling the various stages of gonadal development. See text for abbreviations.

Steroidogenic factor 1 (SF-1). SF-1 is an orphan nuclear receptor originally identified as an adrenal binding protein – adrenal 4-binding protein (Ad4BP). As discussed in Chapter 8, the first target genes for SF-1 action to be identified were steroid hydroxylases and aromatase (Lala *et al.,* 1992, 1995). Subsequently, the recognition sequence of SF-1 (CCAAGGTCA) was found not only in the promoter regions of steroidogenic enzyme genes but also in that of MIS (Shen *et al.,* 1994). SF-1 also has a developmental, and possibly physiological role, centrally. Putative SF-1 response elements have been identified in the DAX1 promoter (see Section 10.2.2), suggesting that these two elements act in concert (Burris *et al.,* 1995). Interestingly, both DAX1 and SF-1 are expressed in the adrenal cortex, gonads, hypothalamus and pituitary gland. Within the hypothalamus these two factors may determine gonadotrophin-releasing hormone (GnRH) secretion whilst in the pituitary gland SF-1 regulates the expression of the glycoprotein α-subunit gene (Barnhart and Mellon, 1994) thus modulating gonadotrophin secretion.

A knockout mouse model has demonstrated an early role for SF-1 in gonadal development as there was complete gonadal and adrenal agenesis in these animals (Luo *et al.,* 1994). Both male and female SF-1 knockout mice possess normal female internal genitalia. The pituitary glands of the SF-1 knockout mouse lacked transcripts for LH-β, follicle-stimulating hormone (FSH)-β and the GnRH receptor while the transcript of the glycoprotein α-subunit gene was also reduced (Ingraham *et al.,* 1994). SF-1, therefore, not only has a function in controlling the hormonal production throughout the reproductive system but also regulates pituitary and gonadal development at a very early stage.

Wilms' tumour gene (WT1). Wilms' tumour of the kidney is a childhood neoplasm, which is familial in less than 1% of cases and is associated with translocations or deletions of chromosome 11p13. The Wilms' tumour locus was eventually narrowed down by the study of the common deleted sequences on 11p and the WT1 gene was identified (Call *et al.*, 1990; Rose *et al.*, 1990). The WT1 gene, comprising 10 exons, encodes a probable transcription factor with four zinc fingers which is highly expressed in the fetal kidney and gonad (Van Heyningen *et al.*, 1990). One of the familial forms of Wilms' tumour is the Denys-Drash syndrome (Wilms' tumour, nephropathy and pseudohermaphroditism). Clinical presentation of the Denys-Drash syndrome might be a female with childhood renal failure, primary amenorrhoea, perhaps with a 46XY karyotype and the later development of a Wilms' tumour or gonadoblastoma (Friedman and Finlay, 1987). In fact, both 46XY and 46XX individuals with Denys-Drash syndrome exhibit gonadal dysgenesis. Various mutations of the WT1 gene have been found to be associated with the Denys-Drash syndrome, particularly involving exons 8 and 9 encoding the zinc fingers (Pelletier *et al.*, 1991a,b). The mutant product of the WT1 gene will inhibit the function of the wild-type protein in the heterozygote; that is, it has a dominant-negative effect. Wilms' tumour tissue from these subjects exhibits loss of heterozygosity (Pelletier *et al.*, 1991a).

WT1 gene mutations have also been identified in the WAGR syndrome (Wilms' tumour and aniridia, genital abnormalities and mental retardation) (Pelletier *et al.*, 1991a,b). This complex of abnormalities comprises a contiguous gene syndrome with separate genes for aniridia and mental retardation (Schmickel, 1986). WT1 gene mutations probably do not cause the 46XY female phenotype in the absence of renal disease (Nordenskjold *et al.*, 1995). WT1 therefore, like SF-1, has an early major role to play in gonadal development (Pritchard-Jones *et al.*, 1990) as well as a postnatal role in renal function and tumour suppression. Ovarian expression of WT1 is found within granulosa cells and decreases with follicular development perhaps through regulation of the inhibin α gene (Hsu *et al.*, 1995).

10.2.2 Genes contributing to testis development

Genes contributing specifically to testis development have been identified through the study of 46XY sex-reversed individuals. The genes fulfilling these criteria so far identified are the sex-determining region Y chromosome gene (SRY), SRY-related, HMG-box, gene 9 (SOX9), DSS–AHC critical region of the X chromosome gene 1 (DAX1) and LH receptor gene.

Sex-determining region Y chromosome gene (SRY). The SRY and SOX gene family encodes members of the high mobility group (HMG) of DNA-binding proteins. This family of proteins is characterized by a conserved HMG-box domain which, in the case of SRY, comprises 80 amino acids (Harley *et al.*, 1992). The HMG-box domain of SRY binds to the minor groove of the DNA double helix and bends the target DNA through 85–130°(Goodfellow and Lovell-Badge, 1993). It is proposed that this SRY-induced DNA bending activates or represses the expression of target genes.

In mice, the SRY gene is transcribed in the hypothalamus, midbrain and testis in adult males (but not females). This male-specific expression occurs appropriately in the embryonic genital ridge prior to differentiation into testes (Koopman *et al.*, 1990). Of the four cell lineages that make up the gonad, it is the supporting cells rather than the germ, steroidogenic or connective tissue cells which express SRY (Bogan and Page, 1994). Thus, undifferentiated supporting cells develop into Sertoli cells in the presence of SRY or granulosa cells in the absence of this gene. The three other cell lineages then follow to form spermatagonia, Leydig cells and peritubular cells in males, and oogonia, theca and stromal cells in females, respectively. SRY action, therefore, occurs after the bipotential gonad has developed and is a major early factor in testis development.

About 80% of 46XX males possess some Y-derived sequence, presumably containing SRY (Guellaen *et al.*, 1984). Conversely, SRY mutations account for about 20% of 46XY females with gonadal dysgenesis (Goodfellow and Lovell-Badge, 1993). These SRY mutations are clustered in the HMG-box domain of the protein which binds to DNA. The 80% of 46XY females with normal SRY sequence and 20% of 46XX males with absent SRY presumably harbour mutations of downstream genes. The first candidate downstream genes to be considered for SRY are those known to be associated with sex reversal (DAX1 and SOX9) as well as sex differentiation (MIS).

SRY-related, HMG-box, gene 9 (SOX9). The SOX9 gene was identified through the study of camptomelic dwarfism (CMD), where subjects have a variety of bony deformities (particularly bowing of the legs) as well as sex reversal (46XY females) (Hovmoller *et al.*, 1977). Subjects with CMD were noted to have disruption of chromosome 17q and various groups undertook cloning of the breakpoints of chromosome 17q translocations. One such breakpoint was found to disrupt a gene designated SOX9 (Foster *et al.*, 1994). Subsequent mutation analysis of SOX9 in other subjects with CMD identified various sequence alterations causing premature stop codons, amino acid substitutions or splice site mutations in one allele only (Foster *et al.*, 1994). In this case the phenotype is not caused by a dominant-negative effect of the mutation but by a lack of wild-type protein (haploinsufficiency).

Insufficiency of the SOX9 gene product prevents testicular development but appears to have no effect on the ovary. That is, CMD females are fertile while genetic males with CMD are sex-reversed with gonadal dysgenesis.

DSS–AHC critical region of the X chromosome gene 1 (DAX1). The DAX1 gene (locus Xp21.3–21.2) comprises two exons which encode a nuclear hormone receptor related to SF-1 (Zanaria *et al.*, 1994). Deletions or mutations of the DAX1 gene cause adrenal hypoplasia congenita (AHC) and hypogonadotrophic hypogonadism (Muscatelli *et al.*, 1994). The relative position of DAX1 in gonadal development presumably occurs after SF-1 as deficiency of the latter prevents both adrenal and gonadal development while in males lacking DAX1 the adrenal gland does not develop beyond the fetal stage but early gonadal development is normal.

Evidence for a role for DAX1 in gonadal development comes from the fact that the dosage-sensitive sex (DSS) reversal gene also maps to Xp21.3 and may, in fact, be identical to DAX1 (hence DSS-AHC critical region of the X chromosome gene

1). The DSS locus is represented by a 160 kb region of Xp21, which causes sex reversal (46XY females) when duplicated, but appears to have no effect when it is deleted (Bardoni *et al.*, 1994). The nature of the DSS reversal phenomenon, however, is complex as duplication of the entire X chromosome, as in the XXY or XXXY karyotype, is associated with a normal male phenotype, while duplication restricted to Xp21 results in sex reversal. Furthermore, the mouse Daxl gene is expressed not only in the adrenal gland and hypothalamus but also in early stages of gonadal differentiation (Swain *et al.*, 1996). Later in development, testis DAX1 gene expression is downregulated while that in the ovary persists. Taken together, these data are consistent with a role for DAX1 in gonadal sex determination as well as adrenal and hypothalamic development.

The LH receptor gene. The development of the Leydig cell component of the testis is dependent on human choriogonadotrophin (hCG) stimulation in the uterus. Complete LH/hCG resistance caused by loss of function mutations of the LH receptor results in Leydig cell hypoplasia (Kremer *et al.*, 1995). The resulting testis exhibits normal Sertoli cell function and produces MIS. The clinical presentation is of a 46XY female with absent Mullerian structures. Normal testicular development is highly dependent, therefore, on LH/hCG but less so on FSH stimulation (see Section 10.3). Even though the effect of blocking the LH axis was identified first in males, affected females are not completely free of gonadal manifestation of these mutations. One female, homozygous for a blocking mutation of the LH receptor, has now been described with amenorrhoea but with relatively normal ovarian morphology (Latronico *et al.*, 1996).

10.2.3 Genes affecting the male phenotype

Once sexual differentiation has been established and testicular development ensured, various genes come into play which ensure normal male reproductive function. Of these, the control of Mullerian duct inhibition and spermatogenesis will be considered here.

MIS and its receptor. The MIS gene (locus 19p13.3–13.2) comprises five exons encoding a hormone of the transcription growth factor group which includes inhibins and activins (Cate *et al.*, 1986; Cohen-Haguenauer *et al.*, 1987). MIS gene expression occurs in Sertoli cells throughout life being greatest at the time of Mullerian duct regression. MIS expression is also found in the granulosa cells of the fetal ovary but, after birth, ovarian expression is less than 1% of that found in the testis (Josso 1986; Musterberg and Lovell Badge, 1991). This persisting expression in the ovary and testis suggests that there may be a role for MIS in gonadal function.

The MIS receptor is a member of the serine/threonine kinase receptor family (Grootegoed *et al.*, 1994). The MIS receptor gene (locus 12q13) is expressed in mesenchymal cells adjacent to Mullerian ducts and in postnatal gonads within granulosa and Sertoli cells (Imbeaud *et al.*, 1995).

Persistent Mullerian duct syndrome (PMDS) is inherited as an autosomal recessive trait and occurs in males with normal external genitalia but with bilateral cryptorchidism and inguinal hernias. At surgery, the gonads are confirmed to be testes but a uterus and Fallopian tubes are also present, usually within the inguinal canal (Behringer, 1994; Lee and Donahoe, 1993). PMDS type I is due to mutations in the MIS gene (Carre-Eusebe *et al.*, 1992; Knebelmann *et al.*, 1991), while PMDS type II is the result of mutations of the MIS receptor gene (Imbeaud *et al.*, 1995). PMDS type I differs from type II only in the detection of circulating MIS.

Factors influencing MIS gene transcription include epidermal growth factor (EGF) and SRY. EGF inhibits MIS secretion (reviewed in Lee and Donahoe, 1993) while induction of MIS transcription by SRY is mediated by unknown intermediates which bind to the proximal 114 bp of the MIS promoter (Haqq *et al.*, 1994). The mechanism by which MIS induces Mullerian duct involution is unclear as the events downstream of MIS have not been characterized.

Genes controlling spermatogenesis. It has generally been assumed that genes on the Y chromosome that are expressed in spermatogenic cells are likely to play a role in spermatogenesis. Early cytogenetic studies identified an azoospermia factor (AZF) within locus Yq11.23 (Tiepolo and Zuffardi, 1976). This region was further characterized and a 200 kb critical area of the Y chromosome defined possibly encompassing several genes essential for spermatogenesis (Kobayashi *et al.*, 1994; Ma *et al.*, 1992; Vogt *et al.*, 1992). Two candidate genes for AZF have been located in this region, RBM (RNA-binding motif; Ma *et al.*, 1993) and DAZ (deleted in azoospermia; Reijo *et al.*, 1995) and both genes are expressed in the testis. DAZ, like RBM, encodes an RNA-binding protein and was originally found to be the only transcription unit consistently deleted in men with azoospermia. However, subsequent screening showed that 11 out of 60 azoospermic men had microdeletions of the Y chromosome of which four were separate from DAZ (Najmabadi *et al.*, 1996).

Genes affecting androgen synthesis and action. For completeness the reader is referred to related topics which are covered elsewhere (Chapters 4 and 8). The remaining causes of male pseudohermaphroditism (46XY females) are accounted for by defects in androgen synthesis or reception. Androgen receptor mutations can cause complete or partial androgen sensitivity with a spectrum from a normal male phenotype through to testicular feminization.

10.2.4 Genes affecting ovarian development

Compared to the progress in understanding testicular development, relatively little is known about the 'default' female pathway. Indeed, it is now clear that several genes contribute specifically to ovarian development with little effect on the testis. Failure of ovarian development manifests clinically as 46XX ovarian dysgenesis or premature ovarian failure (POF), that is, menopause before the age of 40 (Conway *et al.*, 1996).

The ovary differs from the testis in that gamete production, germ cell proliferation through mitosis, is complete before birth. Over three million ova are produced within each ovary *in utero* but, by birth, two-thirds of these have been destroyed

(Baker, 1963) presumably through programmed cell death. Throughout life there is a continued, slow fall out of ova until accelerated depletion is switched on at the age of 40 so that few ova remain at the time of the menopause at the age of 50 (Richardson *et al.*, 1987). Ovarian function will therefore depend on the processes of germ cell migration, proliferation and apoptosis as well as granulosa cell integrity.

In parallel to the characterization of Y chromosome genes that contribute to spermatogenesis, so there is a search for the X chromosome genes that determine ovarian development. Undoubtedly, there are also autosomal genes vital to ovarian function of which the most obvious is the FSH receptor gene.

Ovarian determinants on the X chromosome. Women with Turner's syndrome (45XO) almost universally undergo POF, along with physical stigmata related to Turner's syndrome such as short stature and webbed neck (Ogata and Matsuo, 1995). The fact that an absent X chromosome is so deleterious to ovarian function, as compared to a largely inactivated X chromosome in normal females, suggests two intact alleles are required for the normal function of some genes on the X chromosome and that the candidate loci causing premature ovarian failure must escape X inactivation (Chu and Connor, 1995). A substantial proportion of women with familial POF and a lesser proportion of patients with sporadic POF may therefore harbour mutations or deletions in genes on the X chromosome.

From the study of women with partial deletions of the X chromosome, it is clear that at least three loci are critical for ovarian development (Bates and Howard, 1990; De La Chappelle *et al.*, 1975; Powell *et al.*, 1994). Genes responsible for the stigmata of Turner's syndrome and primary amenorrhoea are located on the short arm of the X chromosome. Particular interest has focused, however, on Xp22, where there are a series of genes that escape X inactivation in humans among which the zinc finger gene (ZFX) is a prime candidate for an association with POF (see below).

The smallest deletion of the X chromosome known to cause POF is Xq26–28 (Krauss *et al.*, 1987, Powell *et al.*, 1994). A putative POF1 locus contains two genes of particular interest, SOX3 and fragile site, mental retardation gene 1 (FMR1). The third area of the X chromosome that is thought to be critical for ovarian development is the interval Xq13–22. This is the least well characterized of the X chromosome's critical regions although several genes in this area escape X inactivation.

Zinc finger gene, X chromosome (ZFX). The ZFX gene (locus Xp22.3–21.2) is the X chromosome homologue of the ZFY gene which was once a candidate for the testis-determining factor on the Y chromosome (Page *et al.*, 1987). ZFY was subsequently shown not to be the testis-determining factor, as several individuals with testes were found to have inherited a Y chromosome fragment which lacked ZFY (Palmer *et al.*, 1989). ZFX and ZFY encode proteins with 13 zinc fingers appropriate for DNA binding and are therefore, probable transcription factors (Schneider-Gadicke *et al.*, 1989b). ZFX fulfils many criteria for an ovarian development gene: it escapes X inactivation (Schneider-Gadicke *et al.*, 1989a), it lies in a critical region for gonadal dysgenesis in Turner's syndrome with the karyotype 45XO (Ogata and Matsuo, 1995) and a preliminary report of a ZFX knockout mouse exhibits ovarian dysgenesis (Page DC presentation to the International

Conference on Sex Differentiation, Cambridge, UK, July 1995). Interestingly, the male ZFX knockout mouse exhibits normal fertility.

Fragile site, mental retardation gene 1 (FMR1). Fragile X syndrome is the most common cause of X-linked mental retardation in males and is caused by disruption of exon 1 of the FMR1 gene (locus Xq27.3) by an expanded trinucleotide repeat (FRAXA) (Jacobs, 1991; Verkerk *et al.*, 1991). The protein encoded by FMR1 possesses two RNA-binding sites (Ashley *et al.*, 1993a,b). Normally, the CGG repeat occurs less than 50 times, a premutation is defined as 50–200 repeats and the full FRAXA mutation carries greater than 200 repeats. The full mutation causes failure of FMR1 transcription by methylation of the promoter, while the premutation was thought, until recently, to have no effect.

Two observations have suggested that FRAXA premutations may be associated with POF. First, premature menopause is more common in female carriers of FRAXA premutations (Fryns, 1986; Schwartz *et al.*, 1994). Second, FRAXA premutations have been identified in two out of nine pedigrees with familial POF (Conway *et al.*, 1995). The mechanism of the association between FRAXA premutations and ovarian development is unclear but two intriguing possibilities arise. Firstly, FMR1 is expressed in the fetal brain, and its lack may contribute to mental retardation (Hinds *et al.*, 1993). FMR1 is also expressed in the fetal gonad (Bachner *et al.*, 1993). Secondly, the SOX3 gene which may have a role in gonadal development (see below) is close to FMR1 and may be affected by FRAXA expansion.

SRY-related, HMG-box, gene 3 (SOX3). Centromeric to the FRAXA locus is the SOX3 gene which has been mapped to Xq26.2–27.2 and also fulfils several criteria as a candidate for POF. This gene, which also escapes X inactivation, is the homologue of the SRY gene that determines testicular development in the Y chromosome (Stevanovic *et al.*, 1993). SOX3 was deleted in one male patient with haemophilia (the gene coding for factor 9 was also deleted), mental retardation and partial testicular failure (Rousseau *et al.*, 1991). No female with deletion of SOX3 has yet been described. This association with infertility, along with its site near the putative POF1 locus, make SOX3 a possible candidate for both ovarian and testicular failure.

Ovarian determinants on autosomes. The FSH receptor gene on chromosome 2p and genes on chromosome 3q23 are implicated in ovarian development. For example, an inactivating mutation of the FSH receptor gene has been shown to cause ovarian dysgenesis in six families in Finland (Aittomäki *et al.*, 1995). Interestingly, some male homozygotes for the FSH receptor-inactivating mutation were fertile, demonstrating that testis development and spermatogenesis had little or no requirement for FSH.

Disruption of the locus 3q23 either by deletion or translocation results in the blepharophimosis, epicanthus inversus and ptosis (BPES) syndrome which is reviewed by Oley and Baraitser (1988). From the reproductive point of view this condition is intriguing. Some families with BPES exhibit infertility in females only (Towns and Muechler, 1979). In fact the presentation is of POF or occasionally ovarian resistance (for discussion on the distinction between these conditions, see Section 10.3). It may be therefore, that the long arm of chromosome 3 contains a gene which contributes to ovarian development. It is possible that BPES and ovarian failure represent a contiguous gene syndrome (Smith *et al.*, 1989) in which case pursuit of the ovarian component will be particularly interesting in the context of the genes discussed here.

10.3 The hormonal control of reproductive function

The reproductive axis from central control of GnRH to the release of the gonadal hormones is depicted in *Figure 10.3*. GnRH is released, under the influence of central neurotransmitters and gonadal feedback, in a pulsatile fashion at 90 min intervals resulting in concomitant pulsatile release of gonadotrophins. The nature of the GnRH pulse generator within the hypothalamus has been the focus of much research in neurophysiology. The clinical manifestation of defects at hypothalamic level would be hypogonadotrophic hypogonadism of which one form associated with adrenal hypoplasia congenita has been defined in molecular terms as caused by mutations in the DAX1 gene (see Section 10.2.3). Interestingly, despite an extensive search, no defect of the GnRH or GnRH receptor genes has been identified at the time of writing. This section starts with discussion of several genes known to affect GnRH secretion.

The gonadotrophins, LH and FSH, are secreted from the anterior pituitary

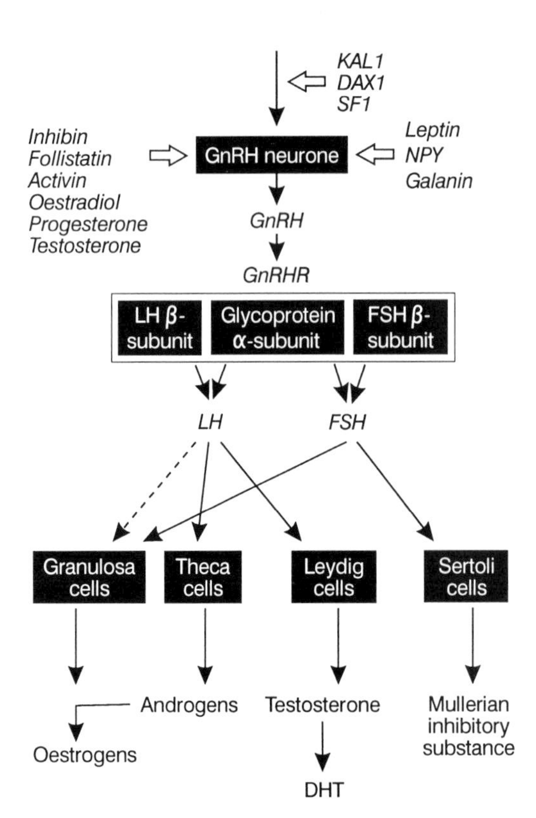

Figure 10.3. Scheme of the hormonal control of reproduction. See text for abbreviations.

gland under the control of GnRH. This gonadotrophin secretion is modulated by the negative feed back from gonadal hormones, both steroidal, oestrogen, testosterone and progesterone, and peptide, for example inhibin and activin. The gonadotrophins are glycoproteins, in common with thyroid-stimulating hormone (TSH) and hCG. The glycoprotein hormones are composed of a common α-subunit coded for by a single gene and unique β-subunits with each hormone derived from a separate β-subunit gene. Defects in the pituitary/gonadal axis include abnormalities of the structure of the hormone or of its receptor and come under the heading of hormone resistance syndromes.

It is over 50 years since the concept of hormone resistance was first proposed (Albright *et al.*, 1942) and in many instances the prediction of structural defects within hormone receptors has been confirmed by molecular genetic analysis. The reproductive axes, however, have lagged behind others in the molecular definition of hormone resistance.

Raised gonadotrophin concentrations may result from primary gonadal failure or from insensitivity to glycoprotein hormones. Hormonal resistance may be caused by an abnormality of hormone structure, for example insulin (Shoelson *et al.*, 1983), or to structural defects in receptors, for example androgen, glucocorticoid, vitamin D, insulin and thyroxine resistance (Arai and Chrousos 1994; Chatterjee and Beck-Peccoz, 1994; Hewison and O'Riordan, 1994; Patterson *et al.*, 1994; Taylor *et al.*, 1990). Thus, in 'ovarian resistance', there may be defects in the genes encoding LH, FSH or their respective receptors. Molecular characterization of gonadotrophin resistance syndromes includes reports of defects in genes encoding all four of these reproductive components.

Within the gonad the two gonadotrophins exert their effect on two endocrine cell types. In the testis, LH acts on the Leydig cell to promote testosterone production and FSH acts on the Sertoli cell to promote spermatogenesis and the synthesis of inhibin. In the ovary, the situation is a little more complex. LH acts on ovarian theca cells to promote androgen synthesis while granulosa cell LH receptors appear only in the pre-ovulatory phase under the influence of FSH stimulation. In granulosa cells LH promotes progesterone synthesis. The action of FSH on granulosa cells, in addition to inducing LH receptors, stimulates the aromatization of androgens of thecal origin to form oestrogen.

10.3.1 Genes affecting GnRH function

The 10 amino acids of GnRH are encoded by a gene (locus 8p21–11.2) comprising four exons and are cleaved from a 92-amino-acid precursor molecule (Hayflick *et al.*, 1989; Seeburg and Adelman 1984). This large precursor comprises not only GnRH but also GnRH-associated peptide which is a prolactin release inhibitory factor. While the hypogonadal mouse is the result of a deletion of part of the GnRH gene (Mason *et al.*, 1986) no human equivalent has yet been described (Layman *et al.*, 1992).

Leptin – the ob gene. An interaction between GnRH secretion and leptin, a helical cytokine of the tumour necrosis factor group, has been proposed primarily as a result of studies on the obese (ob) and diabetic (db) mouse models. Leptin, a

167-amino-acid protein, is encoded by the ob gene which was isolated by positional cloning (Zhang *et al.*, 1994). The ob/ob mouse is infertile and has subnormal gonadotrophin concentrations with an impaired response to castration (Swedloff *et al.*, 1976); that is, they have hypogonadotrophic hypogonadism. Weight loss alone, when forced by dietary restriction upon the ob/ob mouse, does not reverse infertility. Leptin administration to female ob/ob mice, however, results in a prompt return of fertility, presumably through the stimulation of GnRH (Chehab *et al.*, 1996).

The expression of the human ob gene in white fat cells is stimulated by insulin, glucocorticoids, noradrenaline and nutrients (Rohner-Jeanrenaud and Jeanrenaud, 1996; Saladin *et al.*, 1995). Leptin receptors have been identified in various peripheral tissues as well as in the hypothalamus and choroid plexus (Tartaglia *et al.*, 1995). Within the hypothalamus, leptin inhibits neuropeptide Y (NPY) synthesis and release (Stephens *et al.*, 1995). The link between the leptin and the GnRH neurone is, in part, mediated through NPY although there is redundancy in this hypothalamic signalling pathway, as the NPY knockout mouse not only maintains normal body weight but is fertile (Erickson *et al.*, 1996). A novel isoform of the leptin receptor has also been identified in human haemopoietic cells, prostate and ovary (Cioffi *et al.*, 1996) raising the possibility that leptin may have a direct ovarian effect in addition to that routed through the hypothalamus.

Primary leptin deficiency, therefore, is represented by the obese ob/ob mouse whereby a mutation of the ob gene and failure to synthesize leptin causes obesity and hypogonadotrophism. The db mouse, which is indistinguishable from the ob mouse phenotypically, was found to express a mutant leptin receptor and therefore is the product of leptin resistance (Chua *et al.*, 1996). The administration of leptin to the ob mouse (but not the db mouse) results in weight loss both through decreased appetite and increased energy expenditure (Pelleymounter *et al.*, 1995). In wild-type mice, leptin administration at pharmacological doses has only a minor effect on weight loss, making a physiological role for leptin in weight reduction questionable (Halaas *et al.*, 1995; Pelleymounter *et al.*, 1995). Human obesity is not a state of leptin deficiency. Circulating serum leptin concentrations and adipose tissue ob gene expression are raised in obese humans (Considine *et al.*, 1996). Neither is obesity a state of leptin resistance because the raised circulating leptin concentrations are appropriate for the amount of fat tissue and there is no evidence that leptin controls food intake in humans. However, secondary leptin deficiency is represented by weight-related amenorrhoea in humans. Reduced calorie intake, perhaps the result of an eating disorder, causes a deficit of body fat, reduced leptin secretion and hypogonadotrophism. While current knowledge routes the reproductive role for leptin through the GnRH neurone, there is also a possibility that leptin acts directly upon the ovary.

X-linked Kallmann syndrome (KAL1). X-linked Kallmann syndrome comprises agenesis of the olfactory bulbs causing anosmia and hypogonadotrophic hypogonadism in males. Female carriers exhibit only anosmia although there are autosomal recessive (KAL2) and dominant (KAL3) forms of the condition in which females have the complete phenotype. The hypogonadotrophic hypogonadism is the result of GnRH deficiency and is responsive to treatment with pulsatile GnRH.

The KAL1 locus (Xp22.3) was identified through the study of naturally occurring deletions of the short arm of the X chromosome (Meitinger *et al.*, 1990). The KAL1 gene, comprising 14 exons, shares homology with proteins involved with cell adhesion and axonal path-finding (Franco *et al.*, 1991). Such a function of the KAL1 gene product matches the prime defect in Kallmann syndrome whereby the GnRH neurones fail to migrate from the embryonic nasal epithelium to the hypothalamus. Strictly, therefore, the KAL1 gene is involved in the development of the hypothalamus rather than GnRH secretion. Several individuals have been identified in whom the KAL1 gene is not deleted but contains a point mutation causing premature stop codons (Hardelin *et al.*, 1992).

GnRH receptor. The above modulators of GnRH function find a common pathway to the pituitary gland via the GnRH receptor. The GnRH receptor is a member of the seven-transmembrane-domain, G-protein-coupled receptor family (Section 3.2.3) and is localized on the surface of gonadotropes in the anterior pituitary gland (Kakar *et al.*, 1992). The GnRH receptor gene (locus 4q13–21.2) comprises three exons (Fan *et al.*, 1994, 1995). No naturally occurring mutations of the GnRH gene have yet been described although activating mutations might be predicted to cause familial central precocious puberty and inactivating mutations might cause hypogonadotrophic hypogonadism.

The gonadotrophins. The glycoproteins, LH, FSH, TSH and hCG share a common α-subunit and differ by their unique β-subunit. The gene encoding the glycoprotein α-subunit consists of four exons located on chromosome 6 (Fiddles and Goodman, 1981). No naturally occurring mutations of this gene have yet been described although an α-subunit knockout mouse has been created by homologous recombination (Kendall *et al.*, 1995). This mouse, deficient in LH, FSH and TSH, is hypogonadal and hypothyroid but exhibits no abnormality of sexual differentiation or genital development. However, regulation of the glycoprotein α-subunit gene may be a fundamental mechanism for the control of gonadotrophins as sequences responsive to oestrogens, androgens, GnRH, phorbol esters and cAMP are found in the first 350–500 bp of the 5′-flanking region of this gene (Hamernik, 1995; Schoderbek *et al.*, 1993). One group of elements which take part in the pituitary-specific expression of the α-subunit gene are the basic helix–loop–helix proteins (Jackson *et al.*, 1995).

The LHβ and FSHβ genes are similar in structure, each comprising three exons, and are located on chromosomes 19 and 11, respectively (Talmadge *et al.*, 1984; Watkins *et al.*, 1987).

The LHβ gene. The phenotype of isolated LH deficiency was originally predicted to be the 'fertile eunuch' with a 46XY karyotype, male external genitalia, normal size testes with absent Leydig cells and lack of virilization which responded to hCG administration (Rogol *et al.*, 1980; Smals *et al.*, 1978; Williams *et al.*, 1975). In fact, the fertility status described in reports of isolated LH deficiency was very variable, with several males being azoospermic. One case, which has been characterized at a molecular level, shared some features with the 'fertile eunuch' but had raised immunoreactive serum LH concentrations with reduced bioactivity and normal FSH (Beitins *et al.*, 1981). This male presented with delayed puberty and a testicular biopsy showed arrested spermatogenesis and absent Leydig cells. Spermatogenesis and testosterone synthesis were stimulated by the administration of hCG. A mutation (E54R) was identified in exon 3 of the LHβ gene and was shown to prevent binding of the LH molecule to the LH receptor (Weiss *et al.*, 1992). Of his relatives who were demonstrated to be heterozygous for the mutation, three uncles were infertile with raised immunoreactive LH concentrations and variably raised FSH concentrations, while his mother and sister were fertile. His father was not investigated and his heterozygote status was therefore presumed.

Polymorphisms of the LHβ gene were reported in five women with variable presentations such as primary or secondary infertility, menstrual disturbance and polycystic ovary syndrome, who were found to have raised immunoreactive LH which possessed reduced bioactivity (Furui *et al.*, 1994; Suganuma *et al.*, 1995). The two sequence alterations identified in the LHβ gene in these women (W8R and I15T), were also found in a cohort of healthy women confirming that these variations were polymorphisms rather than mutations (Haavisto *et al.*, 1995; Pettersson *et al.*, 1994).

FSHβ gene. The clinical features of isolated FSH deficiency have been described in several reports to be those of primary amenorrhoea in 46XX females in whom fertility could be induced with FSH treatment (Rabin *et al.*, 1972, Rabinowitz *et al.*, 1979). Isolated FSH deficiency has recently been characterized in a female presenting with primary amenorrhoea who was found to be homozygote for a deletion of two nucleotides from codon 61 of the FSH β-subunit gene causing a frameshift and premature stop codon (Matthews *et al.*, 1993). Unlike the LHβ mutant described above, the circulating FSH measurements were low, as the more severe genetic defect did not result in an immunoreactive FSH molecule. Fertility was restored in this individual with the administration of FSH. The mother of the proband was also heterozygous for the FSHβ deletion and had a history of menstrual irregularity, secondary amenorrhoea and primary infertility, before the spontaneous conception of the proband. Screening for FSHβ mutations in 18 women with POF, seven of whom had primary amenorrhoea, failed to identify further mutations (Layman *et al.*, 1993).

The effects of FSH deficiency in men await clarification. The father of the proband above, a presumed heterozygote for the FSHβ deletion even though there was no apparent consanguinity, was deceased, and the genetic and phenotypic details of son have not yet been reported. Reports of sporadic males with isolated FSH deficiency and depressed spermatogenesis have not so far been substantiated by the genetic studies (Al-Ansari *et al.*, 1984; Maroulis *et al.*, 1977).

Gonadotrophin receptors. The genes encoding the receptors for LH/hCG, FSH and TSH form a subgroup of the seven-transmembrane-domain, G-protein-linked receptor family. They are unusual in possessing extended extracellular, ligand-binding domains. The genes encoding the FSH and LH receptors are located on the short arm of chromosome 2 and share an unusual arrangement of nine or 10 small exons making up the extracellular domain and a large exon encoding nearly two-thirds of the receptor including the seven-transmembrane and intracellular domains (*Figure 10.4*) (Heckert *et al.*, 1992; Koo *et al.*, 1991; Minegishi *et al.*, 1990, 1991; Rousseau-Merck *et al.*, 1993).

LH receptor gene. The first LH receptor mutations reported were those which resulted in constitutive receptor activity and presented as precocious puberty in males (Kosugi *et al.*, 1995; Kremer *et al.*, 1993; Latronico *et al.*, 1995; Laue *et al.*, 1995; Shenker *et al.*, 1993). All but one of these reports describe homozygous activating mutations of the LH receptor gene. Interestingly, affected female individuals appear to have no manifestation of gonadal dysfunction suggesting that the ovary is relatively insensitive to the degree of LH activity. The effect of these mutations is to alter the amino acid composition of the third intracellular loop of these receptors which is thought to delay the dissociation of the stimulatory G-protein, $G_s\alpha$, from the receptor so that signal transduction is prolonged (*Figure 10.4*).

It was originally anticipated that LH receptor defects would block function and as such the phenotype would be Leydig cell hypoplasia which occurs in both sporadic (Berthezene *et al.*, 1976; Brown *et al.*, 1978; David *et al.*, 1984; Schwartz *et al.*, 1981) and familial forms (El-Awady *et al.*, 1987; Kremer *et al.*, 1995; Latronico *et al.*, 1996; Perez-Palacios *et al.*, 1981). Leydig cell hypoplasia was initially described in 46XY females presenting with primary amenorrhoea but a milder phenotype with partial virilization *in utero* is also possible. The more markedly severe phenotype observed in Leydig cell hypoplasia, compared with isolated LH deficiency, can now be explained in molecular terms. The fetal testis relies on stimulation by placental hCG for testosterone synthesis and, in turn, the development of external male genitalia. This action of hCG is blocked by inactivating mutations of the LH receptor gene (Leydig cell hypoplasia) while LHβ mutations

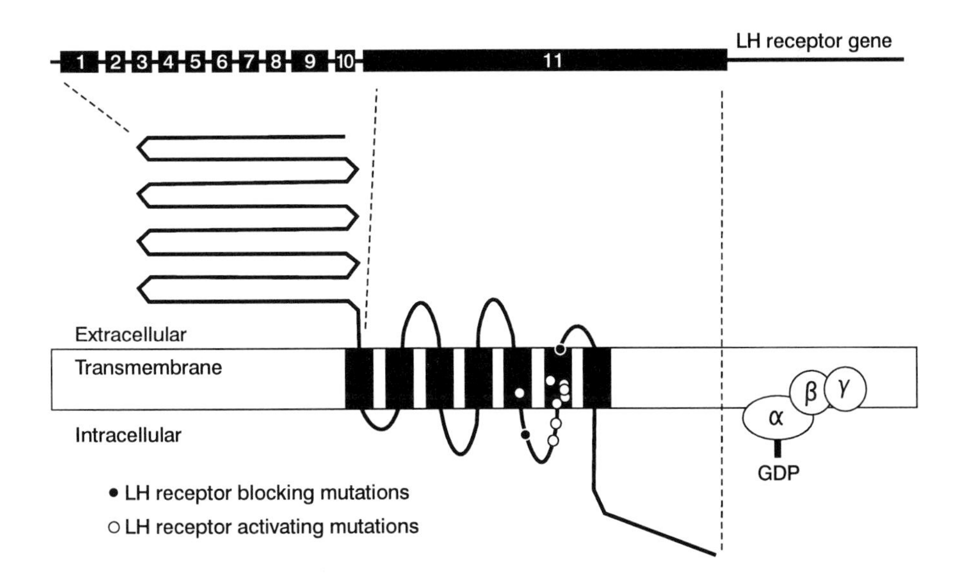

Figure 10.4. The LH receptor gene and the LH receptor showing the site of amino acid changes caused by mutations resulting in loss or gain of function.

become manifest only when testosterone production becomes LH dependent, at puberty (isolated LH deficiency).

Complete LH resistance, caused by a blocking mutation of the LH receptor, was described in two 46XY siblings with Leydig cell hypoplasia presenting with primary amenorrhoea, a female phenotype but with absent Müllerian structures (Kremer *et al.*, 1995). A homozygous mutation within the sixth transmembrane domain (A593P) of the LH receptor gene was found in these siblings (*Figure 10.4*). Serum LH concentrations were raised, FSH concentrations were normal and the histology of the testes showed an absence of Leydig cells. A similar family has recently been reported in whom a mutation (R554X) disrupts the third intracellular loop of the transmembrane domain of the LH receptor (Latronico *et al.*, 1996). This last report is the final piece in the molecular jigsaw of gonadotrophins – the female phenotype. Earlier reports had alluded to amenorrhoeic sisters of 46XY pseudohermaphrodites with Leydig cell hypoplasia (Kremer *et al.*, 1995; Saldanha *et al.*, 1987). The female in the latest pedigree, who was homozygous for the R554X mutation of the LH receptor gene, had normal pubertal development but only a single menstrual bleed at the age of 13 (Latronico *et al.*, 1996). The ultrasonic appearances of the ovaries were described as 'cystic' and of unequal size (1.9 and 7.2 ml). The serum LH concentration was five times greater than normal while the FSH and oestradiol concentrations were normal.

FSH receptor gene. For many years, the phenotype for FSH receptor mutations was thought to be ovarian dysgenesis, POF or resistant ovary syndrome. Ovarian dysgenesis refers to females with primary amenorrhoea and streak gonads, POF refers to women who experience the menopause before the age of 40 and resistant ovary syndrome refers to the combination of hypergonadotrophic amenorrhoea and normal follicular apparatus on ovarian biopsy. Implicit in the original notion of ovarian resistance to gonadotrophin stimulation, was the hope that the resistance might be overcome and fertility regained. In fact, the prospect of fertility for women with 'resistant ovary syndrome' never materialized, despite occasional optimistic case reports, and most investigators concluded that this syndrome was an early form of primary ovarian failure. Now that ovarian ultrasound has replaced ovarian biopsy in the investigation of women with hypergonadotrophic amenorrhoea we realize that the majority of such women have identifiable ovaries with follicles (Conway *et al.*, 1996). The three conditions, ovarian dysgenesis, POF and resistant ovary syndrome, are in fact within the spectrum of one phenomenon – germ cell depletion.

The description of complete FSH resistance caused by mutations of the FSH receptor gene has emerged from the investigation of six Finnish families with early ovarian failure (Aittomäki, 1994). Probands from six families were homozygous for a missense mutation (A189V) of exon 7 of the FSH receptor gene which encodes part of the extracellular ligand-binding domain of the FSH receptor (*Figure 10.5*) (Aittomäki *et al.*, 1995). Expression of this mutation caused reduced ligand-binding capacity with normal binding affinity resulting in reduced signal transduction as measured by cAMP production in transfected MSC-1 cells. The corresponding phenotype, which was also the criterion for ascertainment in this report, was of primary amenorrhoea or secondary amenorrhoea before the age of 20 with ovarian dysgene-

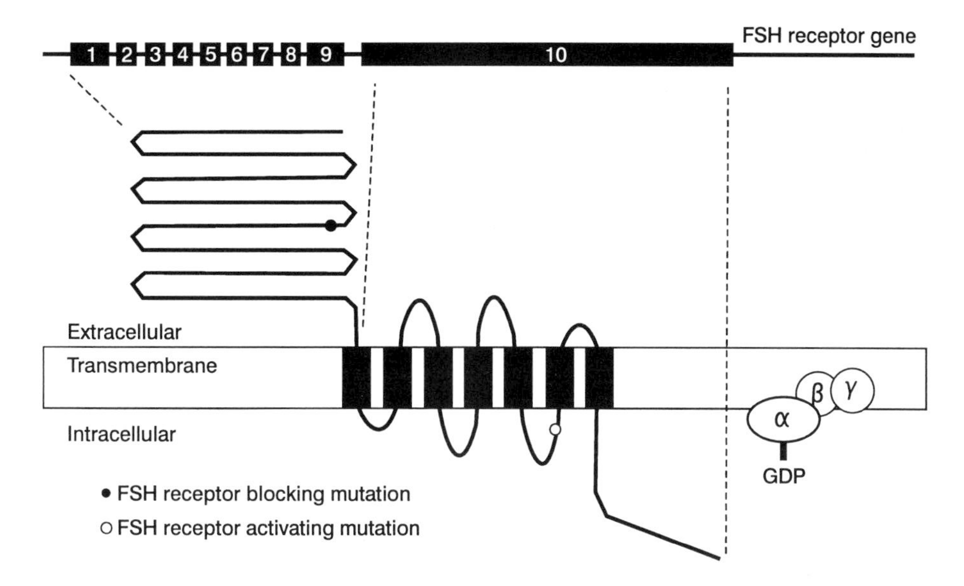

Figure 10.5. The FSH receptor gene and the FSH receptor showing the site of amino acid changes caused by mutations resulting in loss or gain of function.

sis. The obligate heterozygote mothers of the probands exhibited normal fertility and no menstrual disturbance. Interestingly, two of three brothers of probands, who were proven homozygotes for the A189V mutation, were fertile while the third was thought to be infertile. This report raises the question of the wider prevalence of FSH receptor defects in women with POF and of the requirement of FSH in spermatogenesis. An activating mutation of the FSH receptor has recently been described in a male who maintained spermatogenesis and testicular volume despite prior hypophysectomy (Gromoll *et al.,* 1996). This individual is heterozygous for a mutation (D567G) in the third intracytoplasmic loop of the FSH receptor, a site directly analogous to the activating mutations of the LH receptor. Transient expression of this mutation in COS7 cells confirmed a marked increase in basal cAMP production.

In summary, raised activity of the gonadotrophin axis may be due to LHβ or FSHβ gene polymorphisms or activating mutations of LH receptor gene. Reduced gonadotrophin activity may be caused by defects in the LHβ or FSHβ molecule or to inactivating mutations of the respective receptor genes. Complete gonadotrophin resistance is likely to be very rare, however, so what are we likely to find in partial gonadotrophin resistance – might the 'resistant ovary syndrome' correspond with minor FSH receptor mutations? Experience with insulin and androgen resistance syndromes suggests that such a scenario is unlikely. Insulin receptor gene mutations are found in extreme Type A insulin resistance but not in moderate forms of insulin resistance (O'Rahilly *et al.,* 1991). Androgen receptor gene mutations are found in nearly all cases of complete androgen insensitivity but rarely in partial forms (Patterson *et al.,* 1994). Mild resistance to hormone action is rarely detectable in relatives who are heterozygote for receptor mutations which are inherited in a recessive pattern. It seems unlikely therefore, that the suggestions of relatives with reproductive manifestations of heterozygote receptor defects in the pedigrees described above will be substantiated and account for common conditions. Thus, while mutation analysis provides new insights into the gender-specific role of the gonadotrophins, the cause of early ovarian failure in the majority of women remains a mystery.

G-protein mutations in reproduction. For completeness it is appropriate to consider abnormalities of gonadotrophin signal transduction here. As the gonadotrophin receptors are coupled to the stimulatory G-protein, $G_s\alpha$ (Section 3.3), mutations of this signal transduction unit are manifest in the reproductive axis although not in isolation. Loss-of-function mutations of the $Gs\alpha$ gene cause resistance to parathyroid hormone primarily (Albright's hereditary osteodystrophy) and also to gonadotrophins resulting in hypogonadism (Patten *et al.*, 1990; Weinstein *et al.*, 1990). Activating mutations of the $G_s\alpha$ gene cause McCune–Albright syndrome (Weinstein *et al.*, 1991). These gain-of-function mutations cause widespread endocrine constitutive activity affecting the pituitary, thyroid and adrenal gland as well as skin pigment and bone abnormalities. In the gonads, McCune–Albright syndrome is manifest by precocious puberty in girls.

10.4 Conclusions

Naturally occurring defects in the components of the reproductive axis have provided many new insights into sexual function. For example, only now do we realize the degree to which males are LH dependent and females are FSH dependent, both sexes exhibiting relative insensitivity to alterations in the other gonadotrophin. Much more detail of the relationships between structure and function of the gonadotrophins and their receptors can be expected in the future through the study of both naturally occurring and created mutations. In the case of leptin and NPY, animal models have revealed surprising links with the central control of reproduction. Over the past decade the characterization of many of the inherited defects affecting fertility has greatly accelerated our understanding of reproductive endocrinology.

References

Aittomäki K. (1994) The genetics of XX gonadal dysgenesis. *Am. J. Hum. Genet.* **54:** 844–851.

Aittomäki K, Lucena JLD, Pakarinen P *et al.* (1995) Mutation in the follicle-stimulating hormone receptor gene causes hereditary hypergonadotropic ovarian failure. *Cell* **82:** 959–968.

Al-Ansari AAK, Khalil TH, Kelani Y, Mortimer CH. (1984) Isolated follicle-stimulating hormone deficiency in men: success of long-term gonadotrophin therapy. *Fertil. Steril.* **42:** 618–626.

Albright F, Burnett CH, Smith PH, Parson W. (1942) Pseudo-hypoparathyroidism – an example of 'Seabright-Bantam Syndrome'. *Endocrinology* **30:** 922–932.

Arai K, Chrousos GP. (1994) Glucocorticoid resistance. *Baillieres Clin. Endocrinol. Metab.* **8:** 317–332.

Ashley CT, Sutcliffe JS, Kunst CB, Leiner HA, Eichler EE, Nelson DL, Warren ST. (1993a) Human and murine FMR-1 alternative splicing and translational initiation downstream of the CGG repeat. *Nature Genet.* **4:** 244–251.

Ashley CT, Wilkinson KD, Reines D, Warren ST. (1993b) FMR1 protein: conserved RNP family domains and selective RNA binding. *Science* **262:** 563–566.

Baker TG. (1963) A quantitative and cytological study of germ cells in human ovaries. *Proc. R. Soc. Lond.* **158:** 417–433.

Bardoni B, Zanaria E, Guioli S *et al.* (1994) A dosage sensitive locus at chromosome Xp21 is involved in male to female sex reversal. *Nature Genet.* **7:** 497–501.

Barnhart KM, Mellon PL. (1994) The orphan receptor, steroidogenic factor-1, regulates the glycoprotein hormone alpha-subunit gene in pituitary gonadotrophes. *Mol. Endocrinol.* **8:** 878–885.

Bates A, Howard PJ. (1990) Distal long arm deletions of the X chromosome and ovarian failure. *J. Med. Genet.* **27:** 722–723.

Behringer RR. (1994) The in vivo roles of mullerian-inhibiting substance. *Curr. Top. Devel. Biol.* **29:** 171–187.

Beitins IZ, Axelrod L, Ostrea T, Little R, Badger TM. (1981) Hypogonadism in a male with an immunologically active, biologically inactive luteinizing hormone: characterization of the abnormal hormone. *J. Clin. Endocrinol. Metab.* **52:** 1143–1149.

Berthezene F, Forest MG, Grimaud JA, Claustrat B, Mornex R. (1976) Leydig-cell agenesis. A cause of male pseudohermaphroditism. *New Engl. J. Med.* **295:** 969–972.

Bogan JS, Page DC. (1994) Ovary? Testis? – A mammalian dilemma. *Cell* **76:** 603–607.

Bachner D, Steinbach P, Wohrle D, Just W, Vogel W, Hameister H, Manca A, Poustka A. (1993) Enhanced Fmr-1 expression in the testis. *Nature Genet.* **4:** 115–116.

Brown DM, Markland C, Dehner LP. (1978) Leydig cell hypoplasia: a cause of male pseudohermaphroditism. *J. Clin. Endocrinol. Metab.* **46:** 1–7.

Burris TP, Guo W, Le T, McCabe ER. (1995) Identification of a putative steroidogenic factor-1 response element in the DAX-1 promoter. *Biochem. Biophys. Res. Commun.* **214:** 576–581.

Call KM, Glaser T, Ito CY *et al.* (1990) Isolation and characterisation of a zinc finger polypeptide gene at the human chromosome 11 Wilms' tumor locus. *Cell* **60:** 509–520.

Carre-Eusebe D, Imbeaud S, Harbison M, New MI, Josso N, Picard J-Y. (1992) Variants of the anti-Mullerian hormone gene in a compound heterozygote with the persistent Mullerian duct syndrome and his family. *Hum. Genet.* **90:** 389–394.

Cate RL, Mattalinano RJ, Hession C *et al.* (1986) Isolation of the bovine and human genes for mullerian inhibiting substance and expression of the human gene in animal cells. *Cell* **45:** 685–698.

Chatterjee K, Beck-Peccoz P. (1994) Thyroid hormone resistance. *Baillieres Clin. Endocrinol. Metab.* **8:** 267–284.

Chehab FF, Lim ME, Lu R. (1996) Correction of the sterility defect in homozygous obese female mice by treatment with the human recombinant leptin. *Nature Genet.* **12:** 318–320.

Chu CE, Connor JM. (1995) Molecular biology of Turner's Syndrome. *Arch. Dis. Childhood* **72:** 285–286.

Chua SC, Chung WK, Wu-Peng S, Zhang Y, Liu S-M, Tartaglia L, Leibel RL. (1996) Phenotypes of mouse diabetes and rat fatty due to mutations in the OB (leptin) receptor. *Science* **271:** 994–996.

Cioffi JA, Shafer AW, Zupancic TJ, Smith-Gbur J, Mikhail A, Platika D, Snodgrass HR. (1996) Novel B219/OB receptor isoforms: possible role of leptin in hemapoesis and reproduction. *Nature Med.* **2:** 585–589.

Cohen-Haguenauer O, Picard JY, Mattei M-G *et al.* (1987) Mapping of the gene for anti-mullerian hormone to the short arm of chromosome 19. *Cytogenet. Cell Genet.* **44:** 2–6.

Considine RV, Sinha MK, Heiman ML *et al.* (1996) Serum immunoreactive-leptin concentrations in normal-weight and obese humans. *New Engl. J. Med.* **334:** 292–295.

Conway GS, Hettiarachchi S, Murray A, Jacobs PA. (1995) Fragile X premutations in familial premature ovarian failure. *Lancet* **346:** 309–310.

Conway GS, Kaltsas G, Patel A, Davies MC, Jacobs HS. (1996) Characterization of idiopathic premature ovarian failure. *Fertil. Steril.* **65:** 337–341.

David R, Jin Yoon D, Landin L, Lew L, Sklar C, Schinella R, Golimbu M. (1984) A syndrome of gonadotropin resistance possibly due to a luteinzing hormone receptor defect. *J. Clin. Endocrinol. Metab.* **59:** 156–160.

De La Chapelle A, Schroder J, Haahtela T, Aro P. (1975) Deletion mapping of the human X chromosome. *Hereditas* **80:** 113–120.

El-Awady MK, Temtamy SA, Salam MA, Gad YZ. (1987) Familial Leydig cell hypoplasia as a cause of male pseudohermaphroditism. *Hum. Hered.* **37:** 36–40.

Erickson JC, Clegg KE, Palmiter RD. (1996) Sensitivity to leptin and susceptibility to seizures of mice lacking neuropeptide Y. *Nature* **381:** 415–418.

Fan NC, Jeung EB, Peng C, Olofsson JI, Krisinger J, Leung PC. (1994) The human gonadotropin-releasing hormone (GnRH) receptor gene: cloning, genomic organization and chromosomal assignment. *Mol. Cell. Endocrinol.* **103:** R1–6.

Fan NC, Peng C, Krisinger J, Leung PC. (1995) The human gonadotropin-releasing hormone receptor gene: complete structure including multiple promoters, transcription initiation sites, and polyadenylation signals. *Mol. Cell. Endocrinol.* **107:** R1–8.

Fiddles JC, Goodman HM. (1981) The gene encoding the common alpha subunit of the four human glycoprotein hormones. *J. Mol. Appl. Genet.* **1:** 3–18.

Foster JW, Dominguez-Steglich MA, Guioli S *et al.* (1994) Camptomelic dysplasia and autosomal sex reversal caused by mutations in an SRY-related gene. *Nature* **372:** 525–530.

Franco B, Guioli S, Pragliola A *et al.* (1991) A gene deleted in Kallmann's syndrome shares homology with neural cell adhesion and axonal path-finding molecules. *Nature* **353:** 529–536.

Friedman AL, Finlay JL. (1987) The Drash syndrome revisited: diagnosis and follow-up. *Am. J. Med. Genet. Suppl.* **3:** 293–296.

Fryns J-P. (1986) The female and the fragile X: a study of 144 obligate female carriers. *Am. J. Med. Genet.* **23:** 157–169.

Furui K, Suganuma N, Tsukahara S-I, Asada Y, Kikkawa F, Tanaka M, Ozawa T, Tomoda Y. (1994) Identification of two point mutations in the gene coding luteinizing hormone (LH) alpha-subunit, associated with immunologically anomalous LH variants. *J. Clin. Endocrinol. Metab.* **78:** 107–113.

Goodfellow PN, Lovell-Badge R. (1993) SRY and sex determination in mammals. *Ann. Rev. Genet.* **27:** 71–92.

Gromoll J, Simoni M, Nieschlag E. (1996) An activating mutation of the follicle-stimulating hormone receptor autonomously sustains spermatogenesis in a hypophyesctomized man. *J. Clin. Endocrinol. Metab.* **81:** 1367–1370.

Grootegoed JA, Baarends WM, Themmen APN. (1994) Welcome to the family: the anti-mullerian hormone receptor. *Mol. Cell. Endocrinol.* **100:** 29–34.

Gubbay J, Collignon J, Koopman P, Capel B, Economou A, Munsterberg A, Vivian N, Goodfellow P, Lovell-Badge R. (1990) A gene mapping to the sex-determining region of the mouse Y chromosome is a member of a novel family of embryonically expressed genes. *Nature* **346:** 245–250.

Guellaen G, Casanova M, Bishop C, Geldwerth D, Andre G, Fellous M, Weissenbach J. (1984) Human XX males with Y single-copy DNA fragments. *Nature* **307:** 172–173.

Haavisto A-M, Pettersson K, Bergendahl M, Virkamaki A, Huhtaniemi I. (1995) Occurrence and biological properties of a common genetic variant of luteinizing hormone. *J. Clin. Endocrinol. Metab.* **80:** 1257–1263.

Halaas JL, Gajiwala KS, Maffei M, Cohen SL, Chait BT, Rabinowitz D, Lallone RL, Burley SK, Friedman JM. (1995) Weight-reducing effects on the plasma protein encoded by the obese gene. *Science* **269:** 543–546.

Hamernik DL. (1995) Molecular biology of gonadotrophins. *J. Reprod. Fertil. Suppl.* **49:** 257–269.

Haqq CM, King C-Y, Ukiyama E, Falsafi S, Haqq TN, Donahoe PK, Weiss MA. (1994) Molecular basis of mammalian sexual determination: activation of mullerian inhibiting substance gene expression by SRY. *Science* **266:** 1494–1501.

Hardelin J-P, Levilliers J, del Castillo I *et al.* (1992) X chromosome-linked Kallmann syndrome: stop mutations validate the candidate gene. *Proc. Natl Acad. Sci. USA* **89:** 8190–8194.

Harley VR, Jackson DI, Hextall PJ, Hawkins JR, Berkovitz GD, Sockanathan S, Lovell-Badge R; Goodfellow P. (1992) DNA binding activity of recombinant SRY from normal males and XY females. *Science* **255:** 453–456.

Hayflick JS, Adelman JP, Seeburg PH. (1989) The complete nucleotide sequence of the human gonadotropin-releasing hormone gene. *Nucl. Acids Res.* **17:** 6403–6404.

Heckert LL, Daley IJ, Griswold MD. (1992) Structural organisation of the follicle-stimulating hormone receptor gene. *Mol. Endocrinol.* **6:** 70–80.

Hewison M, O'Riordan JLH. (1994) Vitamin D resistance. *Baillieres Clin. Endocrinol. Metab.* **8:** 305–316.

Hinds HL, Ashley CT, Sutcliffe JS, Nelson, DL, Warren ST, Houseman DE, Schalling M. (1993) Tissue specific expression of FMR-1 provides evidence for a functional role in fragile X syndrome. *Nature Genet.* **2:** 197–200.

Hovmoller ML, Osuna A, Eklof O, Fredga K, Hjerpe A, Lindsten J, Ritzen M, Stanecu V, Svenningsen N. (1977) Camptomelic dwarfism. A genetically determined mesenchymal disorder combined with sex reversal. *Hereditas* **86:** 51–62.

Hsu SY, Kubo M, Chun SY, Haluska FG, Housman DE, Hsueh AJ. (1995) Wilms' tumor protein WT1 as an ovarian transcription factor: decreases in expression during follicle development and repression of inhibin-alpha gene promoter. *Mol. Endocrinol.* **9:** 1356–1366.

Imbeaud S, Faure E, Lamarre I *et al.* (1995) Insensitivity to anti-Mullerian hormone due to a mutation in the human anti-Mullerian hormone receptor. *Nature Genet.* **11:** 382–388.

Ingraham HA, Lala DS, Ikeda Y, Luo X, Shen WH, Nachtigal MW, Abbud R, Nilson JH, Parker KLTI. (1994) The nuclear receptor steroidogenic factor 1 acts at multiple levels of the reproductive axis. *Genes Devel.* **8:** 2302–2312.

Jackson SM, Gutierrez-Hartmann A, Hoeffler JP. (1995) Upstream stimulatory factor, a basic-helix-loop-helix-zipper protein, regulates the activity of the alpha-glycoprotein hormone subunit gene in pituitary cells. *Mol. Endocrinol.* **9:** 278–291.

Jacobs PA. (1991) Fragile X syndrome. *J. Med. Genet.* **28:** 809–810.

Josso N. (1986) Anti-Mullerian hormone: new perspectives for a sexist molecule. *Endocr. Rev.* **7:** 421–433.

Jost A. (1953) Studies of sex differentiation in mammals. *Recent Prog. Hormone Res.* **8:** 379–418.

Kakar SS, Musgrove LC, Devor DC, Sellers JC, Neill JD. (1992) Cloning, sequencing, and expression of human gonadotropin releasing hormone (GnRH) receptor. *Biochem. Biophys. Res. Commun.* **189:** 289–295.

Kendall SK, Samuelson LC, Saunders TL, Wood RI, Camper SA. (1995) Targeted disruption of the pituitary hormone alpha-subunit produces hypogonadal and hypothyroid mice. *Genes Devel.* **9:** 2007–2019.

Knebelmann B, Boussin L, Guerrier D, Legeai L, Kahn A, Josso N, Picard J-Y. (1991) Anti-Mullerian hormone Bruxelles: a nonsense mutation associated with the persistent Mullerian duct syndrome. *Proc. Natl Acad. Sci. USA* **88:** 3767–3771.

Kobayashi K, Mizuno K, Hida A *et al.* (1994) PCR analysis of the Y chromosome long arm in azoospermic patients: evidence for a second locus required for spermatogenesis. *Hum. Mol. Genet.* **3:** 1965–1967.

Koo YB, Ji I, Slaughter RG, Ji TH. (1991) Structure of the luteinizing hormone receptor gene and multiple exons of the coding sequence. *Endocrinology* **128:** 2297–2308.

Koopman P, Munsterberg A, Capel B, Vivian N, Lovell-Badge R. (1990) Expression of a candidate sex-determining gene during mouse testis differentiation. *Nature* **348:** 450–452.

Kosugi S, Van Dop C, Geffner M, Rabl ME, Carel J-C, Chaussain J-L, Mori T, Merendino JJ, Shenker A. (1995) Characterization of heterogeneous mutations causing constitutive activation of the luteinizing hormone receptor in familial male precocious puberty. *Hum. Mol. Genet.* **4:** 183–188.

Krauss CM, Turksoy N, Atkins L, McLaghlin C, Brown LG, Page DC. (1987) Familial premature ovarian failure due to an interstitial deletion of the long arm of the X chromosome. *New Engl. J. Med.* **317:** 125–131.

Kremer H, Mariman E, Otten BJ *et al.* (1993) Cosegregation of missense mutations of the luteinizing hormone receptor gene with familial male-limited precocious puberty. *Hum. Mol. Genet.* **2:** 1779–1783.

Kremer H, Kraaj R, Toledo SPA *et al.* (1995) Male pseudohermaphroditism due to a homozygous missense mutation of the luteinizing hormone receptor gene. *Nature Genet.* **9:** 160–164.

Lala DS, Rice DA, Parker KL. (1992) Steroidogenic factor 1, a key regulator of steroidogenic enzyme expression, is the mouse homolog of fushi tarazu-factor 1. *Mol. Endocrinol.* **6:** 1249–1258.

Lala DS, Ikeda Y, Luo X, Baity LA, Meade JC, Parker KL. (1995) A cell specific nuclear receptor regulates the steroid hydroxylases. *Steroids* **60:** 10–14.

Latronico AC, Anasti J, Arnhold IJP, Mendonca BB, Domenice S, Albano MC, Zachman K, Wajchenberg BL, Tsigos C. (1995) A novel mutation of the luteinizing hormone receptor gene causing male gonadotrophin-independent precocious puberty. *J. Clin. Endocrinol. Metab.* **80:** 2490–2494.

Latronico AC, Anasti J, Arnhold IJP, Rapaport R, Mendonca BB, Bloise W, Castro M, Tsigos C, Chrousos GP. (1996) Testicular and ovarian resistance to luteinizing hormone caused by inactivating mutations of the luteinizing hormone receptor gene. *New Engl. J. Med.* **334:** 507–512.

Laue L, Chan W-Y, Hseuh AJW, Kudo M, Hsu SY, Wu S-M, Blomberg L, Cutler GB. (1995) Genetic heterogeneity of constitutively activating mutations of the human luteinizing hormone receptor in familial male-limited precocious puberty. *Proc. Natl Acad. Sci. USA* **92:** 1906–1910.

Layman LC, Wilson JT, Huey LO, Lanclos KD, Plouffe L, McDonough PG. (1992) Gonadotropin-releasing hormone, follicle-stimulating hormone beta, luteinizing hormone beta gene structure in idiopathic hypogonadotrophic hypogonadism. *Fertil. Steril.*: 42–49.

Layman CL, Shelley ME, Huey LO, Wall SW, Tho SPT, McDonough PG. (1993) Follicle-stimulating hormone beta gene structure in premature ovarian failure. *Fertil. Steril.* **60:** 852–857.

Lee MM, Donahoe PK. (1993) Mullerian inhibiting substance: A gonadal hormone with multiple functions. *Endocr. Rev.* **14:** 152–164.

Luo X, Ikeda Y, Parker KL. (1994) A cell specific nuclear receptor is essential for adrenal and gonadal development and sexual differentiation. *Cell* **77:** 481–490.

Ma K, Sharkey A, Kirsch S, Vogt P, Keil R, Hargreave TB, McBeath S, Chandley AC. (1992) Towards the molecular localisation of the AZF locus: mapping of microdeletions in azoospermic men within 14 subintervals of interval 6 of the human Y chromosome. *Hum. Mol. Genet.* **1:** 29–33.

Ma K, Inglis JD, Sharkey A *et al.* (1993) A Y chromosome gene family with RNA-binding protein homology: candidate for the azoospermia factor AZF controlling human spermatogenesis. *Cell* **75:** 1287–1295.

Maroulis GB, Parlow AF, Marshall JR. (1977) Isolated follicle-stimulating hormone deficiency in man. *Fertil. Steril.* **28:** 8188–8122.

Mason AJ, Hayflick JS, Zoeller RT, Young WS, Phillips HS, Nikolics K, Seeburg PH. (1986) A deletion truncating the gonadotropin-releasing hormone gene is responsible for hypogonadism in the 'hpg' mouse. *Science* **234:** 1366–1371.

Matthews CH, Borgato S, Beck-Peccoz P, Adams MTY, Gambino G, Casagrande S, Tedeschini G, Benedetti A, Chatterjee VKK. (1993) Primary amenorrhoea and infertility due to a mutation in the beta-subunit of follicle-stimulating hormone. *Nature Genet.* **5:** 83–86.

Meitinger T, Heye B, Petit C *et al.* (1990) Definitive localization of X-linked Kallmann syndrome (hypogonadotrophic hypogonadism and anosmia) to Xp22.3: close linkage to the hypervariable repeat sequence CRI-S232. *Am. J. Hum. Genet.* **47:** 664–669.

Minegishi T, Nakamura K, Takakura Y, Miyamoto K, Hasegawa Y, Ibuki Y, Igara M. (1990) Cloning and sequencing of human LH/hCG receptor cDNA. *Biochem. Biophys. Res. Commun.* 172: 1049–1054.

Minegishi T, Nakamura K, Takaura Y, Ibuki Y, Igarashi M. (1991) Cloning and sequencing of human FSH receptor cDNA. *Biochem. Biophys. Res. Commun.* **175:** 1125–1130.

Muscatelli F, Strom TM, Walker AP *et al.* (1994) Mutations in the DAX-1 gene give rise to both X-linked adrenal hypoplasia congenita and hypogonadotrophic hypogonadism. *Nature* **372:** 672–676.

Musterberg A, Lovell-Badge R. (1991) Expression of the mouse anti-Mullerian hormone gene suggests a role in both male and female sexual differentiation. *Development* **113:** 613–624.

Najmabadi H, Huang V, Yen P *et al.* (1996) Substantial prevalence of microdeletions of the Y chromosome in infertile men with idiopathic azoospermia and oligospermia detected using a sequence-tagged site-based mapping strategy. *J. Clin. Endocrinol. Metab.* **81:** 1347–1352.

Nordenskjold A, Fricke G, Anvret M. (1995) Absence of mutations in the WT1 gene in patients with XY gonadal dysgenesis. *Hum. Genet.* **96:** 102–104.

Ogata T, Matsuo N. (1995) Turner syndrome and female sex chromosome aberrations: deduction of the principal factors involved in the development of clinical features. *Hum. Genet.* **95:** 607–629.

Oley C, Baraitser M. (1988) Blepharophimosis, ptosis, epicanthus inversus syndrome (BPES syndrome). *J. Med. Genet.* **25:** 47–51.

O'Rahilly S, Choi WH, Patel P, Turner RC, Flier JS, Moller DE. (1991) Detection of mutations in insulin-receptor gene in NIDDM patients by analysis of single-stranded conformation polymorphisms. *Diabetes* **40:** 777–782.

Page DC, Mosher R, Simpson EM, Fisher EM, Mardon G, Pollack J, McGillivray B, de la Chapelle A, Brown LG. (1987) The sex-determining region of the human Y chromosome encodes a finger protein. *Cell* **51:** 1091–1104.

Palmer MS, Sinclair AH, Berta P, Ellis NA, Goodfellow PN, Abbas NE, Fellous M. (1989) Genetic evidence that ZFY is not the testis determining factor. *Nature* **342:** 937–939.

JL, Johns DR, Valle D, Eil C, Gruppuso PA, Steele G, Smallwood PM, Levine MA. (1990) tation in the gene encoding the stimulatory G protein of adenyl cyclase in Albright's hereditary eodystrophy. *New Engl. J. Med.* **322:** 1412–1419.

terson MN, McPhaul MJ, Hughes IA. (1994) Androgen insensitivity syndrome. *Baillieres Clin. Endocrinol. Metab.* **8:** 379–404.

Pelletier J, Bruening W, Kashtan CE *et al.* (1991a) Germline mutations in the Wilms' tumor suppressor gene are associated with abnormal urogenital development in Denys-Drash syndrome. *Cell* **67:** 437–447.

Pelletier J, Bruening W, Li FP, Haber DA, Glaser T, Housman DE. (1991b) WT1 mutations contribute to abnormal genital system development and hereditary Wilms' tumor. *Nature* **353:** 431–434.

Pellymounter MA, Cullen MJ, Baker MB, Hecht R, Winters D, Boone T, Collins F. (1995) Effects of obese gene product on body weight regulation in ob/ob mice. *Science* **269:** 540–542.

Perez-Palacios G, Scaglia HE, Kofman S, Saavedra D, Ochoa S, Larraza O, Perez AE. (1981) Inherited male pseudohermaphroditism due to gonadotrophin unresponsiveness. *Acta Endocrinol.* **98:** 148–155.

Pettersson K, Makela MM, Dahlen P, Lamminen T, Huoponen K, Huhtaniemi I. (1994) Gene polymorphism found in the LH beta gene of an immunologically anomalous variant of luteinizing hormone. *Eur. J. Endocrinol.* **130** (Suppl 2): 65.

Powell CM, Taggart RT, Drumheller TC, Wangsa D, Qian C, Nelson LM, White BJ. (1994) Molecular and cytogenetic studies of an X; autosome translocation in a patient with premature ovarian failure and review of the literature. *Am. J. Med. Genet.* **52:** 19–26.

Pritchard-Jones K, Fleming S, Davidson D *et al.* (1990) The candidate Wilms' tumor gene is involved in genitourinary development. *Nature* **346:** 194–197.

Rabin D, Spitz I, Bercovici B, Bell J, Laufer A, Benveniste R, Polishuk W. (1972) Isolated deficiency of follicle-stimulating hormone. Clinical and laboratory features. *New Engl. J. Med.* **287:** 1313–1317.

Rabinowitz D, Benveniste R, Linder J, Lorber D, Daniell J. (1979) Isolated follicle-stimulating hormone deficiency revisited. *New Engl. J. Med.* **300:** 126–128.

Reijo R, Lee TY, Salo P *et al.* (1995) Diverse spermatogeneic defects in humans caused by Y chromosome deletions encompassing a novel RNA-binding protein gene. *Nature Genet.* **10:** 383–393.

Richardson SJ, Senikas V, Nelson JF. (1987) Follicular depletion during the menopausal transition: evidence for accelerated loss and ultimate exhaustion. *J. Clin. Endocrinol. Metab.* **65:** 1231–1236.

Rogol AD, Mittal KM, White BJ, McGinniss MH, Lieblich JM, Rosen SW. (1980) HLA-compatible paternity in two 'fertile eunuchs' with congenital hypogonadotrophic hypogonadism and anosmia (the Kallman Syndrome). *J. Clin. Endocrinol. Metab.* **51:** 275–279.

Rohner-Jeanrenaud F, Jeanrenaud B. (1996) Obesity, leptin and the brain. *New Engl. J. Med.* **334:** 324–325.

Rose EA, Glaser T, Jones C *et al.* (1990) Complete physical map of the WAGR region of 11p13 localozes a candidate Wilms' tumor gene. *Cell* **60:** 405–408.

Rousseau F, Vincent A, Rivella S *et al.* (1991) Four chromosomal breakpoints and four new probes mark out a 10-cM region encompassing the fragile-X locus (FRAXA). *Am. J. Hum. Genet.* **48:** 108–116.

Rousseau-Merck MF, Atger M, Loosfelt H, Milgrom E, Berger R. (1993) The chromosomal localisation of the follicle-stimulating hormone receptor gene (FSHR) on 2p21-p16 is similar to that of the luteinizing hormone receptor gene. *Genomics* **15:** 222–224.

Saladin R, De Vos P, Guerre-Millo M, Leturque A, Girard J, Staels B, Auwerx J. (1995) Transient increase in obese gene expression after food intake or insulin administration. *Nature* **377:** 527–529.

Saldanha PH, Arnhold IJP, Mendonca BB, Bloise W, Toledo SPA. (1987) A clinico-genetic investigation of Leydig Cell Hypoplasia. *Am. J. Med. Genet.* **26:** 337–344.

Schmickel RD. (1986) Chromosome deletions and enzyme deficiencies. *J. Pediatr.* **108:** 244–246.

Schneider-Gadicke A, Beer-Romero P, Brown LG, Mardon G, Luoh S-W, Page DC. (1989a) Putative transcription activator with alternative isoforms encoded by human ZFX gene. *Nature* **342:** 708–711.

Schneider-Gadicke A, Beer-Romero P, Brown LG, Nussbaum R, Page DC. (1989b) ZFX has a gene structure similar to ZFY, the putative human sex determinant, and escapes X inactivation. *Cell* **57:** 1247–1258.

Schoderbek WE, Roberson MS, Maurer RA. (1993) Two different elements mediate gonadotropin releasing hormone effects on expression of the glycoprotein hormone alpha-subunit gene. *J. Biol. Chem.* **268:** 3903–3910.

Schwartz CE, Dean J, Howard-Peebles PN *et al.* (1994) Obstetrical and gynecological complic in Fragile X carriers: a multicenter study. *Am. J. Med. Genet.* **51**: 400–402.

Schwartz M, Imperato-McGinley J, Peterson RE, Cooper G, Morris PL, MacGillivray M, He. B. (1981) Male pseudohermaphroditism secondary to an abnormality in Leydig cell differentiatic *J. Clin. Endocrinol. Metab.* **53**: 123–127.

Seeburg PH, Adelman JP. (1984) Characterization of cDNA for precursor of human luteinizing hormone releasing hormone. *Nature* **311**: 666–668.

Shen W-H, Moore CCD, Ileda Y, Parker KL, Ingraham HA. (1994) Nuclear receptor steroidogenic factor 1 regulates the Mullerian inhibiting substance gene: a link to the sex determination cascade. *Cell* **77**: 651–661.

Shenker A, Laue L, Kosugi S, Merendino JJ, Minegishi T, Cutler GB. (1993) A constitutively activating mutation of the luteinizing hormone receptor in familial male precocious puberty. *Nature* **365**: 652–654.

Shoelson S, Fickova M, Haneda M, Nahum A, Musso G, Kaiser ET, Rubenstein AH, Tager H. (1983) Identification of a mutant human insulin predicted to contain a serine-for-phenylalanine substitution. *Proc. Natl Acad. Sci. USA* **80**: 7390–7394.

Sinclair AH, Berta P, Palmer MS *et al.* (1990) A gene from the human sex-determining region encodes a protein with homology to a conserved DNA binding motif. *Nature* **346**: 240–244.

Smals AGH, Kloppenborg PWC, van Haelst UJG, Benraad TJ. (1978) Fertile eunuch syndrome versus classic hypogonadotrophic hypogonadism. *Acta Endocrinol.* **87**: 389–399.

Smith A, Fraser IS, Shearman RP, Russell P. (1989) Blepharophimosis plus ovarian failure: a likely candidate for a contiguous gene syndrome. *J. Med. Genet.* **26**: 434–438.

Stephens TW, Basinski M, Bristow PK *et al.* (1995) The role of neuropeptide Y in the antiobesity action of the obese gene product. *Nature* **377**: 530–532.

Stevanovic M, Lovell-Badge R, Collignon J, Goodfellow PN. (1993) SOX3 is an X-linked gene related to SRY. *Hum. Mol. Genet.* **2**: 2013–2018.

Suganuma N, Asada Y, Furui K, Kikkawa F, Furuhashi M, Tomoda Y. (1995) Screening of the mutations in luteinizing hormone alpha-subunit in patients with menstrual disorders. *Fertil. Steril.* **63**: 989–995.

Swain A, Zanaria E, Hacker A, Lovell-Badge R, Camerino G. (1996) Mouse Dax1 gene expression is consistent with a role in sex determination as well as in adrenal and hypothalamus function. *Nature Genet.* **12**: 404–409.

Swedloff RS, Batt RA, Bray GA. (1976) Reproductive hormonal function in the genetically obese (ob/ob) mouse. *Endocrinology* **98**: 1359–1365.

Talmadge K, Vamvakopoulos NC, Fiddles JC. (1984) Evolution of the genes for the beta subunits of human chorionic gonadotrophin and luteinizing hormone. *Nature* **307**: 37–40.

Tartaglia LA, Dembski M, Weng X *et al.* (1995) Identification and expression cloning of a leptin receptor, OB-R. *Cell* **83**: 1263–1271.

Taylor SI, Kadowaki T, Kadowaki H, Accili D, Cama A, McKeon C. (1990) Mutations in the insulin-receptor gene in insulin-resistant patients. *Diabetes Care* **13**: 257–279.

Tiepolo L, Zuffardi O. (1976) Localization of factors controlling spermatogenesis in the nonfluorescent portion of the human Y chromosome long arm. *Hum. Genet.* **34**: 119–124.

Towns PL, Muechler EK. (1979) Blepharophimosis, ptosis, epicanthus inversus and primary amenorrhoea. *Arch. Ophthalmol.* **97**: 1664–1666.

Van Heyningen V, Boyd PA, Seawright A *et al.* (1990) Role for the Wilms tumor gene in genital development? *Proc. Natl Acad. Sci. USA* **87**: 5383–5386.

Verkerk AJ, Pieretti M, Sutcliffe JS *et al.* (1991) Identification of a gene (FMR-1) containing a CGG repeat coincident with a breakpoint cluster region exhibiting length variation in fragile X syndrome. *Cell* **65**: 905–914.

Vogt P, Chandley AC, Hargreave TB, Keil R, Ma K, Sharkey A. (1992) Microdeletions in interval 6 of the Y chromosome of males with idiopathic sterility point to disruption of AZF, a human spermatogenesis gene. *Hum. Genet.* **89**: 491–496.

Watkins PC, Eddy R, Beck AK, Vellucci V, Leverone B, Tanzi RE, Gusella JF, Shows TB. (1987) DNA sequence and regional assignment of the human follicle-stimulating hormone beta-subunit gene to the short arm of human chromosome 11. *DNA* **6**: 205–212.

Weinstein LS, Gejman PV, Friedman E, Kadowaki T, Collins RM, Gershon ES, Spiegel AM. (1990) Mutations of the Gs alpha subunit gene in Albright hereditary osteodystrophy detected by denaturing gradient gell electrophoresis. *Proc. Natl Acad. Sci. USA* **87**: 8287–8290.

LS, Shenker A, Gejman PV, Merino MJ, Friedman E, Spiegel AM. (1991) Activating ons of the stimulatory G protein in the McCune-Albright syndrome. *New Engl. J. Med.* **325:** 1695.

J, Axelrod L, Whitcomb RW, Harris PE, Crowley WF, Jameson JL. (1992) Hypogonadism sed by a single amino acid substitution in the beta-subunit of luteinizing hormone. *New Engl. J. led.* **326:** 179–183.

elshons WJ, Russell LB. (1959) The Y chromosome is the bearer of male determining factors in the mouse. *Proc. Natl Acad. Sci. USA* **45:** 560–566.

Williams C, Wieland RG, Zorn EM, Hallberg MC. (1975) Effect of synthetic gonadotrophic-releasing hormone (GnRH) in a patient with the 'Fertile Eunuch' syndrome. *J. Clin. Endocrinol. Metab.* **41:** 176–179.

Zanaria E, Muscatelli F, Bardoni N *et al.* (1994) An unusual member of the nuclear hormone receptor superfamily responsible for X-linked adrenal hypoplasia congenita. *Nature* **372:** 635–641.

Zhang Y, Proenca R, Maffei M, Barone M, Leopold L, Friedman JM. (1994) Positional cloning of the mouse obese gene and its human homologue. *Nature* **372:** 425–432.

ex

ORDERING DETAILS

Main address for orders

BIOS Scientific Publishers Ltd
9 Newtec Place, Magdalen Road,
Oxford OX4 1RE, UK
Tel: +44 1865 726286
Fax: +44 1865 246823

Australia and New Zealand
DA Information Services
648 Whitehorse Road, Mitcham, Victoria 3132, Australia
Tel: (03) 9210 7777
Fax: (03) 9210 7788

India
Viva Books Private Ltd
4325/3 Ansari Road, Daryaganj, New Delhi 110 002, India
Tel: 11 3283121
Fax: 11 3267224

Singapore and South East Asia
(Brunei, Hong Kong, Indonesia, Korea, Malaysia, the Philippines, Singapore, Taiwan, and Thailand)
Toppan Company (S) PTE Ltd
38 Liu Fang Road, Jurong, Singapore 2262
Tel: (265) 6666
Fax: (261) 7875

USA and Canada
BIOS Scientific Publishers
PO Box 605, Herndon, VA 20172-0605, USA
Tel: (703) 661 1500
Fax: (703) 661 1501

Payment can be made by cheque or credit card (Visa/Mastercard, quoting number and expiry date). Alternatively, a *pro forma* invoice can be sent.

Prepaid orders must include £2.50/US$5.00 to cover postage and packing (two or more books sent post free)